Frenkel / Schiller

Understanding Plastics Additives

Peter Frenkel
Michael Schiller

Understanding Plastics Additives

A Textbook

HANSER

Print-ISBN: 978-1-56990-087-1
E-Book-ISBN: 978-1-56990-955-3

All information, procedures, and illustrations contained in this work have been compiled to the best of our knowledge and is believed to be true and accurate at the time of going to press. Nevertheless, errors and omissions are possible. Neither the authors, editors, nor publisher assume any responsibility for possible consequences of such errors or omissions. The information contained in this work is not associated with any obligation or guarantee of any kind. The authors, editors, and publisher accept no responsibility and do not assume any liability, consequential or otherwise, arising in any way from the use of this information – or any part thereof. Neither do the authors, editors, and publisher guarantee that the described processes, etc., are free of third party intellectual property rights. The reproduction of common names, trade names, product names, etc., in this work, even without special identification, does not justify the assumption that such names are to be considered free in the sense of trademark and brand protection legislation and may therefore be used by anyone.

The final determination of the suitability of any information for the use contemplated for a given application remains the sole responsibility of the user.

Bibliographic information of the German National Library:
The German National Library lists this publication in the German National Bibliography; detailed bibliographic data are available on the Internet at http://dnb.d-nb.de.

Vilshofener Straße 10 | 81679 Munich | info@hanser.de
www.hanserpublications.com
www.hanser-fachbuch.de
Editor: Dr. Mark Smith
Production management: Cornelia Speckmaier
Cover concept: Marc Müller-Bremer, www.rebranding.de, Munich
Cover design: Max Kostopoulos
Cover picture: © gettyimages.de/Adrienne Bresnahan
Typesetting: le-tex publishing services GmbH, Leipzig
Printed and bound by: CPI Books GmbH, Leck
Printed in Germany

In loving memory of my parents Lev and Roza and with gratitude to my family: Elena, Anna, Sasha, Alex, Hannah, and little Naomi for their patience, support and encouragement.

Peter Frenkel

For my beloved family – Annette, Sarah, and Maria, Sebastian, Andreas and my wonderful little ones Felix, Findus Michael and Karolina. With gratitude for your boundless patience, encouragement, and the joy you bring to my life every day.

Michael Schiller

The Authors

Peter Frenkel

Dr. Frenkel has over 40 years in technical management, leading international technical teams, and conducting industrial and basic research on new chemical product & process technologies.

He is Consultant to the PVC industry. Most recently, Dr. Frenkel served as Vice President of Technology at Galata Chemicals. He has over 30 years of experience in the field of polymer additives, focusing on product innovation and advocacy, with an emphasis on PVC. Dr. Frenkel drives innovation that concentrates on sustainable technologies related to bio-based plasticizers, heat stabilizers, polymer modifiers and catalysts. Together with his teams, he conducted fundamental studies and made major contributions to establishing the Universal Mechanism of PVC stabilization with binary heat stabilizer systems, proposing the mechanism of C-PVC stabilization, developing and market introduction of high efficiency primary bio-based plasticizers, modifying reactivity of organic peroxide initiators and accurately determining the mechanism of in-vitro metabolism of selected alkyltin heat stabilizer components that based on these results were discovered to be non-reprotoxic.

Dr. Frenkel has co-authored over 40 patents, several book chapters and dozens of scientific journal and conference publications; played a critical role in numerous new product launches. He has a broad educational background, holds a Ph.D. in Inorganic Chemistry from Belarus Institute of Technology and a Master of Science degree in Organic Chemistry from Belarus University. Dr. Frenkel was named Fellow of The Society of Plastics Engineers. He enjoys listening to the music of The Beatles.

Michael Schiller

Photo: Mahsa Saffaripour

Dr. Michael Schiller, born in 1959, is a German chemist, scientist, author, editor, and inventor. He has more than 30 years of international experience in the PVC industry, with a strong focus on research and development. His expertise includes the development of new PVC stabilizers and additives, investigation of unusual phenomena, sustainability aspects of the PVC industry, and technical product management. However, Michael considers himself more a student than an expert, always eager to learn.

He is the author of the book PVC Additives (Hanser Verlag, two editions, also available in Russian and German). Additionally, he is co-author of "PVC from A to Z" in the PVC eSolution by Hanser Verlag, author and co-author of around 200 scientific papers, co-editor of the Plastic Additives Handbook in English, Russian, and German, co-editor of the 2nd edition of the "PVC Handbook", and inventor on more than 50 patents.

Michael served as Head of Group R&D at Chemson from 1992 to 2009 and as Head of the Innovation and Sustainability Center at Akdeniz Kimya in Izmir, Turkey, from 2009 to 2015. In 2015, he founded HMS Concept e.U., a consulting company specializing in chemical, environmental, and plastics technology, which serves approximately 50 clients across more than a dozen countries on five continents.

Privately, Michael has been happily married for more than 35 years, is the proud father of two adult, well-educated, and accomplished daughters, and is grandfather to two wonderful grandsons and a granddaughter. He enjoys listening to the music of Queen.

Preface

This book is unique in several ways. It does not merely describe additives; it teaches how to understand them. It was designed as an educational, practical and structured guide that helps readers grasp both fundamentals and real-world applications. It explains how the additives function and how their effects can be interpreted and optimized. This textbook is not just something to read; it is a tool for learning, practicing and applying the learnt information about the main classes of plastics additives.

In contrast to the *Plastics Additives Handbook* (a total of six editions of it have already been published by Hanser Publishers over the last 50 years), this textbook was designed to be less of a reference, and more of a resource and learning and teaching tool for on-the-job training of industry professionals, including those who are looking for an in-depth but still exceptionally concise review of main groups of plastics additives and who are focused on better understanding of additives' fundamentals, intricacies and functionalities. It is also for readers who seek a new role in the industry, for college students and graduates, industry practitioners and researchers, lecturers and coaches who are eager to learn more and possibly teach about plastics additives and for the public at large interested to know more about how the additives work, their role, challenges, limitations and available technical solutions for main types of plastics and their applications.

One of the challenges in compiling this information was adapting our writing styles for educational purposes of this book, since compared to other technical publications, writing a textbook demands expressing its content in more simple and clear terms to ensure that even readers without prior subject matter knowledge could grasp the fundamentals. In this regard, we took inspiration from Albert Einstein's advice: "Everything should be made as simple as possible. But not simpler."

Each chapter of the book covers an individual plastics additive group. It includes main literature sources used, definitions of key terms required for reading and comprehending the chapter, mechanisms of action and performance attributes of the additives, chemical compositions of main commercially viable examples of the group, test methods, market trends, and selected health and environmental aspects. Important features of each chapter are that rules are highlighted that characterize and allow prediction of the performance of the additives in polymer systems, a large number of specific examples are provided and each chapter is concluded with a list of questions that help the reader to better understand and retain the material.

The book was compiled in a relatively small format, so the reader can conveniently place it in a briefcase and/or read it on a plane or keep it handy in a laboratory or office environment. Provided information is well organized, cross-referenced and sourced from the latest industry-specific monographs, handbooks and other publications. It intends to provoke thought and even discussion on the topic.

There is a sort of symbiosis among plastics and plastics additives. Most plastics cannot be processed and/or used without the additives, while the main application of most plastics additives is in plastics. This book was designed to demonstrate and celebrate this symbiosis. It is hard not to extend that passion to the additives that contribute substantially to the performance characteristics of plastics.

The authors are fortunate and greatly appreciate Dr. Mark Smith of Carl Hanser Publishers for his commitment to, guidance and unwavering support of this book throughout its preparation. Mark is an extraordinary human being, true professional and gentleman.

Peter Frenkel, Michael Schiller

Foreword

Work in plastics long enough, and you realize that understanding a polymer is really just the beginning. Early on, it can feel like learning the differences between polyethylene and polypropylene is the biggest hurdle. Then you learn about additives, and a new level of material science opens up.

Additives influence processing, durability, appearance, stability, and countless other performance characteristics. They are often the reason a material can meet the requirements of an application in the first place. In many cases, the difference between success and failure comes down to the additive package.

As materials are expected to deliver superior performance, greater efficiency, and enhanced sustainability, the polymer–additive relationship becomes increasingly important. Meeting those demands requires a deeper understanding of how materials can be tailored for specific applications.

This is where a book like *Understanding Plastics Additives* becomes especially valuable. It builds on foundational polymer knowledge and expands it into the critical role additives play in real-world material performance. Dr. Frenkel and Dr. Schiller have built distinguished careers advancing the science and application of additive technologies. Through their technical contributions, publications, patents, and presentations, they helped shape how the industry understands the role of additives in performance and functionality. Their expertise is evident throughout this book, which combines scientific principles with practical insight drawn from decades of experience.

What I appreciate most about this book is that it goes beyond simply listing additives and their functions. It connects additive selection to material performance in a way that tables and datasheets can't quite capture, giving readers a clearer understanding of not just what additives do, but why they matter and how those choices ultimately shape performance.

For students and early-career professionals, this book opens the door to a topic that is often underrepresented early on but quickly becomes central in practice. For experienced professionals, it is a reference you can return to when real-world challenges demand going a level deeper. Whether you are working in formulation, processing, troubleshooting, product development, research, or education, a stronger understanding of additives leads to better material decisions and better outcomes.

Over the years in this industry, I have been fortunate to work alongside people who are not only highly capable but genuinely generous with their knowledge. That willingness to teach, mentor, and elevate others is one of the things I value most about the plastics community. A book like this reflects that same spirit. Dr. Frenkel and Dr. Schiller have created a resource that brings decades of expertise into one place and makes it more accessible to a wide range of plastics professionals. I am honored to introduce *Understanding Plastics Additives*, and I hope readers find in it both the knowledge and perspective needed to continue advancing how we design, process, and understand plastic materials.

Lynzie Nebel

2026 Executive Committee Chair

Society of Plastics Engineers

Contents

1 Properties of Polymers

1.1 Key Terms and Definitions

Molecule	Smallest unit of a chemical substance, composed of two or more atoms held together by chemical bonds, retaining the substance's chemical properties.
Molecular weight	The sum of the relative atomic masses of all atoms in a molecule, indicating its mass on a molecular scale.
Substituent group	An atom or group of atoms that replaces hydrogen in a molecule, influencing its chemical reactivity and physical properties.
Viscosity	A measure of a fluid's resistance to deformation and flow, reflecting internal friction between its molecules.
Co-catalysts	Substances used with a primary catalyst to increase reaction rate, selectivity, or efficiency in chemical processes.
Mechanical properties	Characteristics that describe a material's response to applied forces, such as tensile strength, elasticity, and hardness.

1.2 Introduction to Monomers and Polymers

Let's embark on a journey into the world of plastics additives. To understand these, we must first understand polymers (Figure 1.1). So, let's lean back and start a journey by train – a means of transport that will also serve us as an analogy for polymers. Just as there are different types of trains, such as freight trains, container trains, tank car trains, express trains with first- and second-class compartments, overnight trains, and commuter trains, there are also various polymers for a wide range of applications.

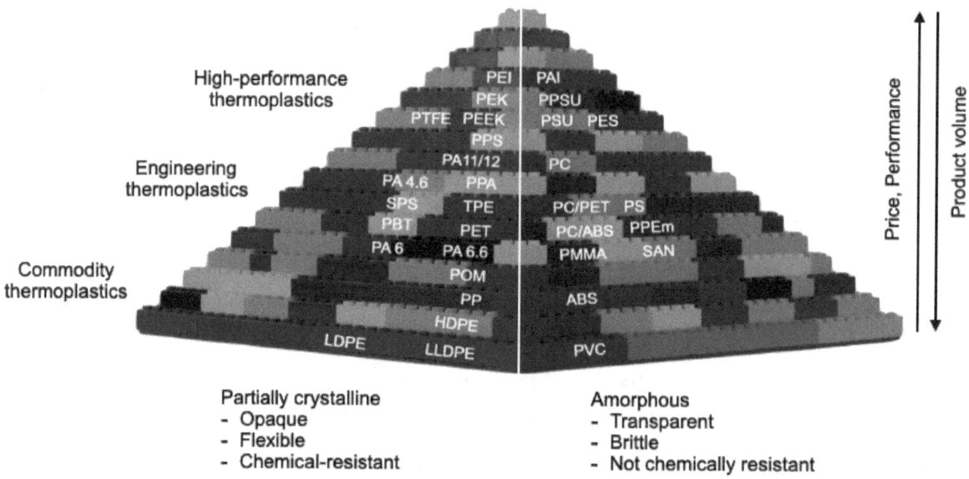

ABS	Acrylonitrile butadiene styrene	PMMA	Polymethyl methacrylate
HDPE	High-density polyethylene	POM	Polyoxymethylene
LDPE	Low-density polyethylene	PP	Polypropylene
LLDPE	Linear low-density polyethylene	PPA	Polyphthalamide
PA	Polyamide	PPEm	Polyphenylene ether, modified
PAI	Polyamide-imide	PPS	Polyphenylene sulfide
PBT	Polybutylene terephthalate	PPSU	Polyphenylsulfone
PC	Polycarbonate	PS	Polystyrene
PE	Polyethylene	PSU	Polysulfone
PEEK	Polyether ether ketone	PTFE	Polyethylene tetrafluoride
PEI	Polyetherimide	PVC	Polyvinyl chloride
PEK	Polyetherketone	SAN	Styrene acrylonitrile
PES	Polyethersulfone	SPS	Syndiotactic polystyrene
PET	Polyethylene terephthalate		

Figure 1.1 The world of plastics in terms of price, performance and production volume. Image credit: Data from Kunststoff-Schweiz (*https://www.kunststoff-schweiz.ch/html/kunststoff.html*); photo and design by Mahsa Saffaripour

A train is assembled from individual carriages with a locomotive, and likewise, polymers are made from individual monomers with a polymerization initiator or a catalyst. Just as the construction of the train cars (Figure 1.2) corresponds to the requirements for the train, the monomer chemistry and the substituent groups of the monomers (Figure 1.3) are among other critical factors that determine properties of the polymer.

Figure 1.2 Various types of railway carriages. Image credit: Mahsa Saffaripour

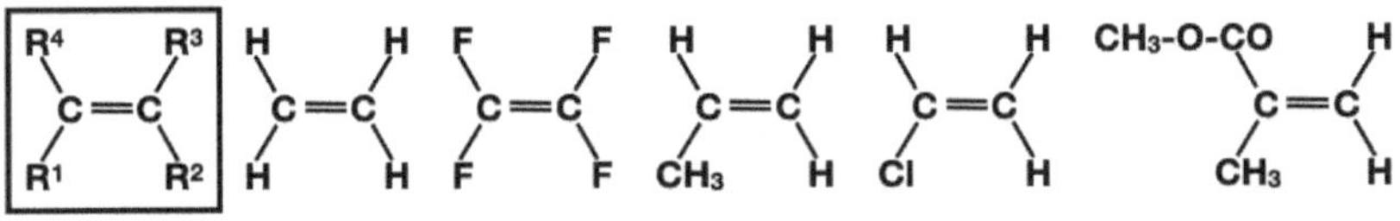

Figure 1.3 General structure of monomers with up to four different substituents (boxed), and examples of common monomers: ethylene, tetrafluoroethylene, propylene, vinyl chloride, methyl methacrylate (from left to right)

If all four substituents from R^1 through R^4 in Figure 1.3 are hydrogen atoms (H), the molecule is ethylene, the monomer for polyethylene (PE), the most widely used plastic of all. Ethylene is one of the few monomers with all four substituent groups being the same. Another one is tetrafluoroethylene, where R^1 through R^4 are fluorine atoms (F). This monomer is used to produce polytetrafluoroethylene (PTFE), also known by the brand name Teflon®.

If R^1 is a methyl group (CH_3) and R^2 through R^4 are hydrogen atoms, it results in propylene, the monomer for polypropylene (PP). When R^1 is a chlorine atom and R^2 through R^4 are H atoms, it gives the vinyl chloride monomer (VCM), from which polyvinyl chloride (PVC) is produced. PE, PP, and PVC are the main technical thermoplastics in Figure 1.1. Table 1.1 lists other important polymers and their selected properties.

An example of a monomer with two different substituents is methyl methacrylate, where $R^1 = CH_3$, $R^4 = C(=O)OCH_3$, R^2 and R^3 are H atoms, and from which poly(methyl methacrylate) (PMMA) is manufactured. PMMA is also known as Plexiglas®.

A locomotive is selected in a way that it can pull a train of specific cars (monomer types), length and load at a required maximum speed. In free radical polymerization, the locomotive can be compared with a free radical initiator, so that an optimal polymer is created under the specific process conditions (Figure 1.4 and Figure 1.5).

Table 1.1 Structure and Properties of Commonly Encountered Monomers and Polymers[1),2)]

Monomer	Molecular weight (g/mol)	Substituents (Figure 1.3)				Polymer (abbreviation)	Density (g/mL)	Hildebrand solubility parameter (HSP) δ (MPa$^{1/2}$)				Cohesive energy E (J/mol)	Moleculars volume V_{mol} (mL/mol)
		R[1)]	R[2)]	R[3)]	R[4)]			Average	van Krevelen	from	to	van Krevelen	
Ethylene	28.0	H	H	H	H	PE	0.851	16.7	17.0	15.8	17.1	9500	32.9
Propylene	42.0	CH_3	H	H	H	PP	0.855	16.2	17.1	16.8	18.8	14,400	49.1
Vinyl chloride	62.5	Cl	H	H	H	PVC	1.383	19.6	19.7	19.2	22.1	17,600	45.2
Vinylidene chloride	97.0	Cl	H	H	Cl	PVDC	1.672	20.6	21.1	20.3	25.0	25,700	58.0
Vinyl acetate[a]	86.0	$O(CO)CH_3$	H	H	H	PVA, PVAc	1.191	19.6	18.7	19.1	22.6	25,300	72.2
Methyl methacrylate	100.0	CH_3	H	H	$O(C=O)CH_3$	PMMA	1.156	19.0	18.9	18.6	26.2	30,800	86.5
Chloroprene	88.5	$C(H)=CH_2$	H	H	Cl	Neoprene (polychloroprene)	1.241	18.1	18.7	16.8	18.9	25,000	71.3
Acrylonitrile[b]	53.0	CN	H	H	H	PAN	1.183	26.2	31.3	25.6	31.5	43,900	44.8

1 D. W. Van Krevelen and K. Te Nijenhuis, *Properties of Polymers*, Elsevier, 2009, pp. 191–222.
2 *https://polymerdatabase.com/polymer%20physics/delta%20Table.html* [accessed 30 June 2023].

Monomer	Molecular weight (g/mol)	Substituents (Figure 1.3)				Polymer (abbreviation)	Density (g/mL)	Hildebrand solubility parameter (HSP) δ (MPa$^{1/2}$)				Cohesive energy E (J/mol)	Moleculars volume V_{mol} (mL/mol)
		R[1)]	R[2)]	R[3)]	R[4)]			Average	van Krevelen	from	to	van Krevelen	
Styrene	104.0	Ph	H	H	H	PS	1.061	18.3	19.8	17.4	19.0	38,300	98.0
Tetrafluoro-ethylene	100.0	F	F	F	F	PTFE	2.020	—	13.3	12.7	—	8700	49.5
1,3-Butadiene	54.0	—				PB, synthetic rubber	0.890	17.0	16.5	16.6	17.6	16,600	60.7
Isoprene	68.1	—				PI	0.900	16.5	17.0	16.2	20.5	21,800	75.7
		—				POM	—	21.1	—	—	—	—	—
Octano-8-prolactam		—				PA 8	—	24.7	23.5	26.0	—	74,800	135.9
Caprolactam		—				PA 6	1.084–1.235[c]	25.5	—	—	—	—	—
		—				PA 6.6	1.13–1.14[3)]	26.1	25.2	27.8	—	132,600	208.3
		—				PET	1.38[4)]	21.2	20.7	19.9	21.9	61,200	143.2

[a] Also a co-monomer of EVA.
[b] Also a co-monomer of styrene acrylonitrile resin (SAN), ABS, nitrile butadiene rubber (NBR), and acrylonitrile styrene acrylate (ASA).
[c] Semi-crystalline phase: 1.130 g/mL; crystalline phase (α-form): 1.235 g/mL; amorphous phase: 1.084 g/mL.

---

3 "Polyamide", Wikipedia, *https://de.wikipedia.org/wiki/Polyamide* [accessed 17 February 2025].

4 "Polyethylenterephthalat", GESTIS-Stoffdatenbank, *https://gestis.dguv.de/data?name=530(?<=\d\d)\,?(?=\d\d\d(?!\d))566* [accessed 17 February 2025].

Figure 1.4 Train before and after assembly. Image credit: Mahsa Saffaripour

R· $H_2C{=}CH_2$ $H_2C{=}CH_2$ $H_2C{=}CH_2$ $H_2C{=}CH_2$ $H_2C{=}CH_2$ $H_2C{=}CH_2$ $H_2C{=}CH_2$ $H_2C{=}CH_2$ $H_2C{=}CH_2$

R- $H_2C\text{-}CH_2\cdot$ $H_2C{=}CH_2$ $H_2C{=}CH_2$ $H_2C{=}CH_2$ $H_2C{=}CH_2$ $H_2C{=}CH_2$ $H_2C{=}CH_2$ $H_2C{=}CH_2$ $H_2C{=}CH_2$

R- $H_2C\text{-}CH_2\text{-}H_2C\text{-}CH_2\text{-}H_2C\text{-}CH_2\text{-}H_2C\text{-}CH_2\text{-}H_2C\text{-}CH_2\text{-}H_2C\text{-}CH_2\text{-}H_2C\text{-}CH_2\text{-}H_2C\text{-}CH_2\cdot$ $H_2C{=}CH_2$

Figure 1.5 Transformation from a monomer to a polymer, as exemplified by ethylene polymerizing to PE

The monomers, initiators or polymerization catalysts, and process conditions, in combination with other parameters, determine properties of the manufactured polymer.

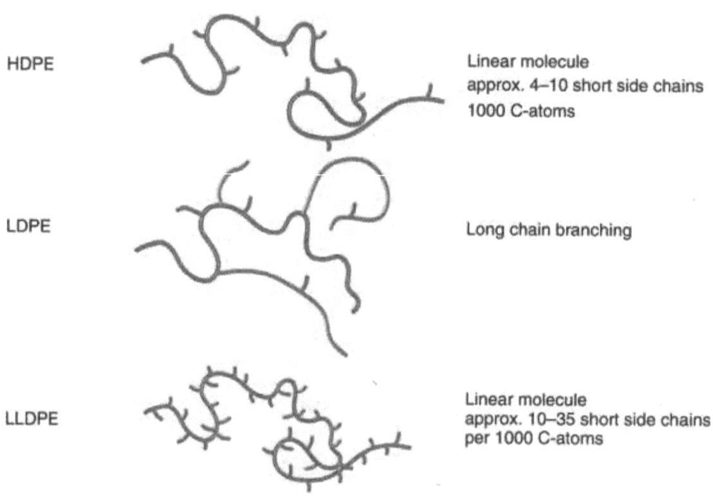

Figure 1.6 Schematic molecular structures of various polyethylene types: HDPE: high-density PE; LDPE: low-density PE; LLDPE: linear low-density PE[5]

[5] E. Baur, T. A. Osswald, and N. Rudolph, *Plastics Handbook: The Resource for Plastics Engineers* (5th edition), Hanser, 2019, p. 21.

Ethylene serves as a raw material not just for producing polyethylene but also for other derivatives, such as polyethylene wax. Depending on variations in the process conditions as well as additives and catalysts used for its polymerization, PE exhibits variations in its properties, such as density, polymer chain length, chain branching, chain length distribution (Figure 1.6), molecular weight distribution, melting point, and viscosity, Chemically, it is still polyethylene, albeit with diverse characteristics that enable its use as anything from a construction material to a wax, and many other things besides.

1.3 Branching and Crystallinity

There is a direct correlation between the linearity (or branching) of polymer chains and the crystallinity of the corresponding plastics:

- Non-branched, linear polymer chains can arrange themselves parallel to one another more easily. This kind of molecular packing promotes material crystallization and leads to higher crystallinity.
- Highly branched polymer chains hinder the well-packed arrangement of the molecules, makes crystallization more difficult, and leads to lower crystallinity.
- The regularity of the placement of side groups on the polymer backbone affects its crystallinity and mechanical properties. For example, isotactic PP crystallizes more easily than atactic PP, which has irregularly distributed side groups.

> The less branched the polymer chains and the more regularly the side groups are arranged on the polymer backbone, the higher its crystallinity is (Figure 1.7).

Terms such as degree of crystallinity and degree of crystallization (usually determined by differential scanning calorimetry (DSC) and expressed in percent) are used synonymously or interchangeably. Crystallinity is a material property that is of particular importance in the case of thermoplastic polymers. Most of the time, it is defined as the mass fraction of the crystalline regions in the total polymer phase that also includes amorphous regions. The crystallinity of thermoplastics is always less than 100%. The number of amorphous thermoplastics with a crystallinity of close to 0% is small, and it includes atactic thermoplastics, such as atactic PMMA and PS. All other thermoplastics can be combined into the large group of semi-crystalline materials, with crystallinity typically in the range 10–80%. In the case of PVC, for example, the crystallinity is 10–15%.

Figure 1.7 Semi-crystalline polymers contain segments (phases) of orderly packed micro-molecules, while amorphous polymers have a completely disorganized structure

1.4 Molecular Weight

There are polymers of low, medium-high, high, very high, and ultra-high molecular weights, corresponding to different lengths of polymer chains. The number of repetitions of the monomer units is called the degree of polymerization P_n. Multiplying the degree of polymerization by the molecular weight of the monomer results in the molecular weight of the polymer. At least, that's the theory; in reality, there is a distinction between the number-average molecular weight M_n (based on the number of monomer units) and the weight-average molecular weight M_w (Figure 1.8).

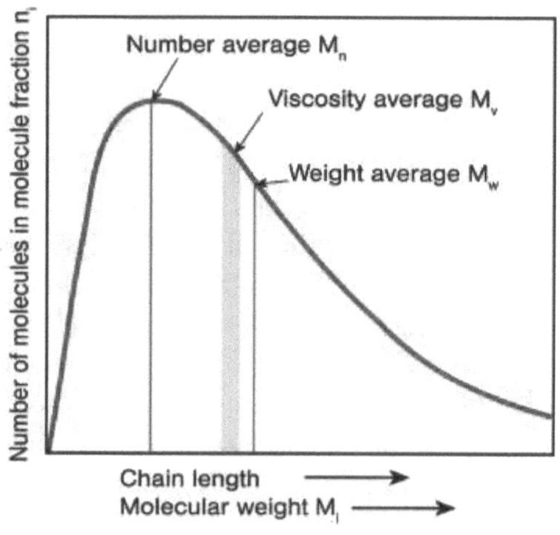

Figure 1.8 Molecular weight distribution of a typical thermoplastic material[6]

[6] E. Baur, T. A. Osswald, and N. Rudolph, *Plastics Handbook: The Resource for Plastics Engineers* (5th edition), Hanser, 2019, p. 22.

M_n is calculated using the following formula:[7]

$$M_\mathrm{n} = \frac{\sum n_i M_i}{\sum n_i} = \frac{\sum m_i}{\sum n_i} \tag{1.1}$$

where:

n_i = number of molecules in class i, each with molecular mass M_i.

M_n rises, for example, during the fractionation of a plastic, involving step-by-step precipitation of macromolecules with increasing chain length.

M_w is defined by the molar mass instead of the number of the mass fraction of the molecules M_i:

$$M_\mathrm{w} = \frac{\sum m_i M_i}{\sum m_i} = \frac{\sum n_i M_i^2}{\sum n_i M_i} \tag{1.2}$$

To describe the mechanical properties of polymers, the z-average molecular weight is often used; this is a higher number that is more sensitive to the high-weight "tail" of the distribution:

$$M_\mathrm{z} = \frac{\sum m_i M_i^2}{\sum m_i M_i} = \frac{\sum n_i M_i^3}{\sum n_i M_i^2} \tag{1.3}$$

The width of the molecular weight distribution, known as dispersity (*Đ*), can be estimated from the quotient of the two mean values M_w and M_n ($Đ = M_w/M_n$) – see Table 1.2. For many polymeric materials, this value fluctuates between 2 and 4. Instead of using the dispersity directly, the non-uniformity parameter u is sometimes used when expressing how much the sample deviates from a perfectly uniform distribution ($Đ = 1$). This parameter can be calculated as:

$$u = Đ - 1 = \frac{M_\mathrm{w}}{M_\mathrm{n}} - 1 \tag{1.4}$$

There is a correlation between the dispersity and the crystallinity of a polymer:

- Low dispersity promotes higher crystallinity: specifically, polymers with a narrow molecular weight distribution (low dispersity) are more likely to crystallize because the more uniform molecular chain lengths allow them to form more compact arrangements.
- High dispersity makes crystallization more difficult: specifically, with a broad molecular weight distribution (high dispersity), the orderly arrangement of the polymer chains is more challenging to achieve, and it leads to lower crystallinity.

[7] W. Kaiser, *Kunststoffchemie für Ingenieure: Von der Synthese bis zur Anwendung* (3rd edition) [in German], Hanser, 2011, pp. 89–90.

- Influence on crystallization kinetics: dispersity also influences the rate of crystallization. More uniform molecular chain lengths lead to faster and more uniform crystal formation.
- Effect on crystallite size: low dispersity may contribute to the formation of larger and more perfectly shaped crystallites, while high dispersity is more likely to result in smaller and less perfect crystallites.

However, crystallinity is not only influenced by dispersity but also by other factors, such as chain linearity, symmetricality of side-group placement on the polymer backbone, and the polymer cooling conditions.

It is important to note that crystallinity of a polymer is a complex phenomenon that depends on many factors, dispersity being only one of them.

The longer the side chain of a polymer, the lower its glass transition temperature T_g. (This is at least true for polyacrylates).[8]

For amorphous thermoplastics, the minimum processing temperature is typically 100 °C above T_g. (This rule does not apply to semi-crystalline polymers, since the energy needed for melting the crystallites must also be provided. Therefore, the difference between the minimum processing temperature and T_g could be substantial. For high-performance polymers, such as PEEK, the latter can be up to 400 °C).

As crystallinity increases, the density of the polymer increases. This is because the molecular chains are more orderly packed in crystalline regions than in amorphous regions.

A higher degree of crystallinity leads to an increase in tensile strength of the polymer. The orderly structure in crystalline areas yields a material with superior mechanical characteristics.

Crystallinity tends to increase T_g. However, this relationship is more complex and less direct than that of density versus tensile strength.

Additional effects of increased crystallinity are:

- Increased stiffness
- Increased melting point
- Increased chemical resistance
- Increased dimensional stability
- Decreased coefficient of thermal expansion.

---

[8] From *https://www.krv.de/artikel/glasuebergangstemperatur* [accessed 17 February 2025].

It is important to note that these correlations represent general trends. The degree at which these effects manifest themselves may vary depending on the specific polymer and other factors, such as molecular structure, processing conditions, and additives.

Table 1.2 Dependence of Dispersity on Polymer Structure and Polymerization Method[9]

Polymer class	Dispersity M_w/M_n
Polymers with consistent molecular weight	1
Polycondensation and polyaddition products	2
Vinyl polymers	2–5
Highly branched polymers	20–50

Another method for characterizing the chain length is to use the average degree of polymerization $\overline{P}$, which is calculated using Equation 1.5:

$$\overline{P} = \frac{\overline{M}}{M_0} \tag{1.5}$$

where:

M_0 = molar mass of the monomer or constitutional repeating unit

$\overline{M}$ can be either M_w or M_n.

So for example, for PE with M_n = 56,000 g/mol and M_0 = 28 g/mol, the average degree of polymerization is $\overline{P}$ = 2000.

The higher the average molecular weight, the superior the mechanical properties and the higher the viscosity of the polymer.

1.5 Tacticity

As long as a substituent group in Figure 1.2 is not hydrogen (H), the *tacticity* of the polymer comes into play and influences its properties. Tacticity is also known as stereoregularity, and describes the recurring arrangement of side-chains on a polymer

[9] E. Baur, T. A. Osswald, and N. Rudolph, *Plastics Handbook: The Resource for Plastics Engineers* (5th edition), Hanser, 2019, p. 24.

backbone. In principle, tacticity can only occur in polymers that are made of asymmetric monomers, such as propylene or styrene, but not ethylene or tetrafluoroethylene. There are three basic types of tacticity (Figure 1.9):

- Syndiotactic, when the substituent groups alternately point forward and backward
- Isotactic, when all substituent groups point in one direction
- Atactic, when the substituent groups are arranged randomly.

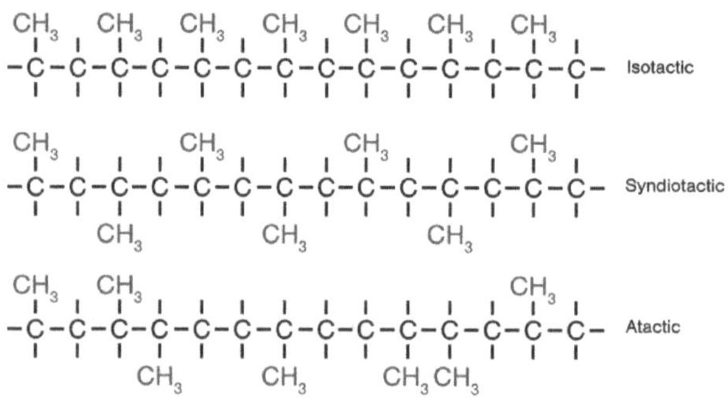

Figure 1.9 Stereoregularity of polypropylene[10]

To take the example of PP:

- Isotactic PP has the highest crystallinity, with a 30–60% degree of crystallization in industrially produced materials. It forms a semi-crystalline structure with a helical arrangement of the molecular chains.
- Syndiotactic PP has lower crystallinity than isotactic PP, with a degree of crystallization of about 30–40%, and this semi-crystalline structure means that it is less orderly than isotactic PP.
- Atactic PP is completely amorphous without crystalline regions.

These structures have effects on the properties. So isotactic PP exhibits the highest mechanical strength, stiffness, and hardness of the three variants. It has an elastic modulus of 1300–1800 N/mm^2. Syndiotactic PP is less stiff than isotactic PP, but it is more transparent and impact-resistant. Atactic PP demonstrates the lowest mechanical strength and stiffness, being sticky and soft at room temperature.

Regarding melting point, isotactic PP has the highest value (185–220 °C for α-crystalline regions) and superior heat resistance. The melting point of syndiotactic PP is lower (170–200 °C), while atactic PP has no defined melting point, resulting in inferior heat resistance.

[10] E. Baur, T. A. Osswald, and N. Rudolph, *Plastics Handbook: The Resource for Plastics Engineers* (5th edition), Hanser, 2019, p. 20.

In terms of density, isotactic PP has the highest value of the three variants, typically being in the range 0.895–0.920 g/mL. Syndiotactic PP has a slightly lower density than isotactic PP, and atactic PP is the least dense.

> The higher the crystallinity of a polymer, the better its mechanical properties, melting point, heat resistance, and density.

1.6 Copolymers and Terpolymers

Let us now review copolymers and terpolymers.

A *copolymer* is a polymer formed from two or more different monomers, in contrast to a *homopolymer*, which is formed from only one type of monomer. Copolymers can themselves be divided into five classes, illustrated below using the case of a binary copolymer derived from two monomers A and B (Figure 1.10):

- *Statistical copolymers* are those in which the distribution of the two monomers in the polymer chains follows a statistical distribution. If the ratio of the monomers in a polymer section corresponds to the molar ratio, this can be described as a *random copolymer* (-AABABBBABAABBBABBABAB-).
- *Gradient copolymers* are similar to random copolymers, but in these the proportion of one monomer gradually increases relative to the other along the polymer chain (-AAAAAABAABBAABABBBAABBBBBB-).
- *Alternating copolymers* are those in which the two monomers alternate regularly (-ABABABABABABABABABAB-).
- *Block copolymers* and *segment copolymers* consist of longer sequences or blocks of each monomer (-AAAAAAAAABBBBBBBBBBBB-). Depending on the number of different blocks on the polymer chain, these might be called diblock copolymers, triblock copolymers, etc. The number of monomers per block is usually up to ten.
- *Graft (or grafted) copolymers* are those in which blocks of one monomer are grafted onto the skeleton (backbone) of another monomer. In other words, the graft copolymer consists of a main polymer chain or backbone (A) covalently bonded to one or more side chains of another polymer (B).

Copolymers that consist of three different monomers are called *terpolymers* (-ABCABCABCABCABCABC-).

Crosslinked polymers are formed as a result of covalent bonding between adjacent polymer backbones, creating an interconnected three-dimensional molecular structure.

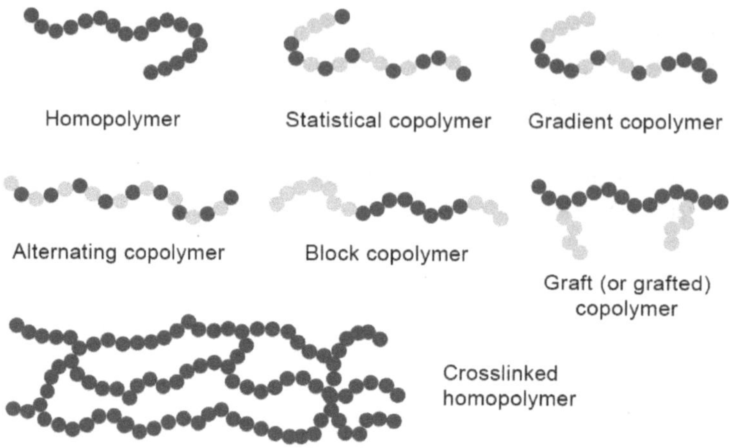

Figure 1.10 Types of copolymers and homopolymers

1.7 Polymerization Processes

As well as polymers produced by free radical polymerization processes, there are several types of plastics that are produced differently. For instance, polyesters like PET (Figure 1.11) and PBT, polycarbonates (Figure 1.12), and polyamides like PA 6.6 (Figure 1.13) are manufactured using the polycondensation reaction. This is a reaction in which two molecules are combined with elimination of low molecular weight by-products, such as water, ammonia, carbon dioxide, hydrogen chloride, and short-chain alcohols. Polycondensation reactions can also take place intramolecularly, and both intermolecular and intramolecular versions are often reversible.

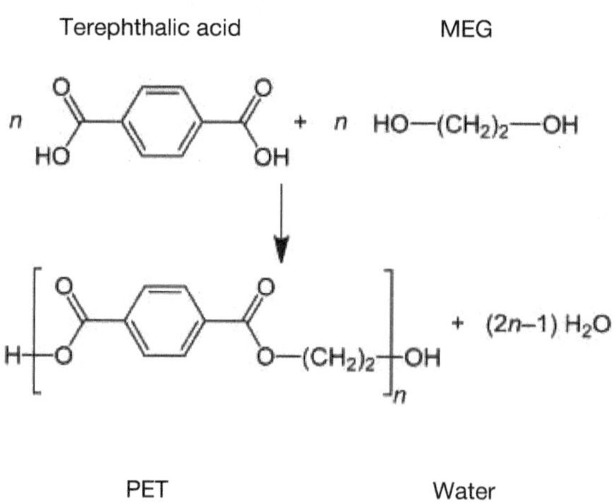

Figure 1.11 Formation of PET by polycondensation of terephthalic acid with monoethylene glycol (MEG)

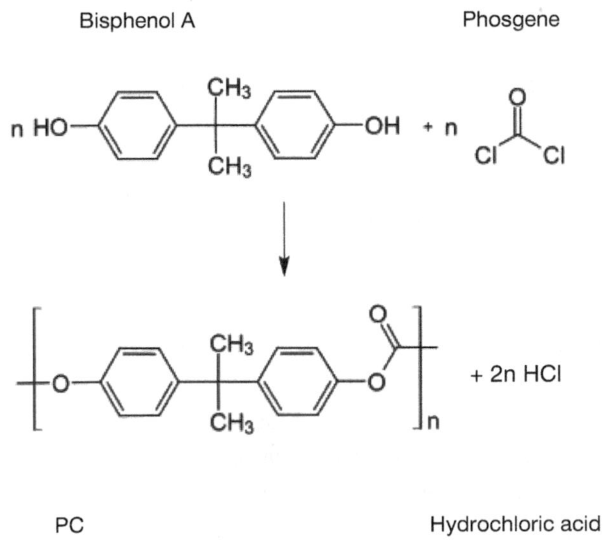

Figure 1.12 Formation of PC by polycondensation of bisphenol A with phosgene

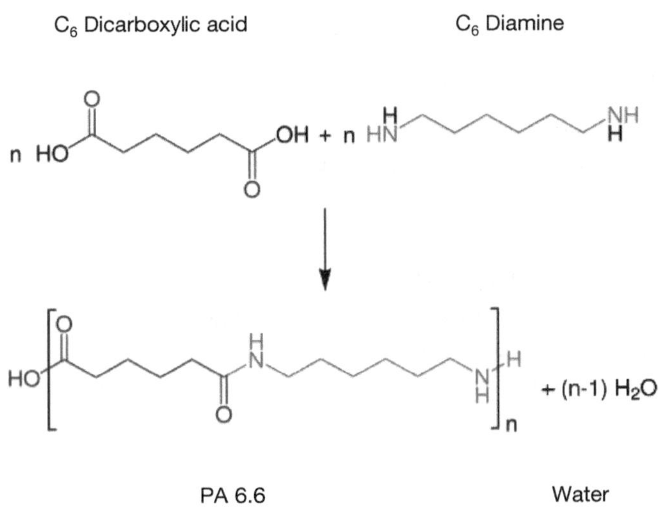

Figure 1.13 Formation of PA 6.6 by polycondensation reaction of C_6 dicarboxylic acid with C_6 diamine

Polyamides are characterized by the presence of repeating amide linkages (-CO–NH-) in their polymer structures. A similar way to form a polyamide (as described in Figure 1.13) is by ring-opening polymerization of caprolactam. The reaction product is known as polyamide PA 6 or Nylon® 6 (Figure 1.14).

Figure 1.14 Formation of PA 6 by ring-opening polymerization of caprolactam

While PA 6 is a homopolymer, PA 6.6 is an alternating copolymer with superior mechanical properties compared to PA 6 at the same molecular weight. The main differences between the two polymers are:

- Strength and stiffness: PA 6.6 has higher strength and stiffness than PA 6. This is due in part to PA 6.6's higher degree of crystallinity.
- Temperature resistance: PA 6.6 shows enhanced thermal stability at elevated temperatures and has a higher melting point compared to PA 6. This makes PA 6.6 a more advantageous material for high-temperature applications.
- Dimensional stability: PA 6.6 outperforms PA 6 in terms of thermal dimensional stability.
- Abrasion resistance: PA 6.6 has outstanding wear-resistance.
- Tensile strength: PA 6.6 has approximately 10% higher tensile strength than PA 6.
- Moisture absorption: PA 6.6 absorbs less moisture compared to PA 6, resulting in extended dimensional stability under humid conditions.

As well as PA 6 and PA 6.6, there are several other polyamides among the engineering plastics (see Figure 1.1).

Polyoxymethylenes (abbreviated as POMs and also known as polyacetals) are high molecular weight thermoplastics. These semi-crystalline polymers are mainly used to produce molded parts via the injection molding process. Due to their high rigidity, low friction, and good dimensional stability, polyoxymethylenes are used to produce precision parts. They are produced from either formaldehyde or trioxane (Figure 1.15).

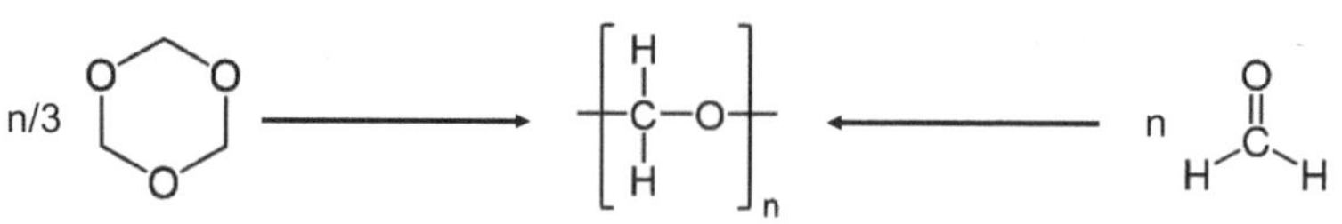

Figure 1.15 Formation of POM by polymerization of formaldehyde or trioxane

Let's take a look at a bioplastic, polylactic acid, also known as poly(lactic acid) or polylactide (PLA). PLA is produced by polycondensation of lactic acid or by ring-opening polymerization of the cyclic lactide formed from two lactic acid molecules (Figure 1.16).

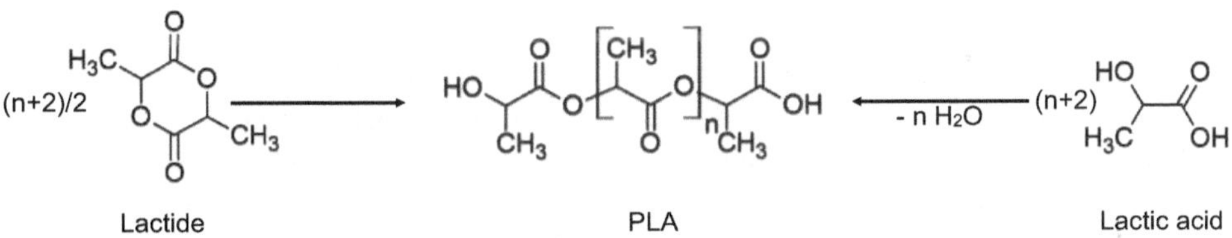

Figure 1.16 Formation of PLA by ring-opening polymerization of cyclic lactide or by polycondensation of lactic acid

Other polymerization methods are:

- *Cationic polymerization* is a type of chain growth polymerization, in which a cationic catalyst transfers charge to a monomer that then becomes reactive. This reactive monomer goes on to react similarly with other monomers to form the polymer. A typical example is the production of POM based on trioxane.
- *Anionic polymerization* is analogous to cationic polymerization, but in which the catalyst is anionic. Production of PA 6 is a typical example.
- *Coordination polymerization* involves the use of catalytic amounts of transition metal salts or complexes to form an active coordination complex between the monomer and the growing polymer chain end. A typical example is the polymerization of HDPE using Ziegler–Natta catalysts, which allows good control over the polymer's molar mass and tacticity.

Ziegler–Natta catalysts enable stereospecific polymerization of α-olefins, such as ethylene or propylene, through the use of mixed organometallic catalysts based on, for example, a titanium chloride and an alkylated aluminium (e.g., $TiCl_4$ + $Al(C_2H_5)_3$). Since the 1950s, this process has revolutionized mass production of polyolefins, and it continues to form the basis for >50% of global plastic production. Today the Ziegler–Natta systems use supported catalysts, such as magnesium chloride-supported titanium complexes, with higher catalytic activity ($>10^6$ g polymer per g catalyst) and more precise control of the polymer architecture. The chiral catalyst structure results in polymer chains with uniform monomer arrangement characterized by increased mechanical strength and thermal stability.[11] This method produces highly crystalline polymers with narrow molecular weight distribution, including HDPE and isotactic PP.

While the Ziegler–Natta catalysts generate stereoregular polymers like isotactic PP, chromium-containing catalysts result in non-stereospecific polymerization of ethene to HDPE. Chromium-catalyzed polymerization (the Phillips process) uses supported chromium oxide on SiO_2/Al_2O_3 and typically requires no co-catalysts such as aluminium alkyls, because ethene itself acts as a reducing agent. Unlike the Ziegler–Natta

[11] "Chiral" means that a molecule or an object cannot be superimposed on its mirror image, similar to how a left hand cannot be superimposed on a right hand. Typically, chiral carbon atoms are attached to four different substituent groups.

systems, this type of catalysis produces broader molecular weight distributions, with a spread of roughly 10–20 mass units around the average, enabling production of HDPE with high environmental stress crack resistance (ESCR). Process temperatures for Cr-containing catalysis are higher (80–150 °C) than that for Ziegler–Natta processes (40–90 °C), favoring gas-phase processes. Use of PP produced by the Phillips process dominates in blow-molding applications (bottles and pipes).

So-called metallocene catalysts enable highly precise control of olefin polymerization, where monomers are sequentially linked at a single active site. A metallocene catalyst refers to a type of organometallic compound, typically consisting of a transition metal (like zirconium or titanium) sandwiched between two cyclopentadienyl rings, which precisely controls the polymerization process. The catalytic activity of the metallocenes is 10–100 times higher than that of traditional Ziegler–Natta systems, with one gram of the zirconium-containing catalyst capable of polymerizing of up to 100 tons of ethylene. By modifying the metallocene complexes, catalysts can be tailored to the targeted co-polymerization of various olefins, allowing a wide range of polymer properties to be produced. The C_2-symmetric *ansa*-metallocenes (in which the two rings are connected by a bridging group) facilitate the production of highly stereoselective polymerized α-olefins, such as isotactic PP. Metallocene-catalyzed polymers exhibit unique properties, including high stiffness, adjustable melting points, and improved dimensional stability compared to the conventionally catalyzed polyolefins.

1.8 Polymer Solubility

The *Hildebrand solubility parameter* (HSP or δ), mentioned in Table 1.1, provides an estimate of the degree of interaction between materials, and can be used as a good indicator of mutual solubility particularly for non-polar materials, such as many polymers. In 1936 J. H. Hildebrand suggested using the square root of the cohesive energy density (CED) to indicate solubility behavior, and this parameter was later given his name. According to the mathematical model, materials with similar solubility parameters interact with each other, resulting in solvency, miscibility, or swelling. Materials with similar values of δ are likely to be miscible. According to the provided definition, δ is calculated as:

$$\sqrt{\mathrm{CED}} = \delta = \sqrt{\frac{\Delta H_\mathrm{V} - RT}{V_\mathrm{m}}} \tag{1.6}$$

where:

ΔH_v = heat of vaporization

V_m = molar volume

R = the gas constant (0.08206 $\mathrm{L{\cdot}atm{\cdot}K^{-1}{\cdot}mol^{-1}}$)

T = absolute temperature.

The principal utility of HSPs is that they predict the phase equilibrium based on a single characteristic that can readily be calculated for most materials. These predictions are often useful for both non-polar and low-polarity systems (with a dipole moment of <2 debyes) without hydrogen bonding. It was found particularly useful for predicting solubility and swelling of polymers in solvents. (More complicated three-dimensional solubility parameters, such as *solubility parameters by Hansen*, were proposed for polar molecules).

The principal limitation of the HSP approach is that it applies only to narrow range of substances, where "like dissolves like". Like any simple predictive theory, it is best used for screening and then cross-referencing the result with additional data. Even then, it is extremely difficult to find information about ΔH_v for most substances and polymers.

Grulke established the theoretical framework for this by determining contributions of different molecular groups to the cohesive energy density (F) (according to van Krevelen) (Table 1.3).[12] Based on these values and structural formulas, the sum of cohesive energy densities ($\sum F$) was calculated. Using the following equation in combination with the molecular weight (M) and density (ρ), Hildebrand solubility parameters (δ) can be calculated (Table 1.3):

$$\delta = \frac{\rho \sum F}{M} = \sqrt{\frac{E}{V_m}} \tag{1.7}$$

where:

$\sum F$ = sum of cohesive energy density F

M = molecular mass of the monomer

ρ = density of (amorphous) polymer

E = cohesive energy

V_m = molecular volume.

V_m can be calculated using the following equation:

$$V_m = M/\rho \tag{1.8}$$

[12] J. Brandrup, E. H. Immergut, and E. A. Grulke, *Polymer Handbook* (4th edition), Wiley, 1999, ch. VII, pp. 675–715.

Table 1.3 Contributions of Molecular Groups to Cohesive Energy Density (F)[12)]

Molecular group	Molecular weight (g/mol)	Contribution to the cohesive energy density F ($MPa^{0.5}$·mL/mol)
H	1.0	140
H	1.0	140
S	32.0	460
F	19.0	164
Cl (primary)	35.5	471
Br (primary)	79.9	695
C	12.0	0
CH_3	15.0	420
CH_2	14.0	280
CH	13.0	140
>$CH(CH_3)$	28.0	560
-$CH(CH_3)_2$	42.0	841
=CH-	13.0	222
=C<	12.0	82
-CH=CH-	26.0	444
-CH=C<	25.0	304
-C_6H_5 (phenyl)	77.0	1520
-C_6H_4- (*p*-phenylene)	76.0	1380
-OH (alkyl)	17.0	754
-O- (ether)	16.0	255
>C(O)	28.0	685
-C(O)–O-	44.0	511
-C(O)–OH	45.0	651
-O–C(O)–O-	60.0	767
-C(O)–NH-	43.0	1290

Molecular group	Molecular weight (g/mol)	Contribution to the cohesive energy density F ($MPa^{0.5}$·mL/mol)
-CH-CN	40.0	1120
-CN	27.0	982

Exercise 1.1

Let us calculate the HSP of PE in two different ways. The molar volume V_m of ethylene is 32.9 mL/mol and its cohesive energy E (according to van Krevelen) is 9500 J/mol. This gives:

$$\text{HSP(PE)} = (E/V_\text{m})^{1/2} = \left(\frac{9500\ \text{J/mol}}{32.9\ \text{mL/mol}}\right)^{1/2} = 16.99\ \text{MPa}^{1/2}$$

According to Table 1.3, ethylene has two CH_2 groups, each with a contribution to the cohesive energy density F of 280 $MPa^{0.5}$·mL/mol. The monomer molecular weight is 28 g/mol and the amorphous polymer density is 0.85 g/mL:

$$\text{HSP(PE)} = \frac{\rho \sum F}{M} = \frac{0.85\ \text{g/mL} \cdot 2 \cdot 280\ \text{MPa}^{0.5}\text{mL/mol}}{28\ \text{g/mol}} = 17.00\ \text{MPa}^{0.5}$$

Exercise 1.2

Repeat Exercise 1.1 for PVC. From Table 1.3, vinyl chloride monomer (VCM) has one CH_2 group (with a contribution to the cohesive energy density F of 280 $MPa^{0.5}$·mL/mol), one CH group (140 $MPa^{0.5}$·mL/mol) and one Cl atom (471 $MPa^{0.5}$·mL/mol). This results in a CED of 891 $MPa^{0.5}$·mL/mol. The monomer molecular weight is 62.5 g/mol. Furthermore, PVC has a density of 1.38 g/mL. This results in an HSP of 19.67 $MPa^{0.5}$.

Calculating the HSP for copolymers and terpolymers is possible if the weight ratio of the monomers is known. The first step is to convert the weight percentages to the mole percentages and calculate the density of the copolymers. HSP can then be determined from the molecular groups using the normal procedure. In Figure 1.17, HSP of EVA was plotted as a function of the vinyl acetate concentration expressed in weight percentages. As can be seen, this is not a perfectly linear relationship but it is close to it based on the established correlation, enabling estimation of HSP of the copolymers to a first approximation. For an EVA with 50% vinyl acetate, the estimated value is 17.76 $MPa^{1/2}$ and the calculated value is 17.71 $MPa^{1/2}$.

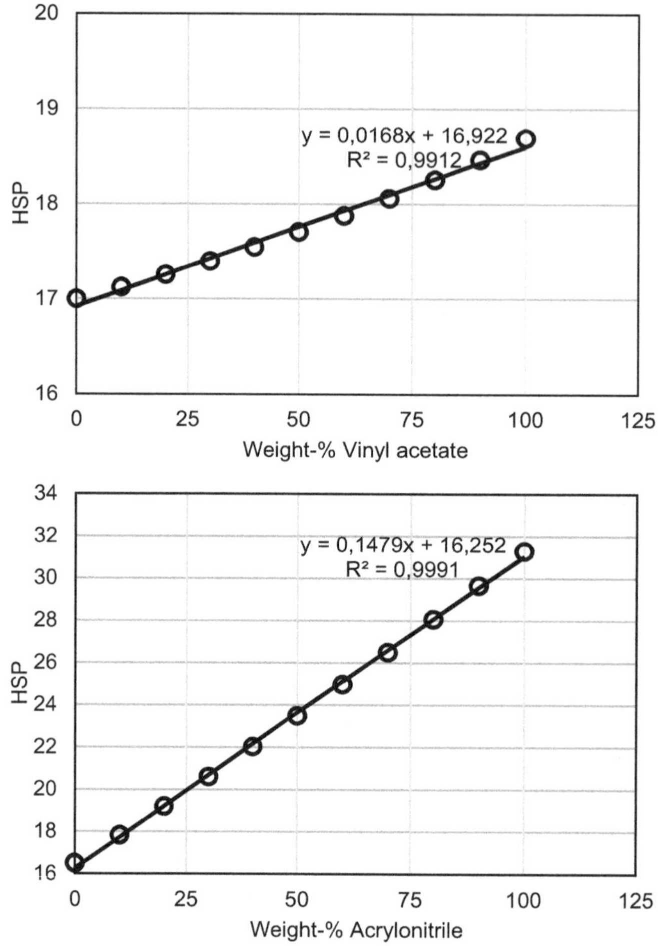

Figure 1.17 Dependency of HSP (expressed in $MPa^{0.5}$·mL/mol) of EVAs on the weight percentage of vinyl acetate (top) and NBRs on the weight percent of acrylonitrile (bottom)

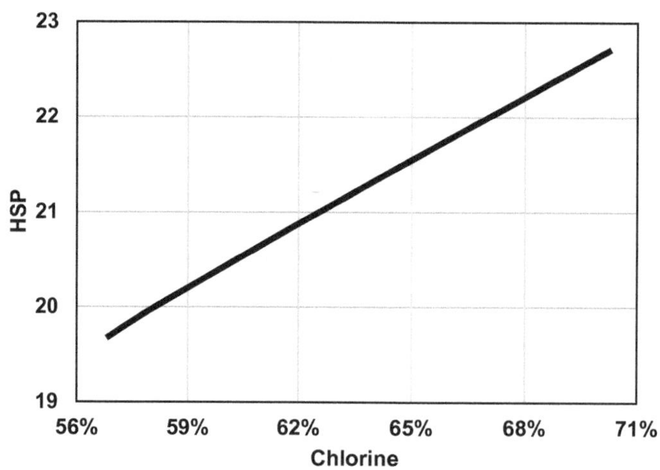

Figure 1.18 Dependence of HSP (in $MPa^{0.5}$·mL/mol) of C-PVC on the chlorine content

Analogous estimates can be made for other copolymers. The situation is more complicated for chlorinated PVC (C-PVC), because many different material grades with different densities and chlorine contents are available on the market. Figure 1.18 shows the dependence of HSP of C-PVC on the chlorine content.

1.9 Polymer Aging

Turning back to our train analogy, another characteristic that our train has in common with a plastic article is that both of them age with use. Most polymers must be mixed with additives from the outset (during pre-processing), so that they can be processed successfully and offer optimum properties during their service life for the application, as well as (hopefully) being recycled or reused. Before looking at possible polymer degradation mechanisms, let's review the methods that are used for converting polymer resins into plastic articles.

1.10 Processing Methods

Let's now review the methods that are used for converting polymer resins into plastic articles.

Extruders are widely adopted in the plastic industry, and have been used in the production of rubber since before 1900. The resin and other components are placed in the "barrel", with a nominal diameter almost equal to the outside diameter of the screw, and pushed through the unit by an extruder screw. The shaping outlet opening (die, tooling, tool) is located at the front end of the screw cylinder. At the rear end of the cylinder there is a drive, in most cases an electric motor with a gear box that ensures rotation of the screw. The materials to be processed are usually fed continuously.

The extruder screw is generally divided into several zones that perform different tasks. The "feeding zone" is located in the rear area of the screw cylinder. In this zone the material to be extruded (a powder or granulate or powder) is fed in via a hopper (funnel). The material is conveyed by the rotation of the screw. In certain machine designs, there is a so-called vacuum zone, in which the polymer mass is neither powdery nor fully fused, and where the necessary removal of gases, moisture, and other volatile components takes place. The material is further compressed in the compression zone. Finally, the discharge zone or metering zone ensures that the homogeneous material flows towards the die.

Extruders may have one, two, or more screws. In both the single-screw extruder and (co-rotating) twin-screw extruder, conveying and pressure build-up are brought about by the friction between the mass rotating with the screw against the standing housing wall (cylinder). This is referred to as friction promotion. The mass that remains in the rotation is pushed from the helical screw flights to the die. The principle of forced conveyance prevails in the counter-rotating twin-screw extruder. For compounding, other extruders might be used, such as a planetary screw extruder, for example.

In the extruders, molders, calenders, and other types of polymer processing equipment, the polymer is subjected to certain in-process shear and elevated temperatures (Table 1.4).

Table 1.4 Typical Processing Temperatures for Selected Polymers[13]

Polymer	Processing temperature (°C)
ABS	200–275
ASA	220–260
CA	180–220
CAB	180–220
CP	190–230
EVA	130–240
HPE	180–250
LDPE	160–220
PA 11	200–250
PA 12	190–270
PA 6	240–290
PA 6.10	220–260
PA 6.6	260–300
PA 6.3 T	250–320
PAI	330–380
PB	220–290
PBT	230–280
PC	270–320
PEEK	350–390
PEI	340–425

[13] E. Baur, T. A. Osswald, and N. Rudolph, *Plastics Handbook: The Resource for Plastics Engineers* (5th edition), Hanser, 2019, pp. 626–627.

Polymer	Processing temperature (°C)
PEK	360–420
PES	320–390
PET	260–300
PFA	380–400
PIB	150–250
PMMA	190–270
PMP	280–310
POM	180–230
PP	200–270
PPS	320–380
pPVC	160–190
PS	170–280
PSU	340–390
PTFE	320–360
PVDF	250–270
SAN	200–260
SB	190–280
SBS	175–250
TPA	170–230
TPC	160–220
TPO	180–200
TPS	180–220
TPU	180–220
uPVC	170–210

1.11 Polymer Degradation

Combinations of the high shear and heat affect polymers and may potentially degrade (or decompose) them, or otherwise cause by-products to be generated, resulting in discoloration and deterioration of their mechanical properties (degradation reactions of selected polymers, such as PVC, chlorinated PVC, and PVA are discussed in other chapters). The polymer chain degradation reaction starts as follows:

$$R–H \xrightarrow{\text{Shear or heat}} R^{\cdot} + H^{\cdot}$$

$$R–R \xrightarrow{\text{Shear or heat}} 2\,R^{\cdot}$$

During melt processing, polymers (which usually contain hydrocarbon groups and are depicted as R–H) can form reactive carbon-centered radicals (so-called macro-alkyl radicals) $R^{\cdot}$ when exposed to heat, with or without shear. Long-term thermal stress can also have the same result.

The initiation phase is followed by the propagation phase, during which olefinic and oxygen-containing by-products are formed, leading to degradation and branching of the polymer chains:

$$R^{\cdot} + {}^{\cdot}O–O^{\cdot} \rightarrow R–O–O^{\cdot}$$

$$R–O–O^{\cdot} + R–H \rightarrow R–O–O–H + R^{\cdot}$$

$$R–O–O–H \rightarrow R–O^{\cdot} + {}^{\cdot}O–H$$

$$R–O^{\cdot} + R–H \rightarrow R–O–H + R^{\cdot}$$

$${}^{\cdot}O–H + R–H \rightarrow H–O–H + R^{\cdot}$$

$$R^{\cdot} + \text{-}CH_2{=}CH_2\text{-} \rightarrow R–CH_2{=}CH_2^{\cdot}$$

$$2\,R–O–O–H \rightarrow R–O^{\cdot} + R–O–O^{\cdot} + H–O–H$$

The alkyl radicals $R^{\cdot}$ react with molecular oxygen dissolved in the polymer to form peroxy radicals $R–O–O^{\cdot}$. Proton abstraction by the peroxy radicals results in hydroperoxides. The decomposition of hydroperoxides R–O–O–H into alkoxy ($R–O^{\cdot}$) and hydroxy (${}^{\cdot}O–H$) radicals requires considerable activation energy, and these types of reactions can also be catalyzed by UV light or metal ions. The $R–O^{\cdot}$ and ${}^{\cdot}O–H$ radicals are extremely reactive and can in turn react with hydrogen atoms of polymers (R–H).

Chain growth and chain branching are followed by chain termination reactions, usually via recombination of two radicals:

$$R^{\cdot} + R–O–O^{\cdot} \rightarrow R–O–O–R$$

$$2\,R–O^{\cdot} \rightarrow R–O–O–R$$

$$2\,R^{\cdot} \rightarrow R–R$$

$$R^{\cdot} + R–O^{\cdot} \rightarrow R–O–R$$

$$2\,R^{\cdot} \rightarrow R–H + \text{olefin}$$

There are several other types of reactions that can generally take place as a result of polymer degradation.

Primary and secondary peroxy radicals are more effective than the analogous tertiary peroxy radicals with regard to their hydrogen abstraction activity.

The oxidative autocatalytic degradation reaction is slow at the start (during the induction period), and it accelerates as the concentration of the resulting hydroperoxides increases (Figure 1.19).

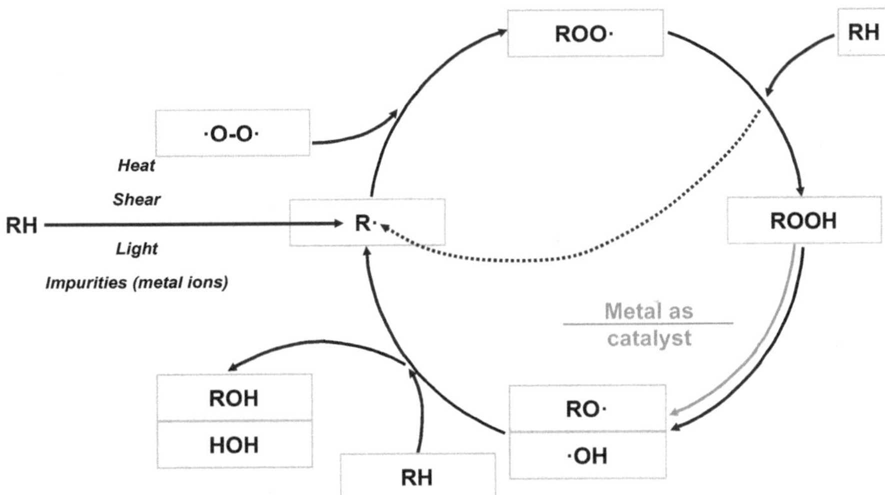

Figure 1.19 Free radical mechanism of polymer degradation (autoxidation)

The key chain-propagating step responsible for backbone cleavage in oxidized polymers is the β-scission of an alkoxy radical on the macromolecule, resulting in the formation of an alkyl radical:

$$R'\text{–}C(O^{\bullet})R''\text{–}R''' \xrightarrow{\beta-\text{scission}} R'\text{–}C(O)\text{–}R''' + R''^{\bullet}$$

During melt processing operations, such as extrusion, dissolved oxygen in the polymer rapidly reacts, so that under the resulting low-oxygen conditions the concentration of alkyl radicals $R^{\bullet}$ typically exceeds that of peroxy radicals $ROO^{\bullet}$. Alkyl radicals may also be generated mechanically by the high shear stresses imparted by the processing equipment.

In addition to β-scission, fragmentation of macroalkyl radicals provides another route by which the average molecular weight of the polymer can be decreased. By contrast, disproportionation of two small alkyl radicals leaves the molecular weight of the chains unchanged, whereas recombination of two macroalkyl radicals bonds the chains together and therefore increases polymer molecular weight.

The ability of a polymer to withstand thermo-oxidative deterioration is generally controlled by preventing or minimizing the peroxide formation. Thus, the oxidative stability of the polymer is linked to the rate of hydrogen abstraction from a C-atom in either the main chain (polymer backbone) or its side group (branch).

The highest temperature at which a particular polymer can be processed is limited by its thermal degradation.

We have seen that radicals (more precisely, free radicals) are atoms, molecules or ions that have an unpaired valence electron. Typical examples are:

- Oxygen, which is a biradical: $^{\bullet}O{-}O^{\bullet}$
- Alkyl radicals, e.g. polymer fragments formed by shear, temperature, and UV light: $R^{\bullet}$
- Alkylperoxy radicals, e.g. formed by reaction of $R^{\bullet}$ with oxygen: $R{-}O{-}O^{\bullet}$
- Peroxy hydroxyl radicals: $H{-}O{-}O^{\bullet}$
- Alkoxy radicals: $R{-}O^{\bullet}$
- Hydroxyl radicals: $H{-}O^{\bullet}$.

The discussed mechanism of polymer degradation, also called autoxidation, can be represented as a cyclic process (Figure 1.19). With each subsequent cycle, additional free radicals are generated that can in turn lead to further degradation of the polymer.

In order to delay and/or slow down the polymer degradation processes or, in the best case, to prevent them, the incorporation of appropriate polymer additives is necessary (Figure 1.20). Examples of these additives include:

- Antioxidants that capture (scavenge) and deactivate free radicals
- Substances that decompose organic peroxides
- Absorbers of UV light that prevent generation of free radicals under irradiation
- Metal deactivators that form complexes with metal ions.

All of these types of additives are discussed in detail in the following chapters.

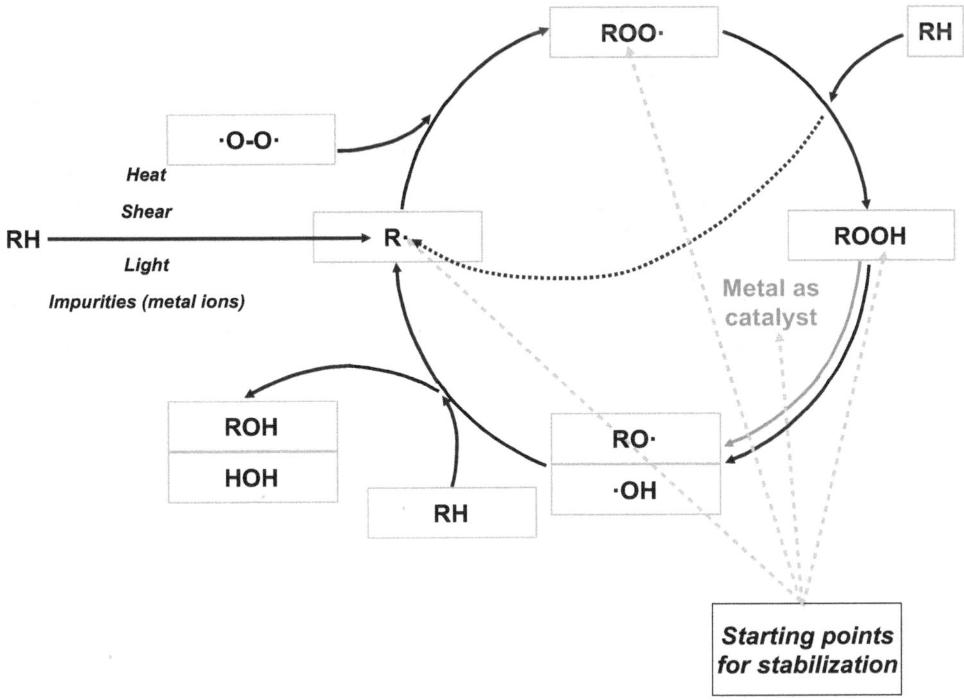

Figure 1.20 Options for stabilizing thermo-oxidative radical degradation of polymers

Plastic materials can also be degraded by the environment during their post-production service life. Free radicals often play a role in this. This type of polymer degradation usually takes place in the absence of polymer-stabilizing additives via free radical mechanisms, and can be initiated by various factors, such as radiation (e.g., UV light), solar heat, air oxygen, moisture of humid air, and mechanical stress (stretching, compression, bending, etc.) (Figure 1.20). The reaction pathways and whether the free radicals react with the polymer or with the polymer-stabilizing additives depend on the energy of competing reactions. According to the Second Law of Thermodynamics, chemical reactions occur if energy is released during the process. Using information included in Table 1.5, the reaction energy can be calculated as:

Reaction energy = Energy of bond cleavage – Energy of bond formation

Table 1.5 Energies (in kJ/mol) of Homolytic Bond Cleavage or Formation[14)]

	H	CH_3	CH_2CH_3	$CH(CH_3)_2$	$C(CH_3)_3$	C_6H_5	$C_6H_5CH_2$	NH_2	$NHCH_3$	$N(CH_3)_2$	OH	F	Cl	Br	Source
H	436	435	410	398	385	461	356	460	431	398	498	569	432	366	17a
CH_3	435	369	355	353	343	417	301	363	350	326	381	456	350	292	17a
CH_3CH_2	410	355	343	337	326	404	289	353	345	320	381	448	338	283	17a
$(CH_3)_2CH$	398	352	337	328	311	397	284	356	348	323	384	442	337	285	17a
$(CH_3)_3C$	385	343	325	311	288	379	272	350	338	314	382	454	336	277	17a
F	569	456	448	442	454	—	—	—	—	—	—	159	255	—	17a
Cl	432	350	338	337	336	—	—	—	—	—	251	255	240	—	17a
Br	366	292	283	285	277	—	—	—	—	—	234	—	—	—	17a
HO	498	381	381	384	382	—	—	—	—	—	213	—	251	—	17a
C_6H_5	461	417	404	397	379	494	329	—	—	—	459	519	396	234	17a
$C_6H_5CH_2$	356	301	289	284	272	329	—	—	—	—	326	403	291	332	17a
CH_3O	440	341	338	341	338	—	—	—	—	—	181	—	—	229	17a
H_2N	460	363	353	356	350	—	—	251	—	—	—	—	—	—	17a
CH_3NH	431	350	345	348	338	—	—	235	193	—	—	—	—	—	17a
$(CH_3)_2N$	398	326	320	323	314	—	—	—	—	180	—	—	—	—	17a

[14] (a) *Photochemie* (2nd edition), VEB Deutscher Verlag der Wissenschaften, 1975, p. 495; (b) M. Lucarini, P. Pedrielli, G. F. Pedulli, S. Cabiddu, and C. Fattuoni, Bond dissociation energies of O–H bonds in substituted phenols from equilibration studies, *The Journal of Organic Chemistry*, 1996, 61: 9259–9263 (Figure 2 also available at *https://www.researchgate.net/figure/Experimental-bond-dissociation-energy-values-for-substituted-phenols_fig1_280780699*).

	H	CH_3	CH_2CH_3	$CH(CH_3)_2$	$C(CH_3)_3$	C_6H_5	$C_6H_5CH_2$	NH_2	$NHCH_3$	$N(CH_3)_2$	OH	F	Cl	Br	Source
CH_3CO	368	343	—	—	—	—	—	—	—	—	456	498	352	—	17a
Cl_3C	396	—	—	—	—	—	—	—	—	—	—	—	302	—	17a
Cl_2CH	414	—	—	—	—	—	—	—	—	—	—	—	332	—	17a
RO	430	—	—	—	—	—	—	—	—	—	—	—	—	—	17a
RS	370	—	—	—	—	—	—	—	—	—	—	—	—	—	17a
R_2N	360	—	—	—	—	—	—	—	—	—	—	—	—	—	17a
C_6H_5O	370	—	—	—	—	—	—	—	—	—	—	—	—	—	17b
***p*-CH_3–C_6H_4O**	361	—	—	—	—	—	—	—	—	—	—	—	—	—	17b
***p*-$(CH_3)_3C$–C_6H_4O**	357	—	—	—	—	—	—	—	—	—	—	—	—	—	17b
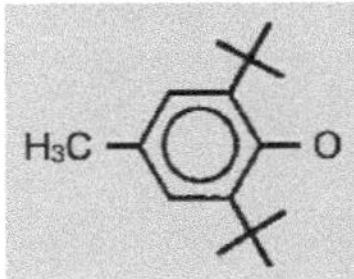	339	—	—	—	—	—	—	—	—	—	—	—	—	—	17b

As Figure 1.19 shows, free radicals can be generated not only by the in-process shear and heat, but also by UV light. The prerequisite for this is that the light is absorbed according to the Grotthus–Draper law (see Section 3.2). Additionally, free radicals can also be formed by absorption of compounds containing a C=C or C=O bond on titanium dioxide. This phenomenon often takes place during the service life of a plastic article.

Exercise 1.3

In Figure 1.21, under what conditions does the bond dissociation proceed in PP? Let us find out which H atom of PP is able to be attacked by the RO˙ radical. If RO˙ abstracts an H atom, 430 kJ/mol is released. So, it is unlikely that the H would be abstracted from the methyl group CH_3, because this would require 435 kJ/mol. The most probable H abstraction would be from the isopropyl ($(CH_3)_2CH$) group, and that would release 32 kJ/mol. H abstraction is also possible from the ethyl group (CH_3CH_2).

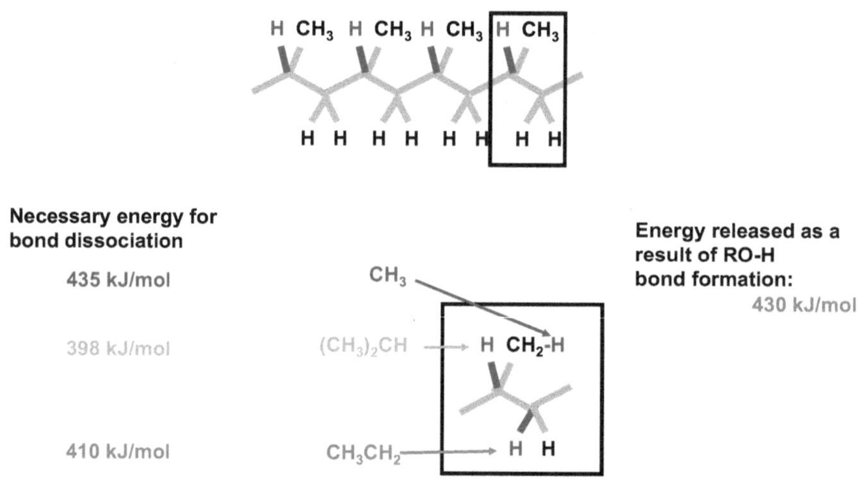

Figure 1.21 H abstraction from PP

Exercise 1.4

What rearrangement is likely in PS (Figure 1.22).Let's get back to what happens in the extruder during processing. As can be seen from Table 1.5, significantly less energy is needed to cleave a C–C single bond than a C–H bond. Therefore, during extrusion, mainly polymer backbone carbon chains are cleaved. Presence of the shorter polymer chains translates into a reduction in viscosity of the polymer melt in most cases.

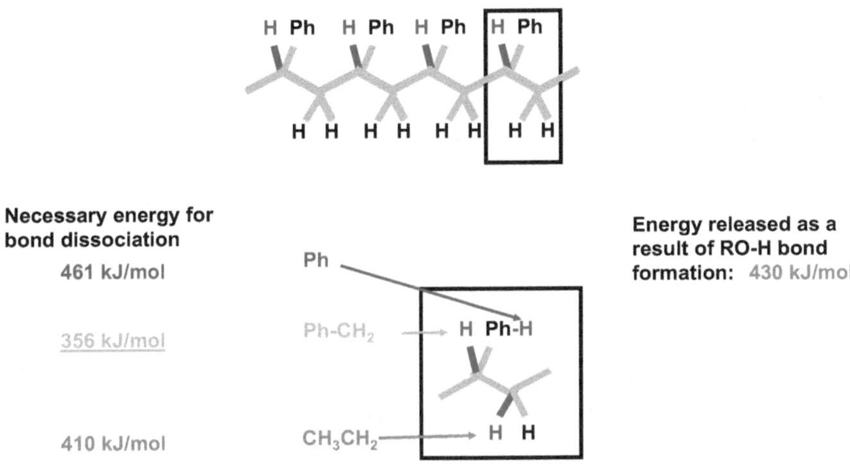

Figure 1.22 H abstraction from PS

As the temperature increases, the MFI increases (Figure 1.23), and thus both the average molecular weight and the viscosity of the polymer decrease.

In addition to the free radical degradation mechanisms, ionic or combined ionic–free radical mechanisms can operate during the degradation of polymers including PVC, chlorinated PE, chlorinated PVC, polyvinylidene chloride (PVDC), and PVA, under high heat and in-process shear. If polymerization of PVC progressed ideally and resulted in the stereo-regular linear polymer chains without any defects, processing of PVC resins would not require the addition of any heat stabilizers. In reality, during polymerization, defects are formed on the polymer backbone and reduce its thermal stability. The most critical structural defects formed on PVC are tertiary chlorine atoms that result from polymer branching, and allylic chlorines that are formed in part via the termination of the polymerization reaction (Figure 1.24). The actual PVC degradation mechanism is more complex.

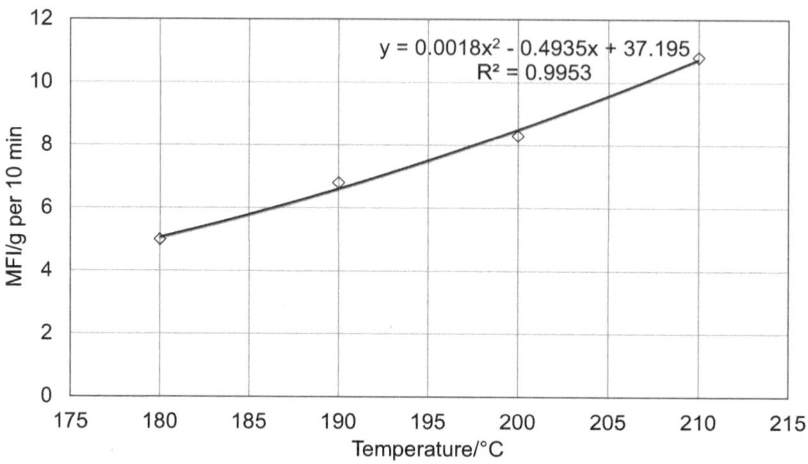

Figure 1.23 The melt flow index (MFI)[15] of untreated LDPE at different temperatures[16]

If a stabilizer is not added, the degradation once started follows the zipper-chain elimination mechanism (Figure 1.25).

All other irreversible processes that do not lead to any material changes are covered under the term *physical aging*. These include certain changes in the polymer structure, the molecular order/packing, and the concentration ratio of the components (in multi-component systems). Typical physical aging processes are:

- Post-crystallization, accompanied by lowering of T_g (volume contraction)
- Reduction of internal stress and disorientation
- Plasticizer loss, migration, and extraction
- Agglomerations, such as blooming and exudation.

Our journey into the world of plastics additives has begun, and will continue through the following chapters. The authors hope that readers are now well-positioned to appreciate how even small amounts of additives can substantially improve the performance of polymers in a variety of processes and applications.

[15] MFI is a measure of the ease with which the thermoplastic polymer melt flows. It is defined as the mass of polymer expressed in grams that flows within 10 min through a capillary of a specific diameter and length by pressure applied using the defined weights at the defined temperature. MFI is an indirect measure of molecular weight, with a high MFI corresponding to low molecular weight. At the same time, MFI is a measure of a material melt's ability to flow under pressure. It is inversely proportional to the viscosity of the melt at the conditions of the test; it should be kept in mind that the viscosity for any such material depends on the applied force: MFI $\propto 1/\eta$.

[16] S. Husseinsyah, B. Y. Lim, and P. L. Teh, Rheological and thermal properties of palm kernel shell-filled low-density polyethylene composites with acrylic acid. *Journal of Thermoplastic Composite Materials*, 2013, 26: 1155–1167.

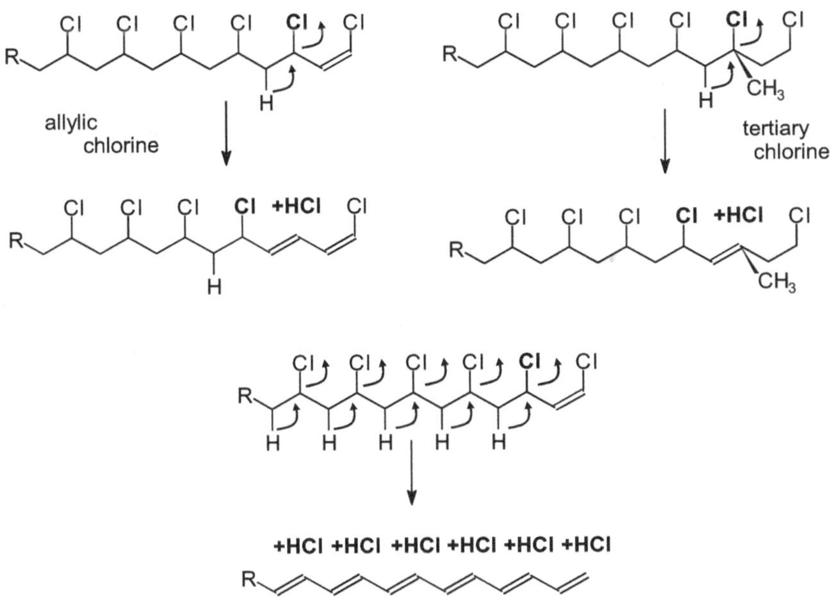

Figure 1.24 Degradation of PVC at the allylic chlorine and tertiary carbon atom defects due to exposure to heat and/or shear

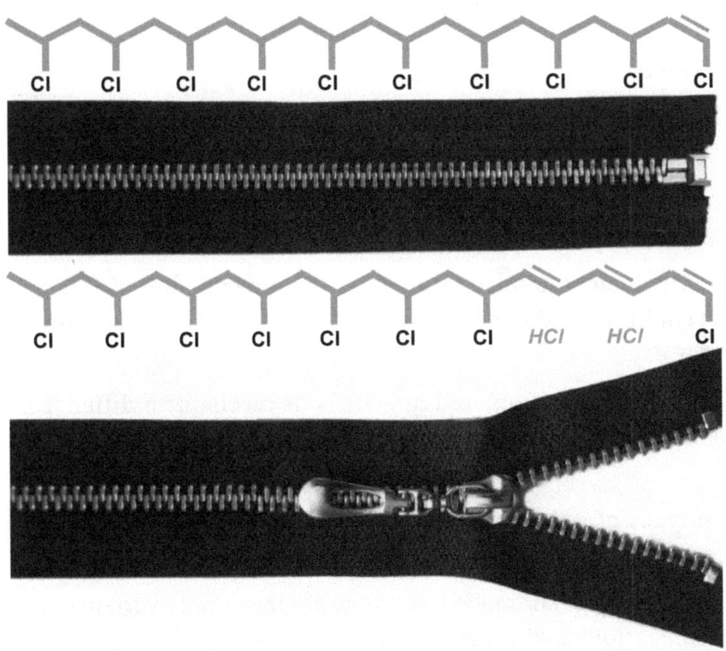

Figure 1.25 "Zipper-chain" elimination mechanism of PVC degradation. Image credit: Mahsa Saffaripour

1.12 Questions

1. What is the definition of a monomer in polymer chemistry, and what structural elements must a molecule possess to act as a monomer?
2. What is the definition of a polymer? Provide examples.
3. How are polymer molecular weights defined and measured, and why do polymers exhibit a range of molecular weights rather than a single value?
4. What is polymer crystallinity, and what structural and processing factors influence the degree of crystallinity in semicrystalline polymers?
5. What does polymer dispersity describe, and how is it quantified by the the parameter Đ?
6. How does the glass transition temperature T_g of a polymer influence its mechanical properties and processing characteristics?
7. What factors determine the minimum processing temperature of a thermoplastic polymer, and how is this temperature related to its glass transition and melting temperatures?
8. How does the average molecular weight of a polymer correlate with its mechanical properties, and what characteristics are typically improved with increasing chain length of polymers?
9. What are the three types of polymer tacticity, and how does tacticity affect physical properties of a polymer and its ability to crystallize?
10. How are copolymers defined? How do the composition and arrangement of different monomer units in a copolymer affect material properties?
11. How are crosslinked polymers defined? How does the introduction of covalent crosslinks between chains alter the mechanical and thermal properties of linear or branched polymers?
12. What is a polycondensation reaction, and how does its mechanism differ from chain-growth polymerization in terms of reaction kinetics and by-product formation?
13. What are the main types of polymerization catalysts, and how do they control polymer microstructure and molecular weight distribution?
14. What are Hildebrand solubility parameters, and how are they used to predict polymer solubility and compatibility with solvents or other polymers?
15. What are free radicals in the context of polymer chemistry, and what role do they play in polymerization and polymer degradation processes?

16. What methods are commonly used for generating free radicals used for initiating radical polymerization reactions? How does the choice of initiators and process conditions influence the characteristics of the resulting polymer?
17. How are free radicals formed during polymer degradation, and what factors promote their generation?
18. How does the melt flow index (MFI) correlate with polymer molecular weight, and what is the reason for this correlation?
19. How does the MFI relate to polymer melt viscosity, and what does a change in MFI indicate about the processability of a polymer melt?

2 Antioxidants

2.1 Terms and Definitions

Free radicals	Atoms or molecules with unpaired electrons, making them highly reactive and capable of initiating chain chemical reactions.
Thermal degradation of polymers	Breakdown of polymer chains due to heat, often resulting in reduced molecular weight, discoloration, and property loss.
Thermal stabilization of polymers	Use of stabilizers or additives to protect polymers from heat-induced degradation during processing or use.
Melt processing of polymers	Forming and shaping polymers by heating them to a molten state for extrusion, molding, or other processes.
Oven aging of polymers	Accelerated testing method in which polymers are exposed to elevated temperatures over time to assess long-term stability.

2.2 Introduction

When polymers undergo oxidation, their strength and toughness decrease, and the articles made from these materials often show surface roughness and changes in color. Oxidative processes may take place at any step of the polymer production, including resin manufacturing and storage, melt processing and shaping, and later during service life of the finished parts. Because chemical structure varies widely among different plastics, their intrinsic resistance to oxidative attack can differ greatly.

In material science, oxidation of polymers is commonly treated as a form of aging, and the accompanying changes in molecular structure are grouped under the term "degradation". Such aging and degradation processes can be slowed down, and this chapter focuses on approaches that are used to delay these reactions.

One possible strategy is to alter the polymer backbone or its composition to make it less reactive toward oxygen, but in industrial practice the more usual approach for reducing the polymer oxidation is to incorporate additives that are classified as antioxidants. These substances interrupt or suppress the reaction pathways responsible for thermo-oxidative damage, so that the polymer is "better equipped" to withstand heat and oxygen during its processing and long-term use.

Accordingly, this chapter examines how plastics behave when exposed to combined thermal and mechanical loads, as well as during long-term exposure to heat and oxygen, typically described as thermo-oxidative or long-term heat aging.

2.3 Oxidative Degradation and Autoxidation

During thermo-oxidative breakdown of polymers, a variety of oxygen-containing groups can be formed on the polymer chains, including hydroperoxides, alcohols, ketones, aldehydes, and carboxylic acids and related derivatives (such as peracids, peresters, and lactones). Higher temperatures, irradiation, and the presence of catalytic metal ions may significantly accelerate these oxidation reactions, affecting polymers that contain structural units that are especially vulnerable to such attacks.

Thermoplastics are obtained via chain-growth polymerization, polyaddition, or polycondensation (Chapter 1) and are then transformed into articles during melt processing operations. During these steps the material is exposed to heat and shear, and the resulting articles are often intended for long-term use with as little change as possible in mechanical performance, appearance, and other attributes. Throughout their service life, however, plastic articles are often continuously subjected to oxygen, heat, UV radiation, and moisture, so that chain scission, branching, and crosslinking reactions can occur through the oxidation reactions of the macromolecules. Knowledge of these oxidative degradation pathways and strategies to suppress them is therefore crucial for ensuring the durability of plastic materials.

Aging of polymers can also be affected by the presence of other additives, such as fillers, pigments, and reinforcing fibers.

The spontaneous reaction of organic substances with molecular oxygen under ambient conditions is termed *autoxidation*, reflecting the fact that it often proceeds without external source of heating, radiation, or ignition. Autoxidation can be slowed or interrupted by incorporation of specifically designed stabilizers called antioxidants. On the molecular level it follows a free-radical chain mechanism with distinct phases of initiation, propagation, and termination, during which radicals are formed, consumed, and finally deactivated. Details about the chemical mechanisms of oxidation processes and typical processing temperatures of plastics (Table 1.4) can be found in Chapter 1.

Alkyl radical formation during polymer processing can also result from high shear forces in the processing machines (Figure 2.1).

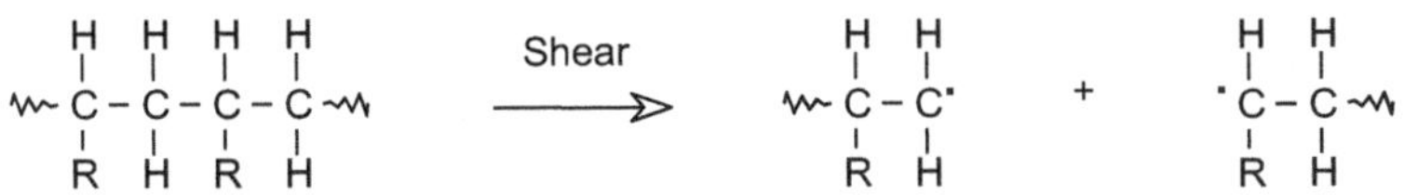

Figure 2.1 Alkyl radical formation during polymer processing

2.4 Inhibited Autoxidation

Thermo-oxidative breakdown of polymers can be limited by incorporating suitable stabilizing additives, known as antioxidants, at low loadings, generally not exceeding about 2 wt%.

Using radical scavengers for trapping free carbon-centered radicals is the main commercial option for suppressing the autoxidation. Molecular oxygen reacts very rapidly with carbon-centered radicals ($R^{\cdot}$). Under conditions where oxygen is scarce, species that trap these carbon-centered radicals act as highly effective antioxidants. There are currently no well-established methods for directly scavenging $RO^{\cdot}$ or $HO^{\cdot}$ radicals, whereas peroxy radicals ($ROO^{\cdot}$) can be intercepted by hydrogen-donor molecules to give comparatively stable hydroperoxides (see Section 2.5.1).

Radical-trapping species are commonly described as chain-breaking acceptors (CB-As), whereas hydrogen-donating species are termed chain-breaking donors (CB-Ds). To prevent further chain growth via branching reactions that arise from thermally generated alkoxy and hydroxyl radicals, so-called hydroperoxide decomposers (HDs) are employed, which convert hydroperoxides into non-reactive moieties (see Section 2.5.2). The CB-A- and CB-D-type stabilizers are usually grouped under the term *primary antioxidants*, while hydroperoxide decomposers are classified as *secondary antioxidants*.

Commonly, polymers that are in direct contact with metals, e.g., copper in cable insulation or fittings, degrade more readily. In these applications, the so-called metal deactivators (MDs) may significantly improve the service life of the end products (see Section 2.5.4).

Antioxidants are formulated to safeguard plastics against oxidative damage during processing, shaping, and service life under the relevant environmental conditions. To maintain this protective effect throughout the article's lifetime, any physical loss of the stabilizer from the polymer matrix should be kept as low as possible.

Antioxidants used for long-term heat exposure applications generally have molecular weights >700 g/mol. How well an antioxidant performs during processing depends on its ability to diffuse through the viscous polymer melt and reach sites that are under the oxidative attack in the presence of the formed free radicals. Their key performance characteristics include good compatibility and solubility in the solid polymer, low volatility, and resistance to being extracted into the surrounding environment.

2.5 Antioxidants: Mechanisms of Action

2.5.1 H-donors

In the classical autoxidation sequence, the slow, rate-controlling step is the abstraction of a hydrogen atom from the polymer backbone by a peroxy radical. It results in generation of a relatively stable, polymer-bound hydroperoxide. If an added hydrogen donor, InH, provides a more easily abstractable hydrogen, the peroxy radical preferentially reacts with InH, so that hydrogen abstraction from the polymer chain is effectively suppressed or even prevented until the external donor is fully consumed ($k_2 \gg k_1$, Figure 2.2). Suitable H-donors do not react further by abstracting a hydrogen atom from the polymer backbone ($k_3 \ll k_2$).

$$\text{R–O–O}^{\cdot} + \text{R–H} \xrightarrow{k_1} \text{R–O–O–H} + R^{\cdot}$$
$$\text{R–O–O}^{\cdot} + \text{In–H} \xrightarrow{k_2} \text{R–O–O–H} + \text{In}^{\cdot}$$
$$\text{In}^{\cdot} + \text{R–H} \xrightarrow{k_3} \text{In–H} + \text{R}^{\cdot}$$

Figure 2.2
Reactions of H-donors

2.5.1.1 Aromatic Amines

Secondary aromatic amines and sterically hindered phenols act as very effective hydrogen-donor antioxidants and have a long history of use as such. Secondary aromatic amines are known to discolor and stain the end articles, and so their use in plastics is largely confined to dark, carbon-black-filled rubber compositions and certain polyurethane systems. In addition, most aromatic amine derivatives are not acceptable for indirect food-contact applications, such as packaging materials. Their function as hydrogen-donor antioxidants is outlined schematically in Figure 2.3.

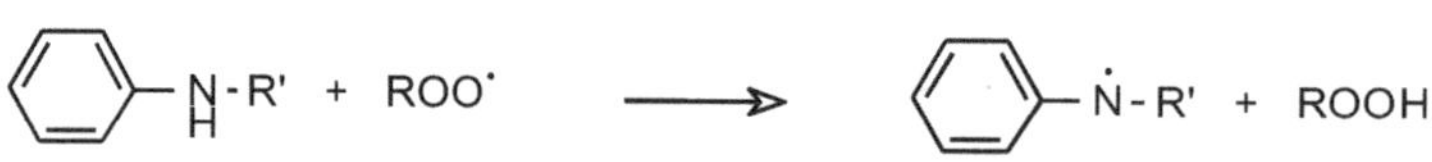

Figure 2.3 Mechanism of stabilization with aromatic amines

Hydrogen abstraction in these systems produces aminyl radicals and related resonance-stabilized carbon-centered radicals, which can then undergo N–N, N–C, or C–C coupling, as shown in Figure 2.4. These coupled by-products retain antioxidant activity and can be converted into nitroxides or benzoquinone monoimides, both of which can further participate in the trapping of carbon-centered radicals.

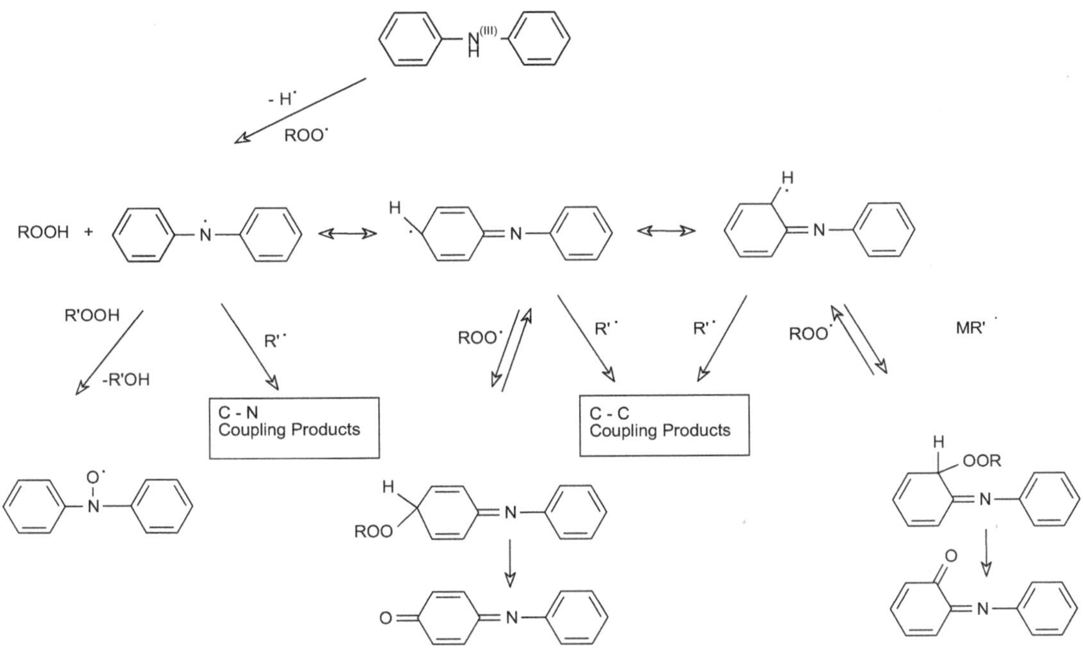

Figure 2.4 Reactions of aromatic amines

2.5.1.2 Phenols

Phenolic antioxidants that act as hydrogen donors are among the most effective and commonly used stabilizers in polymeric materials. In their key functionality, they transfer a hydrogen atom from the phenolic hydroxyl group to a radical species, thereby generating a hydroperoxide and a corresponding phenoxy radical, as shown in Figure 2.5. The stability of the phenoxy radical is enabled by steric hindrance from the substituents in the 2,6-positions on the aromatic ring.

> Hydrogen abstraction by phenolics takes place at elevated temperatures. At ambient temperatures, phenoxy radicals do not abstract hydrogen from the polymer backbone.

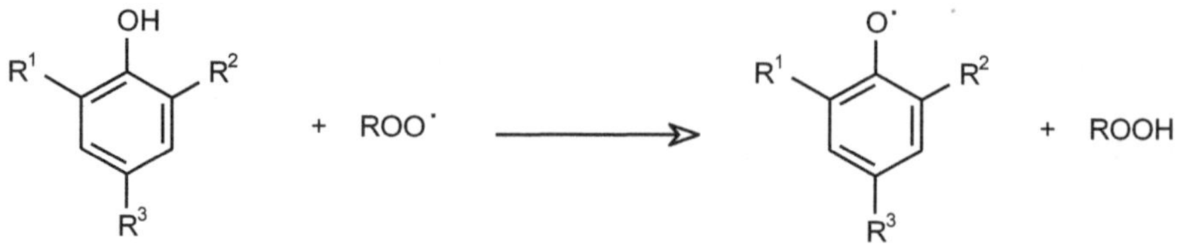

Figure 2.5 Sterically hindered phenolic antioxidants: mechanism of action

Long-term heat-stabilizing performance of the sterically hindered phenolic antioxidants in polymers becomes particularly effective at elevated temperatures (above 120–150 °C). These phenolic antioxidants represent the most widely used group of commercial stabilizers, as they provide both long-term thermal stability during service life of and protection to the polymer during melt processing.

Since they generate chromophoric moieties, oxidation by-products formed in the presence of certain phenolic structures can, however, lead to discoloration in the polymer. Such discoloration can be reduced by choosing phenolic antioxidants with appropriate molecular structures and by combining them with suitable co-stabilizers.

2.5.2 Hydroperoxide Decomposers (HDs)

Suitable hydroperoxide decomposers (HDs) convert hydroperoxides, ROOH, into non-radical, non-reactive, and thermally stable derivatives. To be effective, the HD must compete successfully with the thermal decomposition of the hydroperoxide group, thereby preventing chain-branching via alkoxy radical and hydroxyl radical formation in the autoxidation cycle. In this process, the hydroperoxide group ROOH is reduced to an alcohol, ROH, while the decomposer itself is stoichiometrically oxidized (Figure 2.6, first equation).

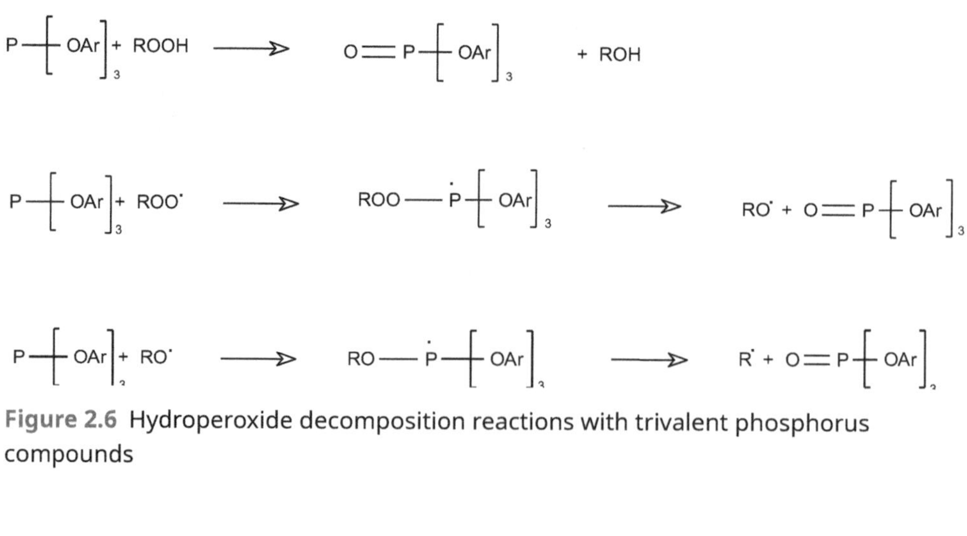

Figure 2.6 Hydroperoxide decomposition reactions with trivalent phosphorus compounds

In practice, hydroperoxide decomposers are typically used in combination with hydrogen-donor stabilizers such as phenolic antioxidants. Common examples of HD additives include organic trivalent phosphorus compounds (phosphites and phosphonites), organic sulfur compounds, such as sulfides, and metal salts of dialkyldithiocarbamates and dithiophosphates.

2.5.2.1 Phosphites/Phosphonites

> Especially in polymer melt processing, trivalent phosphorus compounds are the preferred hydroperoxide decomposers. They are mostly used in blends together with sterically hindered phenols.

Trivalent phosphorus stabilizers, such as phosphites and phosphonites, are converted to the corresponding phosphates, as shown in Figure 2.6. Through this redox process, peroxy radicals ($ROO^{\bullet}$) and alkoxy radicals ($RO^{\bullet}$) are effectively reduced by the phosphorus(III) species. Under oxygen-deficient conditions, the remaining alkyl radicals ($R^{\bullet}$) can still participate in disproportionation or fragmentation reactions. When appropriate alkyl radical scavengers are employed in combination with such HDs, they make a significant contribution to maintaining melt stability of polymers during processing. However, since phosphites and phosphonites react stoichiometrically with hydroperoxides, their effect on the long-term thermal stability of polymers is relatively limited.

> The tendency of phosphites and phosphonites to undergo hydrolysis with the formation of acidic by-products, can lead to corrosion of metal components in processing equipment. To reduce this risk, phosphites bearing alkyl-substituted aromatic groups are generally favored, as they are typically more resistant to hydrolytic degradation than aliphatic alkyl phosphites.

2.5.2.2 Thiosynergists

Among sulfur-based hydroperoxide decomposers, esters of 3,3-thiodipropionic acid are commercially viable. In their primary reaction pathway, the initially generated sulfoxide group undergoes thermal decomposition to yield a sulfenic acid, and the acidic species formed in this sequence are capable of decomposing hydroperoxides, ROOH, in an effectively over-stoichiometric fashion. Figure 2.7 shows main reactions of the thiodipropionic acid ester thiosynergists.

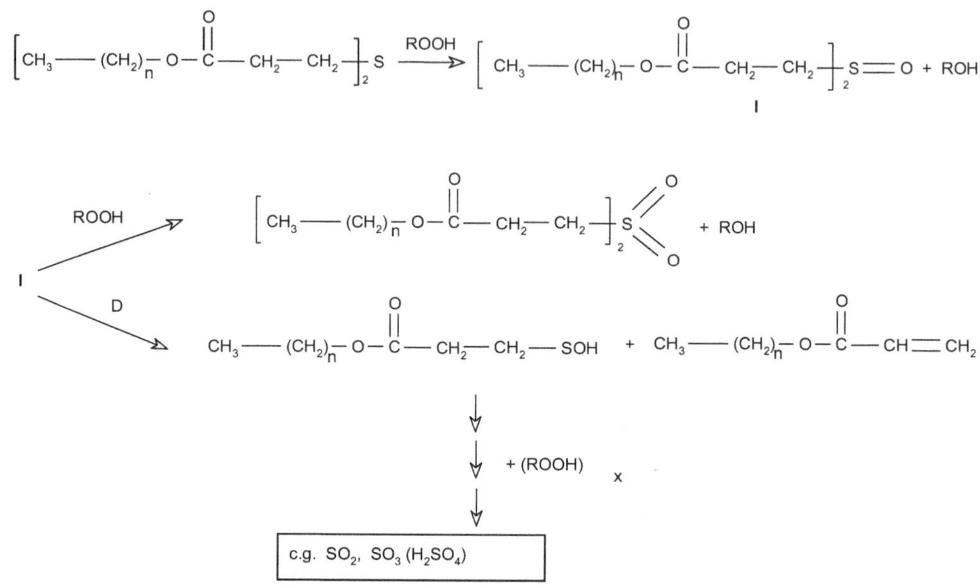

Figure 2.7 Hydroperoxide decomposition reactions in the presence of thiodipropionic acid esters

These HDs do not improve the melt stability of polymers during polymer melt processing. In contrast to phosphorus(III)-based hydroperoxide decomposers, thiodipropionate-type additives are highly effective for improving the long-term thermal aging resistance of polymers in the temperature range of about 100–150 °C. Sulfur-containing HDs of this kind are most commonly applied in combination with sterically hindered phenolic antioxidants.

2.5.3 Sterically Hindered Amine Stabilizers (HAS)

Trapping alkyl radicals (R$^{\cdot}$) is an effective way to bring the autoxidation of polymers to a halt. Sterically hindered amines (often abbreviated as HAS or HALS) are based on tetramethylpiperidine structures and are readily oxidized by either peroxy radicals to give nitroxyl radicals, or by peracids to form the corresponding hydroxylamines, as illustrated in Figure 2.8. Oxidation of the polymer and subsequent formation of peroxy radicals result in the formation of nitroxyl radicals.

Generally, the mechanism of stabilization with hindered amines is explained by the reaction of alkyl radicals present on the oxidatively degraded polymer with the nitroxyl radicals.

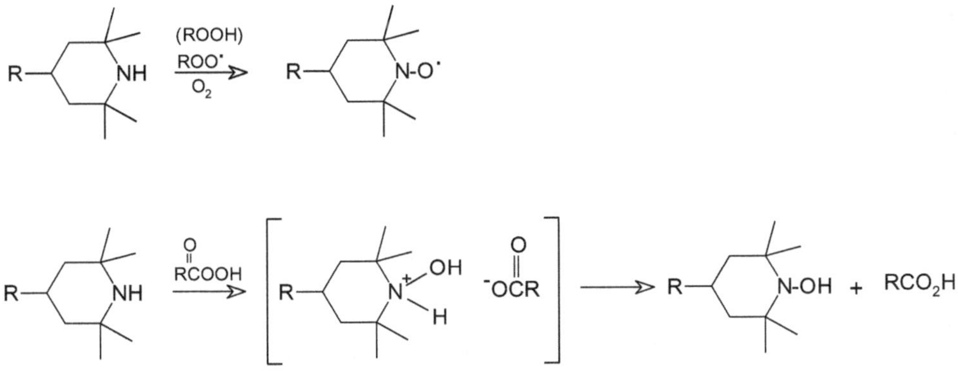

Figure 2.8 Formation of nitroxyl radicals and hydroxylamines

The reaction of nitroxyl radicals with alkyl radicals occurs at a rate only slightly lower than the addition of alkyl radicals to oxygen. This reaction operates in a catalytic cycle in which the nitroxyl moieties are continuously regenerated until they are eventually consumed in side-reactions.

In addition to trapping alkyl radicals, nitroxyl radicals can also react with peroxy radicals and acylperoxy radicals (paths A, B, and C), which further enhances their stabilizing action in plastics, as shown in Figure 2.9.

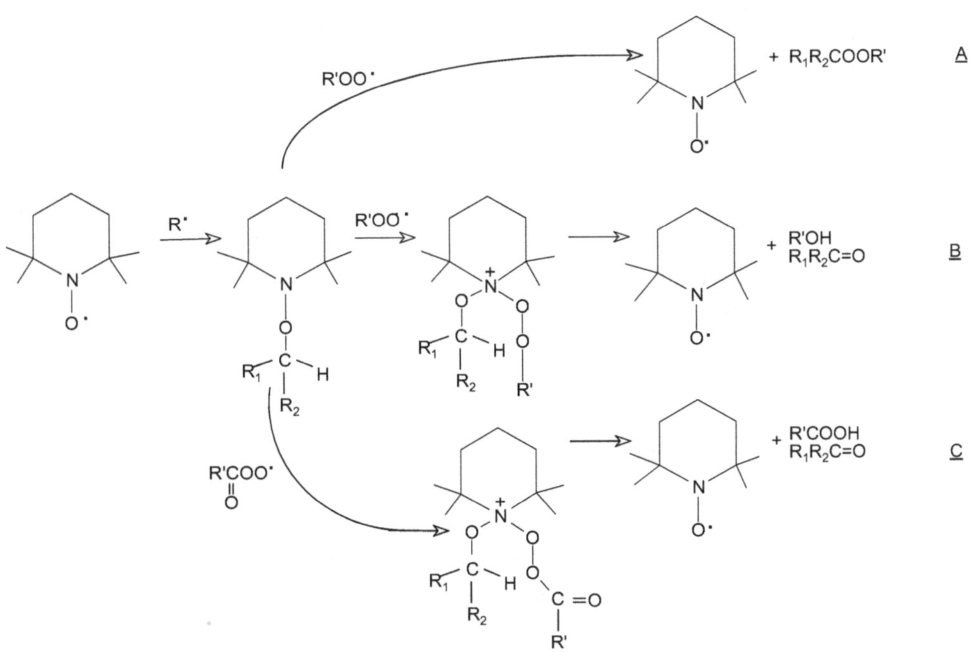

Figure 2.9 Scavenging of alkyl radicals of the oxidatively degraded polymer by nitroxyl radicals

2.5.4 Metal Deactivators

Metal ions catalyze the decomposition of hydroperoxides in the degrading polymers and form reactive radicals as shown in Figure 2.10.

$$M^{n+} + ROOH \rightarrow M^{(n+1)} + RO^{\cdot} + HO^{-}$$
$$M^{(n+1)} + ROOH \rightarrow M^{(n+)} + ROO^{\cdot} + H^{+}$$

Figure 2.10 Decomposition of hydroperoxides catalyzed by metal ions

Metal deactivators form stable coordination complexes with metal ions, particularly with copper ions, as illustrated in Figure 2.11. By tying up these catalytically active metal species, they significantly enhance the thermal stability of polymers used, for example, in cable insulation that is in continuous contact with copper wires.

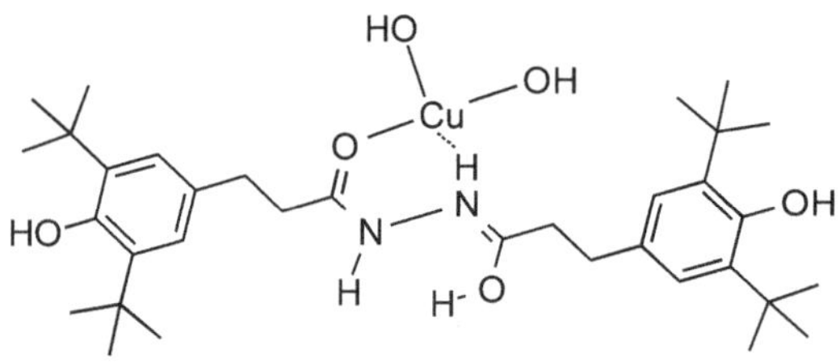

Figure 2.11 Stable complex of a metal deactivator with a copper ion

2.5.5 Multifunctional Stabilizers

Stabilizers that incorporate multiple protective functions within a single molecule (for example, species that are capable of both donating hydrogen atoms and decomposing hydroperoxides) are widely employed in industrial applications. Sterically hindered phenolic antioxidants bearing sulfur-containing substituent groups are among the best-studied representatives of this type.

2.5.6 Blends of Stabilizers

Mixtures of antioxidant components that act through different stabilization pathways in polymers are produced and applied on the commercial scale. Such systems often combine phenolic hydrogen-donor antioxidants with trivalent phosphorus-based hydroperoxide decomposers or sulfur-containing thiosynergists, and these combinations are capable of exhibiting synergistic behavior, meaning that their overall protective effect exceeds the sum of the individual contributions from each component used alone.

2.6 Testing of Antioxidants

2.6.1 General Remarks

During resin isolation, compounding and processing, macromolecules may undergo solvent removal, drying, and extrusion, for example, together with the associated thermal and mechanical (shear) stresses and exposure to oxygen. In the service phase of the finished article, polymers are then subjected to various combinations of physical and chemical stresses.

These factors trigger structural modifications on both the side groups and along the polymer backbone. Particularly critical are bond scissions on the main polymer chains that may result in the formation of radical species capable of initiating further chain reactions. Competing depolymerization pathways and intra- or intermolecular radical transfer processes lead in parallel to the formation of low-molar weight fragments, oligomers and new molecular components, thereby altering the composition, average molar mass, and the molar-mass distribution. Methods used to assess the effectiveness of antioxidants are discussed in Section 2.6.4.

2.6.2 Failure Mechanisms in Polymers

2.6.2.1 Amorphous Polymers

Amorphous polymers, which lack long-range ordered morphology and can be viewed structurally as supercooled liquids, are susceptible to various aging processes. Contributing factors include residual molding stresses, surface defects, absorbed moisture, chemical degradation, and general physical deterioration, promoting environmental stress cracking particularly with in-mold stresses and surface flaws.

These factors relax only slowly over time and are accompanied by physical aging that changes the mechanical response of the material. Moisture-induced chemical degradation, for example hydrolysis in susceptible polymers, can reduce molar mass and thus impair mechanical properties, such as strength and toughness.

Amorphous polymers eventually become brittle when subjected to stress as characterized by the elongation at break.

2.6.2.2 Semi-crystalline Polymers

Semi-crystalline polymers consist of macromolecules organized into regions of ordered crystals interspersed with amorphous segments. Compared with fully amorphous polymers, this mixed morphology allows for a broader spectrum of aging phenomena.

Over time, relatively small molecular rearrangements can yield more perfect crystalline structures, thereby reducing crystal defects and changing key performance characteristics of the material. Crystallites may continue to develop even after molding in a process known as secondary crystallization. Its kinetics are influenced by factors such as moisture uptake in polyamides and exposure to temperatures above the glass transition that may promote crystallization of amorphous materials.

Secondary crystallization leads to localized densification and shrinkage in crystalline regions that impose stresses on the surrounding amorphous domains. These stressed amorphous areas become preferred sites for the onset of thermo-oxidative degradation and can act as initiation points for brittle fracture.

2.6.3 Sample Preparation

For reliable testing, particularly when screening multiple additives, antioxidants must be distributed uniformly throughout the polymer matrix, while avoiding excessive shear and high processing temperatures. Poor dispersion leads to large scatter in the resulting data. Specimens prepared for heat-aging studies must also have comparable surface morphologies.

Since thermo-oxidative degradation can occur at every stage of processing and service life, it is a common practice to introduce a certain amount of the stabilizer as early as possible in the production train. In some systems, small amounts of antioxidants are added even during the polymerization process; if this is not feasible, then the stabilizer should be incorporated before the polymer's first contact with air, for example before resin drying.

When antioxidants are dosed during polymerization, they must be soluble in the monomer phase, and in emulsion polymerization processes they need to be supplied in an emulsion form.

Typically, antioxidants are added during dry-blending and/or compounding steps. The complete antioxidant package is usually in a powder form, masterbatch, or non-dusting blends.

2.6.4 Overview of Test Methods

Primary and secondary antioxidants are consumed during processing as a result of their exposure to oxygen at elevated temperatures.

In general, it is easier to monitor relevant polymer properties and correlate them with the activity of the added antioxidants than measuring the activity itself. An overview of useful test methods for assessing the effectiveness of antioxidants in plastics is in Table 2.1.

Table 2.1 Test Methods and Material Properties for Assessing Processability and Thermo-oxidative Stability of Polymers

Melt stability	Long-term thermal stability (heat aging)

▪ Melt processing
▪ Melt mass flow rate (MFR)
▪ Melt volume flow rate (MVR)
▪ Torque increase
▪ Viscosity
  ▪ Low-shear/high-shear viscosity (η)
  ▪ Solution viscosity
▪ Dynamic moduli
  ▪ Storage modulus
  ▪ Loss modulus
▪ GPC
  ▪ M_w, M_n, M_z, polydispersity
▪ Color (Yellowness Index YI)
	▪ Oven aging
	▪ Tensile test
	  ▪ Tensile strength
	  ▪ Elongation at break
	▪ Impact resistance test
	  ▪ Drop Dart impact test
	  ▪ Charpy impact test
	  ▪ Izod impact test
	▪ Optical properties
	  ▪ Color (Y.I.)
	  ▪ Gloss
	  ▪ Haze
	▪ Surface properties
	  ▪ Surface roughness
	  ▪ Craze formation
	▪ Oxygen uptake
	▪ Oxygen induction time (OIT)
	▪ Carbonyl index
	▪ GPC
	  ▪ M_w, M_n, M_z, polydispersity
	▪ Weight loss (TGA)

2.6.4.1 Melt Processing Stability

Polymers are transformed into finished articles by processes, such as extrusion, injection molding, and calendering, and in all of these operations the main drivers of degradation are elevated temperature, high shear stresses, and the unavoidable presence of oxygen. Even modest degradation during processing can already produce noticeable aging defects, not only impairing the physical properties but also compromising the long-term stability of the material in service. At the molecular level, these defects are associated with the formation of oxygen-containing groups (for example carbonyl and peroxy functions), the introduction of unsaturations, and changes in molar mass through chain scission and/or crosslinking.

Processing stability is commonly evaluated using torque rheometers, such as Brabender-type batch mixers, where the time-dependent torque under defined temperature and oxygen conditions is related to the resistance of the polymer to degradation. More sophisticated approaches employ single- or twin-screw extruders, where the material is subjected to repeatable extrusion cycles, and the melt behavior is characterized by multiple parameters, such as melt mass-flow rate (MFR) or melt volume-flow rate (MVR). Newer methods make the use of bypass and in-line capillary rheometers, in which a small side stream of melt is continuously diverted from the extruder and pushed through a capillary to obtain viscosity data in real time.

Key extrusion variables such as melt temperature, imposed shear, oxygen concentration, and the residence time have a strong impact on the measured stability of the polymer. For meaningful results, the processing conditions selected for each polymer–stabilizer system in the test extruders should be tuned to model and reproduce the full-scale production conditions as closely as possible.

2.6.4.2 Thermal Analysis

Thermal analysis techniques commonly used to study oxidation of polymers and the effectiveness of antioxidants include differential thermal analysis (DTA), differential scanning calorimetry (DSC), and thermogravimetric analysis (TGA). TGA measures mass loss as a function of temperature or time under a controlled atmosphere and is frequently applied to determine moisture content or volatility, or the polymer decomposition characteristics, including oxidative stability and the effect of non-antioxidant additives on polymer formulations.

DSC and DTA tests for oxidative stability are usually carried out isothermally at temperatures well above the polymer's melting point. After establishing an isothermal baseline under an inert gas such as nitrogen, the atmosphere is switched to oxygen or air and the induction time to the onset of the exothermic oxidation peak is recorded; standardized procedures include ASTM D3895[1)] for polyolefins and ASTM D4565 for cable insulation and jackets.

The oxidative induction time (OIT) measured by DSC or DTA is widely used as a quality control parameter for primary antioxidants, due to a roughly linear dependence of OIT on the concentration of phenolic antioxidants in polyolefins.

1 ISO 11357-6:2018 is nearly equivalent.

2.6.4.3 Long-term Thermal Stability Testing

2.6.4.3.1 Oven Aging Techniques

A straightforward way to assess the effectiveness of antioxidants is to subject stabilized polymer formulations to accelerated aging in a hot-air oven at defined temperatures with periodic examination of the specimens. Over the course of the exposure, changes in appearance, mechanical performance, and chemical structure are monitored to track the progress of degradation; when non-destructive analytical methods are used, the same samples can be returned to the oven, so that it helps to reduce the total number of specimens needed for testing each formulation.

Despite its simplicity, oven aging has critical limitations. Forced-air ovens can promote volatilization and loss of antioxidants before significant oxidative breakdown of the polymer occurs, especially at high test temperatures. Additionally, semi-crystalline materials such as polypropylene may experience substantial shifts in their crystallinity during annealing near the melting range (for instance, an increase of several percentage points in crystallinity can be observed within a few hours at about 135 °C, leading to marked changes in mechanical properties that are not solely attributed to oxidation). Loss of mechanical performance during such tests generally correlates with the reduction in average molar mass of the polymer due to chain scission.

In parallel with mechanical testing, chemical indicators of thermo-oxidative damage can be followed analytically. Infrared spectroscopy is frequently used to monitor the growth of carbonyl absorption bands, providing a measure of oxidation that often correlates well with changes in polymer properties and can reveal damage at relatively early stages, even before the pronounced losses in parameters such as yield stress are observed.

2.6.4.3.2 Testing under External Stress

Antioxidant testing in plastics generally aims to identify additives that extend the time to failure of the polymer. In many screening test protocols, specimens are not simultaneously subjected to several different mechanical loads to investigate changes in chemical and physical properties under controlled conditions.

In pressure pipe applications, failure typically manifests itself as brittle cracking and pinhole leaks, with the time to failure being strongly determined by the applied stress. Microcracks originate where material degradation occurs on the pipe's inner surface, with crack propagation occurring under internal pressure, and ultimately compromising the structural integrity.

2.6.4.4 Service Life Extrapolation

Accurately estimating the service life of plastic parts on the basis of accelerated aging experiments remains a key challenge and a focus of ongoing research. Meaningful lifetime predictions for a given material–application combination require that the entire temperature window relevant to thermo-oxidative failure be characterized. Furthermore, such extrapolations are only valid if additional factors – such as concurrent chemical reactions, applied mechanical stresses or ultraviolet radiation – can be assumed as negligible. In situation where taking into consideration these factors is important, they must either be accounted for by suitable correction schemes or addressed through the addition direct test protocols.

2.6.5 Stabilization of Selected Polymeric Substrates

Examples of specific antioxidant (AO) systems for selected polymeric materials are shown in Table 2.2. More details about AOs used in PVC compounds are in Section 4.3.4.

For most plastics, most antioxidant systems consist of phenols and/or combinations of phenols with organic phosphites.

Table 2.2 Examples of Antioxidant Systems Used for Selected Polymers

Polymer	Polymer Abbreviation	AO type	Examples
Polypropylene	PP, PP cable	Phenol	AO3, AO18
		Phenol/phosphite	AO18/PS2, AO18/PS4
		Phenol/thiosynergist	AO18/TS1
		Phenol/phosphite/ hindered amine	AO18/PS2/HAS4
		Phenol/phosphite/ hindered amine	AO3/PS2/HAS2
High-density polyethylene	HDPE, HDPE cable	Phenol	AO18
		Phenol/phosphite	AO18/PS2, AO3/PS2
		Phenol/hindered amine	AO18/HAS2
		Metal deactivator	MD1
		Metal deactivator/phenol	MD1/AO18
Poly(vinyl chloride)	u-PVC, p-PVC	Phenol	AO2, AO3
		Phenol/phosphite	AO3/PS1, AO18/PS1
Crystal polystyrene	CPS	Phenol	AO3

Polymer	Polymer Abbreviation	AO type	Examples
High-impact polystyrene	HIPS	Phenol	AO3, AO10
Styrene–acrylonitrile copolymer	SAN	Phenol/phosphite	AO3/PS2
Acrylonitrile–butadiene–styrene copolymer	ABS	Phenol	AO3, AO10
		Phenol/thiosynergist	AO3/TS2, AO10/TS2
Methyl methacrylate- butadiene–styrene copolymer	MBS	Phenol/thiosynergist	AP10/TS2, AO21/TS2
Polyamide	PA	Phenol	AO7
		Phenol/phosphite	AO7/PS2
Polyacetal	POM	Phenol	AO7, AO10, AO18
Poly(butylene terephthalate)	PBT	Phenol	AO10, AO18
		Phenol/phosphite	AO10/PS2
Polycarbonate	PC	Phenol/phosphite	AO3/PS2
		Phosphite	PS2
Polyurethane	PUR	Phenol	AO4/AO34/AO40
		Phenol/phosphite	AO4/PS1

Table 2.3 Chemical Compositions and Designated Codes of Selected AO Types and Components

Code	Structure	CAS No.	Abbreviation
Primary antioxidants			
AO-2	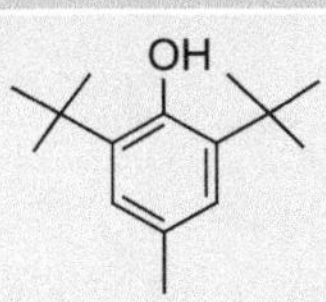	128-37-0	BHT
AO-3	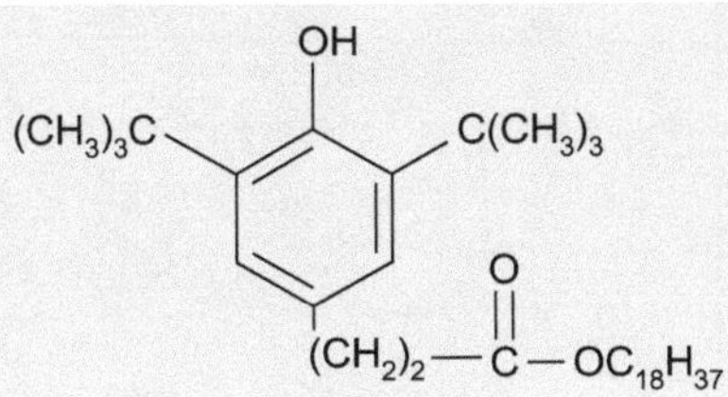	2082-79-3	AO 1076

Table 2.3 Chemical Compositions and Designated Codes of Selected AO Types and Components (*continued*)

Code	Structure	CAS No.	Abbreviation
AO-4	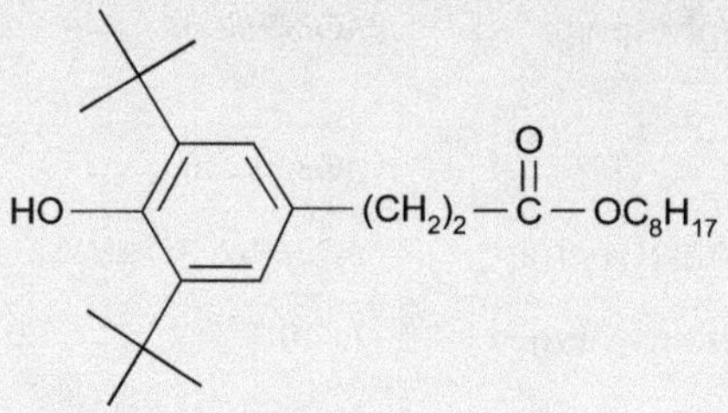	12643-61-0	-
AO-7	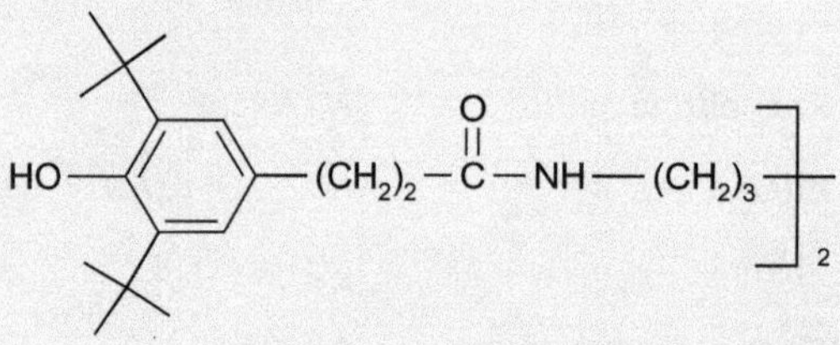	23128-74-7	-
AO-10	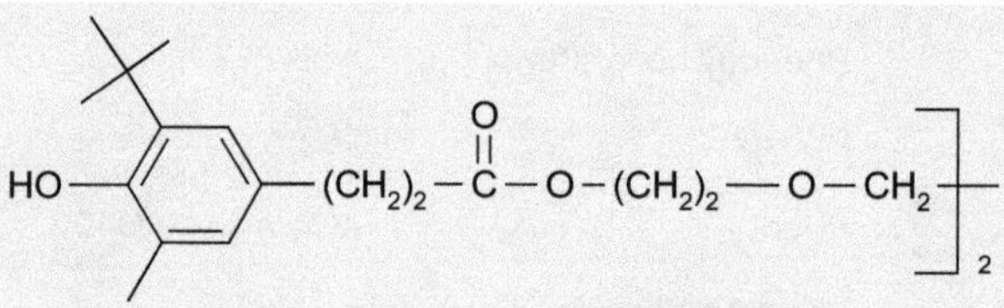	36443-68-2	-
AO-14	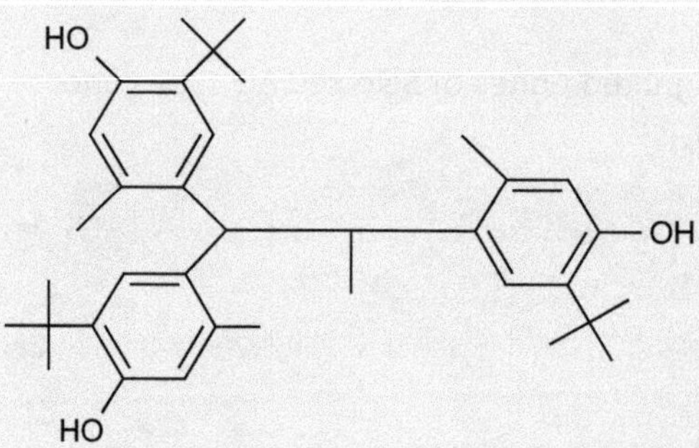	1843-03-4	-
AO-18	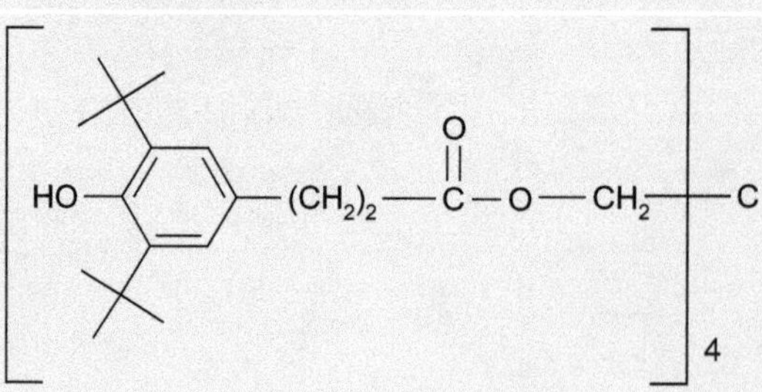	6683-19-8	AO 1010

Code	Structure	CAS No.	Abbreviation
AO-23	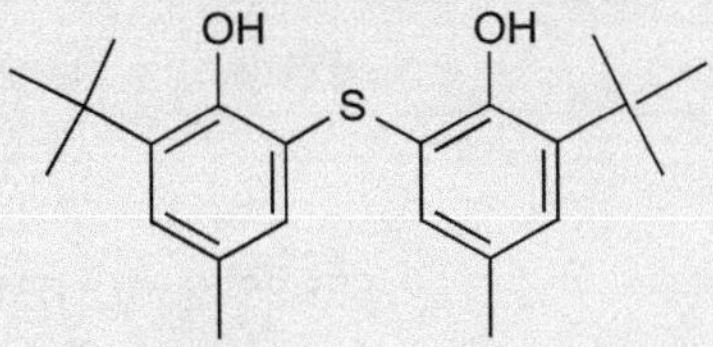	90-66-4	-
AO-34	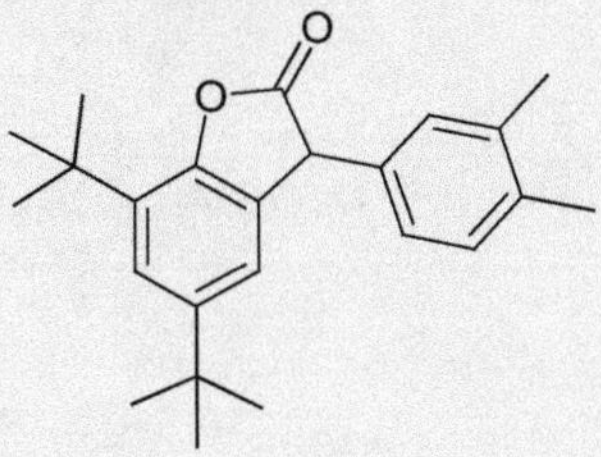	181314-48-7	-
AO-40	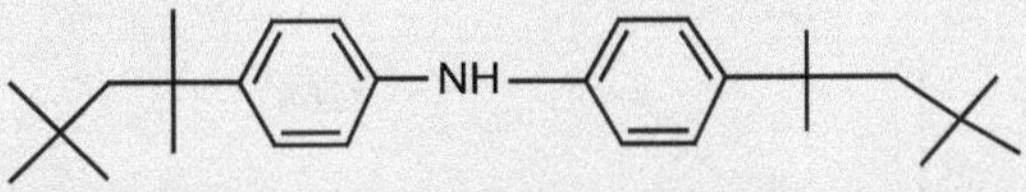	68411-46-1	-
Secondary antioxidants: Phosphites/phosphonites			
PS-1	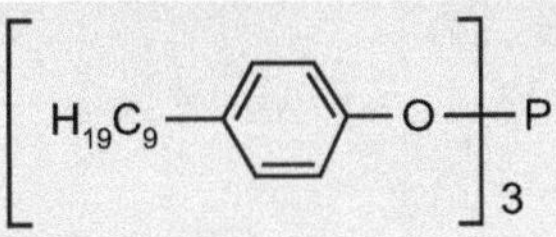	26523-78-4	TNPP
PS-2	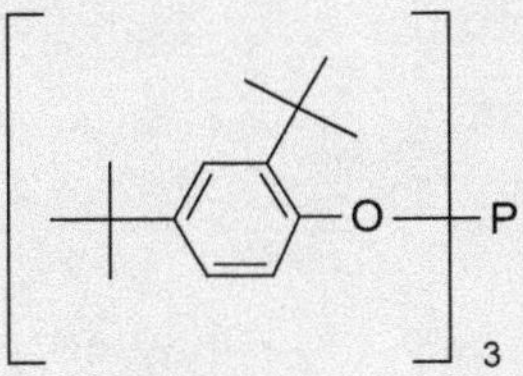	31570-04-4	AO 168
PS-4	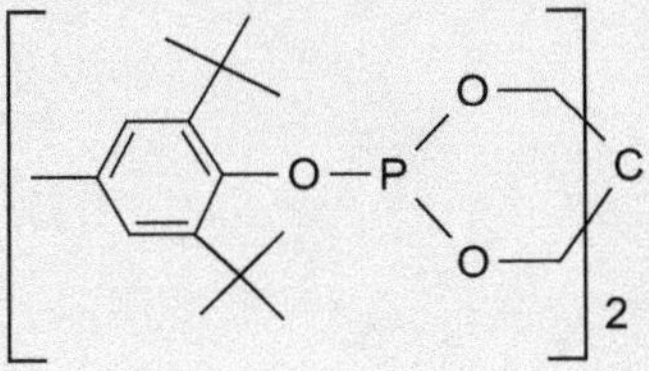	26523-78-4	-

Table 2.3 Chemical Compositions and Designated Codes of Selected AO Types and Components (*continued*)

Code	Structure	CAS No.	Abbreviation
Secondary antioxidants: Sulfur-containing compounds			
TS-1	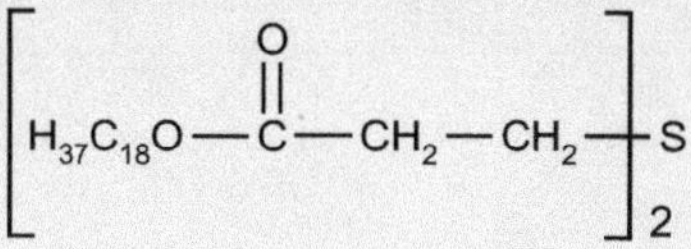	693-36-7	DSTDP
TS-2	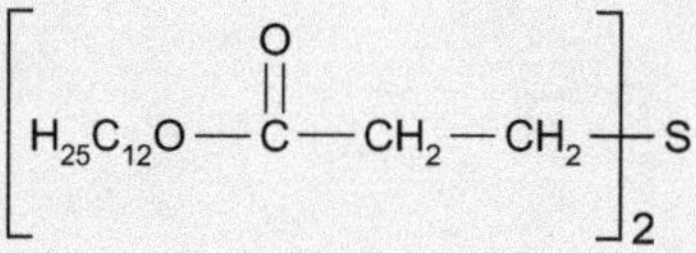	123-28-4	DLTDP
Metal deactivators			
MD-1	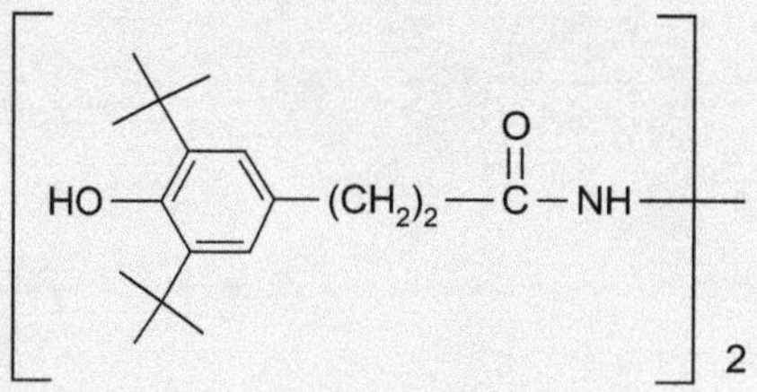	32687-78-8	-
MD-2	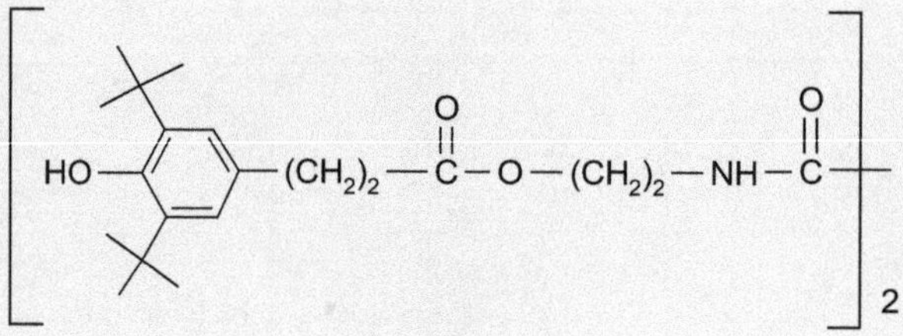	70331-94-1	-

Code	Structure	CAS No.	Abbreviation
Hindered amine stabilizers (HA(L)S)			
HA(L)S-2	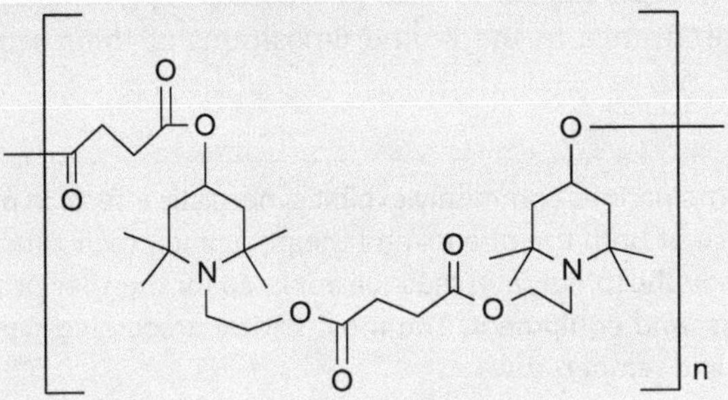	65447-77-0	-
HA(L)S-3	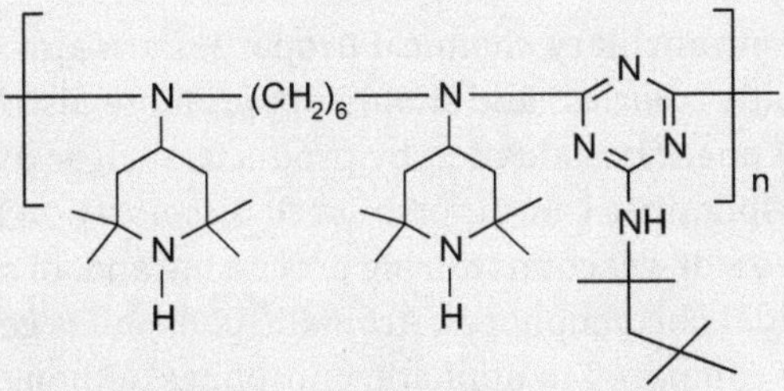	71878-19-8	-
HA(L)S-4	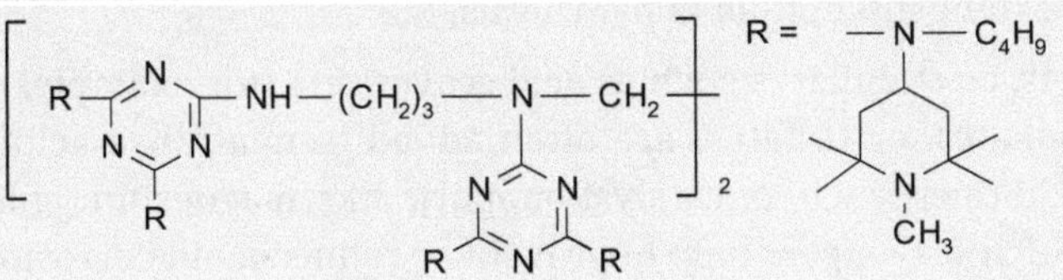	106990-43-6	-

2.6.5.1 Polyolefins

This section outlines how polyolefins are stabilized during melt processing and protected against thermo-oxidative damage throughout their service life. To assess processing stability and changes in material behavior, parameters such as melt flow or viscosity, molecular weight distribution, and color are typically measured before and after processing, for example after passage through an extruder.

Thermo-oxidative stability in service is often evaluated by aging the test specimens or parts at elevated temperature in a circulating-air oven and monitoring mechanical, chemical, and physical properties as a function of exposure time at a chosen temperature. In addition to such oven tests, thermo-analytical techniques like differential scanning calorimetry (DSC) are widely employed to characterize the resistance of polymer formulations to thermo-oxidative degradation.

2.6.5.1.1 Melt Processing Stabilization

Sterically hindered phenols are effective antioxidants for the melt processing of polyolefins. Their effectiveness depends on the number of phenolic groups capable of serving as H-donors and on steric hindrance in the 2- and 6-positions of their aromatic ring structures.

> Phenol/phosphite combinations commonly exhibit synergistic effects in polyolefin processing. The choice of both the phenol and the phosphite, their total concentration, and the phenol/phosphite ratio must be adjusted for the specific polymer, processing conditions, and equipment. The most critical processing parameters are in-process shear and temperature.

When selecting antioxidants, their ancillary chemical properties are also important. The stabilizer should be essentially colorless and should not promote discoloration of the polymer that can arise from phenolic oxidation by-products, such as quinoid species. Suitable phosphites or phosphonites can suppress such discoloration by protecting the phenol from the oxidative side-reactions during processing and, in some cases, by reacting directly with quinoidal chromophores. Aromatic phosphites generally offer superior hydrolytic stability compared to aliphatic phosphites, although all phosphites are to some extent susceptible to hydrolysis.

Besides antioxidants, co-stabilizers such as acid scavengers (for example, metal stearates, oxides, carbonates, or lactates) are often added to neutralize acidic residues caused by other additives, such as catalyst supports like magnesium chloride, that might be present. Such scavengers help to prevent or reduce mold corrosion but typically have little direct effect on melt-processing stability. In many polyolefin systems, blends of hindered phenols with organic phosphites provide superior melt stabilization compared with phenols alone; for instance, phenolic AO-18 in combination with phosphite PS-2 at a 1:2 ratio was identified as particularly effective for polypropylene extrusions, while PE grades are well-protected by analogous phenol/phosphite formulations with the phosphite largely consumed during processing.

HDPE produced with chromium catalysts may undergo both crosslinking and chain scission during melt processing, which significantly alters the molecular weight distribution and can degrade mechanical properties; this can be mitigated by using a phenol AO-18/phosphite PS-2 blend with a ratio of about 1:4 to stabilize the melt against such changes. In contrast, titanium-catalyzed HDPE more typically shows a reduction in molecular weight under processing, for which a roughly 1:1 AO-18/PS-2 blend affords optimal stabilization, while linear low-density polyethylene (LLDPE) can be melt-stabilized in a similar manner, provided that phenol/phosphite levels are kept low enough to maintain good compatibility and avoid blooming.

Examples of stabilizing polyolefins by selected antioxidants during melt processing are provided in Table 2.4.

Table 2.4 Stabilizing Effect of Selected Antioxidants on the Melt Flow Rate (MFR) (g/10 min) During Melt Processing

Polyolefin	Antioxidant	No. of extrusion cycles	
		1	5
PP homopolymer at 270 °C[a]	None	7.0	>30
	0.075% AO2	5.2	8.0
	0.075% AO1	4.7	7.0
	0.075% AO3	6.8	25
HDPE homopolymer (Cr-type catalyst) at 260 °C[b]	None	40	65
	0.05% AO18	24	30
	0.05% AO18/0.05% PS2	23	25
	0.05% AO18/0.2% PS2	21	21

[a] All samples contained 0.075% Ca stearate.
[b] All samples contained 0.02% Ca stearate.

The principles described for melt-processing stabilization of polyethylene and polypropylene homopolymers apply the same way to most of their copolymers. Sterically hindered phenolic antioxidants, phosphites, or phosphonites, and carbon-centered radical scavengers (particularly when used in suitable combinations) provide effective protection for these materials during processing and subsequent long-term thermal exposure.

2.6.5.1.2 Stabilization for Long-Term Thermal Exposure

Most plastic materials are exposed to elevated temperatures, light, moisture, and atmospheric oxygen during their use, allowing thermo-oxidative degradation (autoxidation under excess oxygen) to proceed readily. The following section focuses on the long-term thermal stabilization of polyolefins, commonly referred to as long-term thermal stability (LTTS), which is typically assessed by aging specimens at elevated temperature in a circulating-air oven.

Mechanical properties such as bend embrittlement, elongation at break, and tensile strength are recorded as a function of exposure time at a chosen test temperature, and the oven-aging lifetime is usually reported in days. The results depend on several parameters, including the type and concentration of stabilizers employed, specimen thickness, aging temperature, and other test conditions, all of which must be controlled in order to compare performance of the polymer formulations properly.

Sterically hindered phenols that are used either alone or in combinations with thiosynergists as hydroperoxide decomposers are effective stabilizers for long-term thermal stability of polyolefins. Sterically hindered amines (HAS/HALS) are increasingly used for this purpose as well.

The effects of various sterically hindered phenols on the LTTS of a PP homopolymer at 135 °C and 150 °C are shown in Table 2.5.

Table 2.5 Effect of Phenolic Antioxidants on the Time to Embrittlement (in Days) of PP Homopolymer[a]

Phenol	Oven aging temperature	
	135 °C	150 °C
None	<1	<1
AO-1	10	<2
AO-3	41	11
AO-13	130	33
AO-18	160	50
AO-19	45	10

[a] All samples contained 0.1% Ca stearate, and the phenols were added at 0.2%, using 1 mm compression-molded plaques in a circulating-air oven.

The effect of phenolic antioxidants on oven-aging behavior is largely governed by their molecular weight and chemical structures.

In practice, low-molecular-mass phenols with molecular weights below about 500 g/mol tend to be too volatile at typical test temperatures and are rapidly lost from the polymer specimens, limiting their long-term protective effect.

The dependence of a polypropylene homopolymer's long-term thermal stability on phenol loading and aging temperature is illustrated in plots shown in Figure 2.12.

It should be noted that at lower aging temperatures, the ranking of efficiencies of the phenols can be the inverse of that found at higher aging temperatures.

Extrapolating high-temperature LTTS data to lower temperatures usually results in an over-estimation of the time-to-failure.

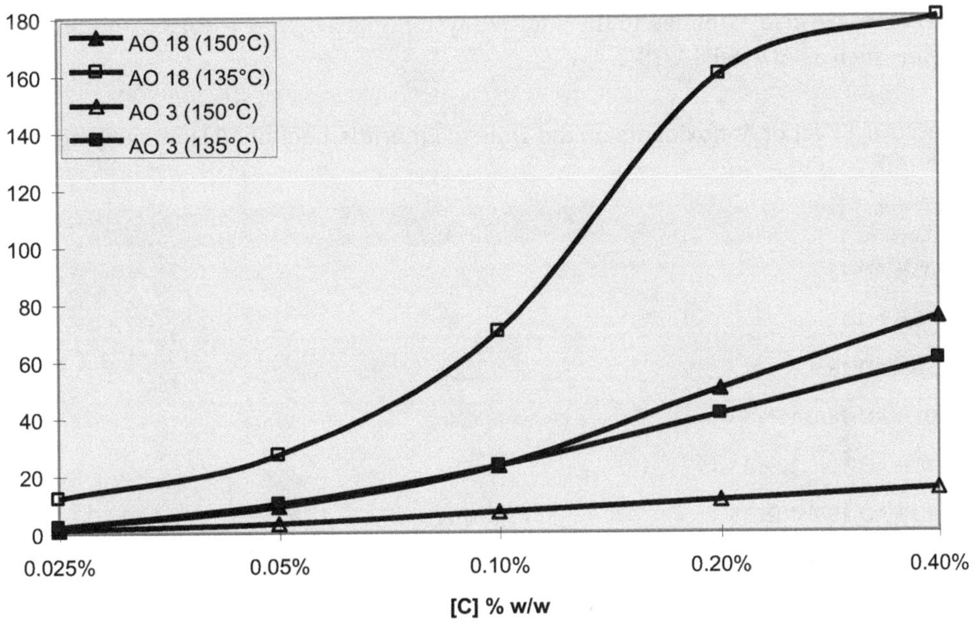

Figure 2.12 Effect of stabilizer concentration on the LTTS of PP homopolymer at 135 and 150 °C. All samples contained 0.1% Ca stearate, and 1 mm compression-molded plaques were used in a circulating-air oven

The most effective protection against thermo-oxidative degradation of polyolefins during their melt processing is achieved with the phenol/phosphite (or phosphonite) combinations. Phosphorus-containing hydroperoxide decomposers are primarily used to shield phenolic antioxidants during melt processing so that the phenols remain largely intact and can continue to provide long-term thermal stability.

The phosphites themselves make little direct contribution to LTTS; rather, LTTS is governed by the fraction of "phosphite-protected" phenol that survives the processing steps. To further enhance long-term stability, thioether-type hydroperoxide decomposers (thiosynergists) are often combined with phenolic antioxidants. The oven-aging data for polypropylene homopolymer stabilized with a phenol such as AO-18 in combination with a thiosynergist like TS-1 illustrates this beneficial effect, as shown in Table 2.6.

The positive impact of thiosynergists on LTTS is evident, and these additives also tend to reduce the discoloration that often develops during oven aging, particularly when they are used in combination with phenolic antioxidants at ratios of about 1:3. The phenol/thiosynergist packages are therefore widely favored for stabilizing HDPE and LLDPE, as well as in poly(1-butene) pipe applications. High-efficiency performance attributes are also extended to the additive systems based on AO-18/PS-2 and ternary blends such as AO-18/PS-2/TS-2.

Table 2.6 Effect of Antioxidants on the Time to Embrittlement of PP Homopolymer at 150 °C[a]

Phenol/thiosynergist	Time to embrittlement (days)
0.05% AO-18	9
0.10% AO-18	23
0.20% AO-18	50
0.05% AO-18/0.15% TS-1	54
0.10% AO-18/0.30% TS-1	78
0.20% AO-18/0.40% TS-1	101

[a] All samples contain 0.1% PS-2 and 0.1% Ca stearate, and 1.0 mm compression-molded plaques were used in a circulating-air oven.

To minimize color formation, many producers adopt phenol-free stabilization approaches in sensitive uses like fibers, where phenols are replaced by benzofuranone/phosphite systems or hydroxylamine/phosphite/HAS packages. HASs act mainly as oxidation retarders and are particularly efficient at lower oven-aging temperatures. They can be a substitute for phenols in an effort to improve LTTS while simultaneously providing light stability to the polyolefin substrates.

Phenol-free systems also provide superior initial color and stability during the reprocessing of PP resins.

When HASs are protonated by acidic residues that may be present in polymer matrix, they lose their effectiveness. In applications where mineral acids may be generated (such as greenhouse films exposed to sulfur- or halogen-containing agrochemicals, or plastics formulated with halogenated flame retardants that can form acid *in situ*), it is therefore necessary to use low-basicity or essentially neutral HAS derivatives.

For applications in which plastic parts are likely to be exposed to water, such as washing machines, household appliances or automotive components, the stabilizer system must also show good resistance to hydrolysis and leaching in order to maintain long-term thermal stability.

2.6.5.1.3 Influence of Fillers, Pigments, and Reinforcing Additives

Global PP usage includes a substantial fraction of filled or reinforced material grades, with estimates suggesting that roughly one quarter of PP consumption involves formulations containing mineral fillers or fibers. Talc and calcium carbonate are the predominant fillers in these systems, though smaller amounts of mica and wollastonite are also employed. Many mineral fillers are known to adversely affect the long-term thermal and light stability of polyolefins by interacting with the stabilizers or modifying degradation pathways.

Typical oven-aging results for two types of filled, stabilized PP homopolymer are presented in Figure 2.13. As can be seen, both fillers significantly affect the LTTS of the polymer. In talc-filled PP, the overall LTTS remains inferior to that of unfilled PP, even when a phenol/thiosynergist antioxidant package is added, whereas formulations filled with coated calcium carbonate show no severe reduction in LTTS.

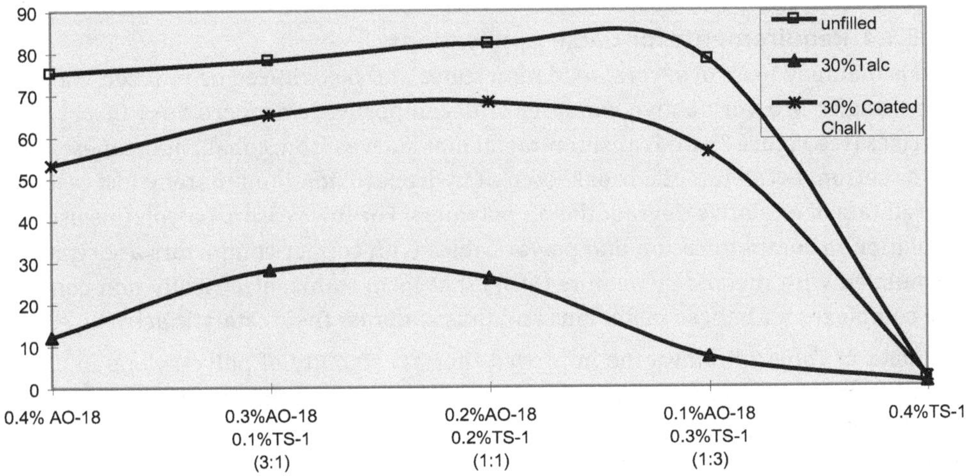

Figure 2.13 LTTS at 150 °C of PP homopolymer filled with talc and coated calcium carbonate and stabilized with the phenolic antioxidant AO-18, both alone and in combination with the thiosynergist TS-1, added at a total concentration of 0.4%. All samples contained 0.1% PS-2 and 0.1% Ca stearate, and 0.5 mm compression-molded plaques were used in a circulating-air oven

Many commercial filled PP grades for interior, exterior, and under-the-hood automotive parts are additionally colored with carbon black, which can markedly shorten LTTS; however, both talc-filled and carbon black-pigmented PP can be effectively stabilized with the use of HASs.

Rutile (titanium dioxide) is generally beneficial for light stability and, unlike many other pigments, does not significantly impair the long-term thermal stability of polymers. When its surface is coated with silicone and/or inorganic modifiers such as zinc or aluminium compounds, it can greatly enhance the weathering resistance of pigmented plastics. In contrast, inorganic pigments containing transition metals, while offering good thermal stability compared to organic pigments, may have a detrimental effect on LTTS due to catalytic effects imparted by metal-containing moieties or impurities.

As with melt-processing stabilization, the approaches described for long-term thermal stabilization of PE and PP homopolymers apply broadly to their copolymers, and systems based on either sterically hindered phenols in combination with thiosynergists, or HASs, provide effective protection of these materials during extended heat aging when appropriately formulated.

Since the crystallinity of ethylene copolymers is lower than that of PE homopolymer, the antioxidant concentrations required for adequate service life of the copolymers may need to be slightly higher.

2.6.5.1.4 Requirements for Cable Applications

Metals that may exist in several oxidation states and be oxidized or reduced via electron transfer are very active catalysts of decomposing hydroperoxides in polymer matrixes (see Figure 2.10). Transition-metal ions such as iron, cobalt, manganese, copper or cerium accelerate the breakdown of hydroperoxides and thereby increase the overall rate of oxidative degradation in polymers. For this reason, polyolefins used as insulation in communication and power cables with copper conductors are typically formulated with metal deactivators (MDs) that form stable, electrically non-conductive complexes with these metal ions and thus suppress their catalytic activity.

The data in Table 2.7 shows the improved thermal stability of polyethylene in direct contact with the copper conductor. Since MD-1 combines both the phenolic antioxidant functionality and metal-complexing capability, it substantially improves the oven aging performance of the PE insulation.

Table 2.7 Effect of Stabilizers on the Oven Aging Time (Days) of PE Cable Insulation[a]

Stabilizer system	Aging temperature	
	90 °C	110 °C
0.05% AO-18	32	5
0.05% MD-1	119	29
0.1% MD-1	527	275
0.05% AO-18/0.1% MD-1	410	136

[a] The copper conductor was 0.5 mm diameter, and the PE insulation thickness was 0.2 mm with a density of 0.927 g/cm^3. A circulating-air oven was used with pigtail specimens in accordance with VDE 0209 and FTZ 72 TZV1.

In communication cables, the conductors are often surrounded by petrolatum (petroleum jelly) to enhance resistance to water ingress, and foamed (cellular) PS insulation is widely used to offer superior dielectric performance. Both the presence of petrolatum and the cellular polymer morphology, however, tend to reduce the oxidative stability of the cable insulation. Aging tests have shown that exposure to petrolatum can lower the LTTS of HDPE insulation by about 40%, while a cellular insulation structure alone can decrease LTTS by about 25–55%.

These detrimental effects can be addressed by increasing the level of metal deactivators in the polymer formulation. With sufficiently high MD loadings, even the most demanding cable configurations (cellular HDPE insulation filled with petrolatum) can be stabilized, so that the expected LTTS is effectively maintained.

2.6.5.1.5 Crosslinked PE

Crosslinked PE (XLPE) is increasingly selected for cable insulation applications, since its structure provides significantly improved thermal stability and allows production operations at higher temperatures compared to thermoplastic PE. The stabilizer package must be suitable for the chosen PE crosslinking route (which is typically either peroxide curing or silane grafting), so that it does not unduly inhibit the crosslinking reaction or reduce the crosslink density.

In peroxide-cured crosslinked LDPE systems, conventional phenolic antioxidants and multifunctional phenols, such as AO-23 (which also contains a thioether group and acts as a hydroperoxide decomposer) can deliver excellent LTTS. For silane-cured LDPE cable insulation, high-molecular-weight, sterically hindered amines, such as HAS-2 and HAS-3, provide particularly favorable oven-aging performance, combining good retention of mechanical properties with effective resistance to thermo-oxidative degradation.

2.6.5.1.6 Irradiation Effects

A wide range of polymeric materials are exposed to ionizing radiation and then placed into service. These include, for example, radiation-cured rubbers and plastics, and components for medical devices sterilized by irradiation. Gamma irradiation is one of the principal sterilization methods, in addition to ethylene oxide treatment. It is notable for its high penetration depth at relatively low doses, whereas electron-beam (beta) irradiation delivers high doses of electrons with more limited penetration.

In such applications, radiation-induced degradation plays a crucial role in determining the usable lifetime of the irradiated plastics. The high-energy radiation field generates carbon-centered radicals in the polymer that then react with oxygen, accelerating autoxidation and chain scission. Many of these formed radicals are trapped in the solid matrix and can persist there for long periods of time, propagating polymer degradation even well after the irradiation treatment is completed. The primary strategy for mitigating post-irradiation aging is the incorporation of suitable antioxidants and radical scavengers that shorten radical lifetimes and inhibit the associated oxidative reaction cascades.

Effects of antioxidants in PP following beta- and gamma-irradiation are shown in Table 2.8 and Table 2.9, respectively. The best performance in terms of oven aging and color retention of the post-irradiated PP was achieved with the use of a ternary stabilizer blend, consisting of a hindered phenol AO-3 or AO-18, a phosphite (PS-2), and a hindered amine (HAS-2).

Table 2.8 Effect of Antioxidants on Oven Aging and Discoloration of PP after γ-Irradiation[a]

Stabilizer system	Irradiation dose (Mrad)					
	0	2.5	5.0	0	2.5	5.0
	Time to embrittlement (days)[b]			Discoloration (Y.I.)[c]		
0.015% AO-3/0.06% PS-2	40	4	2	0.80	1.20	1.60
0.015% AO-3/0.06% PS-2/0.05% HAS-2	125	70	70	0.85	0.75	0.85

[a] 1 mm compression-molded plaques.
[b] Oven aging at 120 °C.
[c] Yellowness Index determined according to ASTM 1925-77.

Table 2.9 Effect of Antioxidants on Oven Aging and Discoloration of PP before and after β-Irradiation[a]

Stabilizer system	Time to embrittlement (days)[b]		Discoloration (Y.I.)[c]	
	Before	After	Before	After
0.01% AO-1/0.09% PS-2/0.05% HAS-2	130	70	2.10	1.80
0.015% AO-3/0.06% PS-2	40	2	0.80	0.60
0.015% AO-3/0.06% PS-2/0.05% HAS-2	130	80	0.85	0.65

[a] 1 mm compression-molded plaques.
[b] Oven aging at 120 °C.
[c] Yellowness Index determined according to ASTM 1925-77.

Antioxidants are partially consumed during the irradiation step, and this effect is dependent on radiation dose.

The service lifetime of irradiated specimens is shorter than for non-irradiated samples.

In the case of β-irradiated plastic samples, further discoloration occurs upon storage, increasing their color by approximately 2 Y.I. units compared to the color of the specimens immediately after the irradiation.

2.6.5.2 Polystyrene and Styrenic Copolymers

2.6.5.2.1 Crystal Polystyrene (CPS)

Under oxidative conditions, unmodified polystyrene such as CPS is a comparatively stable thermoplastic, so that many PS applications do not require the use of antioxidants. However, foaming and thermoforming processes generate significant amounts of scrap, making recycling of PS increasingly important. Successful reprocessing of PS foam demands that the material retain a sufficiently high molar mass (on the order of about M_w 280,000 g/mol) to avoid foam collapse. In this context, antioxidant stabilizers play a critical role of reducing degradation of CPS during processing and service, thereby enabling to preserve its molecular weight and recyclability.

Maximum stabilizing efficiency is achieved when the antioxidant is added prior to the polymerization of the styrene monomer to CPS in a continuous bulk polymerization process.

Suspension polymerization is the principal route for producing expandable polystyrene (EPS). Phenolic antioxidants can be introduced either prior to or during peroxide-initiated polymerization without significantly affecting the polymerization rate, whereas phosphites or phosphonites are typically incorporated only after completing the polymerization and right before the subsequent melt-processing steps.

2.6.5.2.2 Styrene–Acrylonitrile Copolymer (SAN)

SAN represents a range of non-elastomeric styrenic copolymers, and it commonly serves as the rigid core phase in ABS. It is also processed on its own and used for manufacturing kitchenware, housings for electrical appliances and electronics, automotive interior parts, medical devices, and packaging components. SAN grades are processed using the standard thermoplastic processing methods, with injection molding being particularly suitable at melt temperatures typically about 220–240 °C, depending on grade and equipment.

Degradation of SAN is strongly influenced by the acrylonitrile units, which are usually present at about 24–35 wt%, and often impart yellowing during processing or storage of SAN at elevated temperatures. Since this discoloration is attributed primarily to non-oxidative reactions of the acrylonitrile component rather than from classical thermo-oxidative pathways, conventional antioxidants exhibit only limited effectiveness in preventing color development in SAN.

2.6.5.2.3 High-impact Polystyrene (HIPS)

As with crystal PS, HIPS is predominantly produced via continuous bulk polymerization. Since styrene–butadiene copolymers are more prone to oxidative attacks compared to unmodified PS, they are susceptible to yellowing and loss of mechanical performance, making the incorporation of suitable antioxidants essential.

Antioxidants can, in principle, be dosed into the reaction mass before, during or after polymerization, typically as concentrated solutions whose good solubility in styrene monomer is crucial. The optimal point of addition depends on the initiation method: for thermally initiated styrene polymerization, stabilizers may be introduced at almost any stage, whereas in peroxide-initiated systems only phenolic antioxidants are normally added during or before polymerization, since phosphite antioxidants tend to interfere with the radical polymerization kinetics and are therefore deferred to later process steps if used at all.

2.6.5.2.4 Acrylonitrile–Butadiene–Styrene (ABS)

ABS is a graft copolymer of acrylonitrile, butadiene and styrene, produced predominantly via emulsion polymerization, although bulk (mass) polymerization processes are also used and generally yield material with somewhat higher inherent stability. To protect emulsion-made ABS against thermo-oxidative degradation during the drying step, an emulsified antioxidant is usually introduced into the latex before coagulation. Since the resulting coagulated powder has very high specific surface area, drying of the material is conducted in a continuously renewed air stream at 110–150 °C and can lead to self-heating and even ignition. As a result, drying of ABS is recommended in a closed-system under nitrogen.

The oxidative stability of emulsion-polymerized ABS depends on the initiator system used in the polymerization process and on the details of the coagulation process. Since typical ABS contains roughly 25% polybutadiene, it is effectively melt-stabilized by phenolic antioxidants such as AO-10 in combination with thiosynergists like TS-2 that not only protects the co-polymer against thermo-oxidative degradation but also significantly reduces its yellowing in oven-aging tests.

Antioxidant packages may also be incorporated into the copolymer during compounding, which is common practice for both mass- and emulsion-polymerized ABS grades. Phosphite stabilizers such as PS-1 can be added to the phenol/thiosynergist systems to further enhance their oxidative stability. In addition to heat stabilizers, most applications of HIPS, ABS and related styrenic graft copolymers also require light stabilizers and flame retardants. Therefore, care must be taken to avoid additive combinations with components that might react with one another.

2.6.5.2.5 Methyl Methacrylate–Butadiene–Styrene (MBS)

MBS is a range of graft copolymers of methyl methacrylate, styrene, and butadiene rubber; they are mainly used as impact modifiers for PVC (see Section 6.3.2.1), polycarbonate, and polycarbonate/polybutylene terephthalate blends.

Its production process is similar to that of the emulsion-polymerized ABS. Therefore, antioxidants must be added prior to the coagulation step to avoid oxidation during the drying step. Effects of antioxidants on the oxidative stability of MBS at 180 °C are shown in Table 2.10.

Table 2.10 Effects of Antioxidants on Thermo-oxidative Stability of MBS

Antioxidant system	Time to exothermic effect (min)[a]
0.1% AO-3/0.6% TS-2	20
0.2% AO-10/0.6% TS-2	175
0.3% AO-10/0.6% TS-2	350
0.1% AO-21/0.6% TS-2	425

[a] Measured at 180 °C by DSC.

2.6.5.3 Polyamides (PAs)

2.6.5.3.1 Aliphatic Polyamides

A broad range of polyamides is produced commercially, with PA 6, PA 6.6, PA 4,6, PA 11, and PA 12 among the highest-volume grades. Their susceptibility to oxidative degradation depends strongly on overall crystallinity, and the chemical structure and density of the amorphous segments. Aliphatic polyamides have traditionally been stabilized for withstanding long-term thermal exposure with very small amounts of copper salts (typically up to about 50 ppm) in combination with halide ions, such as iodide or bromide. This approach is somewhat counterintuitive, since copper ions are known to act as pro-oxidants in many polyolefin systems.

Aromatic amines are classic long-term heat stabilizers for polyamides but tend to cause pronounced discoloration, so their use is often restricted to black or dark-colored grades, whereas typical phenolic antioxidants can improve the initial color of only aliphatic PAs if added before the end of the polycondensation step. In general, aliphatic polyamides do not require dedicated melt-processing stabilizers. Comparative LTTS data for copper iodide-stabilized PA 6 versus systems containing phenolic antioxidants such as AO-7, alone and also combined with phosphites like PS-2, are summarized in Table 2.11.

Table 2.11 Effects of Stabilizer Systems on LTTS of PA 6 at Different Temperatures

Stabilizer	Aging time (h)[a]			
	100 °C	120 °C	150 °C	165 °C
None	480	24	6	3
0.003% Cu/0.14% I	5520	624	324	44
0.5% AO-7	6840	600	36	15
0.25% AO-7/0.25% PS-2	6480	576	43	12

[a] Time to 50% retained elongation, at specified temperature. 1 mm injection molded dumbbells measured in an air-circulating oven.

At aging temperatures of above 150 °C, copper iodide heat stabilizer systems provide the most effective long-term protection for aliphatic polyamides, whereas at lower temperatures, the phenolic antioxidants, either alone or in combination with phosphites, tend to perform better. Aliphatic polyamides are widely used as fibers for textiles, carpets, and related applications, where long-term thermal stability is a key performance attribute. Representative oven-aging data comparing non-stabilized and AO-7-stabilized PA 6 fibers are summarized in Table 2.12.

Table 2.12 Effects of AO7 on LTTS of PA 6 Fibers at 140 °C and 160 °C

Stabilizer	Aging time (h)[a]	
	140 °C	160 °C
None	0	10
0.5% AO-7	1260	470

[a] Time to 50% retention of tensile strength, at specified temperature. Measured in an air-circulating oven.

Stabilizing PA fibers with a phenolic antioxidant such as AO-7, either alone or in combination with a phosphite like PS-2, provides excellent long-term thermal stability, especially at temperatures below about 150 °C. The copper halide-containing stabilizer systems (for example copper iodide or copper bromide) are less suitable for use in fiber applications, since the metal salts can leach out significantly when the fibers are exposed to moisture, undermining both stability and appearance.

Glass fiber-reinforced PA 6.6 is widely used for under-the-hood and structural automotive components, such as engine fans, radiator tanks and grilles, fluid reservoirs, valve covers, wheel trims and head-rest shells. In these applications, the phenolic antioxidants and their combinations with phosphites markedly extend long-term thermal stability as reflected in oven-aging data set summarized in Table 2.13.

Table 2.13 Effects of Antioxidants on LTTS of PA 6.6 (33% GF) at 120 °C

Stabilizer	Oven aging (days)[a]
None	12
0.5% AO-7	92
0.25% AO-7/0.25% PS-2	150

[a] Time to 50% retained impact strength, at 120 °C. Test bars (4 × 6 × 50 mm) measured in an circulating-air oven.

2.6.5.3.2 Aromatic Polyamides

Aromatic PAs are inherently oxidatively and thermally stable, since they are produced at high processing temperatures. As a result, using antioxidants for processing the aromatic PAs is not necessary.

2.6.5.4 Polyesters

2.6.5.4.1 Poly(ethylene terephthalate) (PET)

Usually, PET is oxidatively stable, so it is processed and used without the addition of antioxidants.

2.6.5.4.2 Poly(butylene terephthalate) (PBT)

PBT is somewhat less stable under long-term heat exposure than PET. The effects of different antioxidants on impact strength after aging in an air-circulating oven at 160 °C are shown in Table 2.14.

Table 2.14 Effects of Antioxidants on Impact Strength of Oven-aged PBT at 160 °C

Stabilizer system	Aging time (h)[a]
None	60
0.02% AO-10	450
0.1% AO-10	570
0.05% AO-10/0.1% PS-2	560

[a] Time to impact strength of 80 kJ/m^2 at 160 °C. Test bars (4 × 6 × 50 mm) measured in a circulating-air oven.

2.6.5.5 Polyacetals

Oxidative degradation of polyacetals, such as polyoxymethylene (POM), generates volatile by-products, leading to measurable weight loss and visible discoloration of the material. Among the stabilizers evaluated for this class of polymers, phenolic antioxidants of the AO-10 type provide particularly good performance, combining strong long-term thermal stability with a significant reduction in color formation during the material storage at elevated temperatures.

Since polyacetals are highly sensitive to acid-catalyzed chain scission, even trace amounts of acidic moieties must be avoided in their formulations. For this reason, phosphite-type hydroperoxide decomposers are generally not recommended in these polymer systems, since they can undergo hydrolysis and yield acidic residues.

2.6.5.6 Polycarbonate (PC)

PC is prone to form discoloration as a result of thermal autoxidation that adversely affects its transparency, a key performance attribute of this polymer. PC also undergoes hydrolytic degradation in the presence of moisture, so even small amounts of acidic moieties must be avoided, since they catalyze chain scission.

To suppress yellowing and maintain transparency characteristics, hydrolytically stable phosphites or phosphonites are recommended for use in PC in combination with suitable acid scavengers. Any stabilizer selected for it must withstand the high processing temperatures of PC without volatilizing or decomposing. A hydrolysis-resistant phosphite such as PS-2 affords effective protection, while the combination of a phenolic antioxidant like AO-3 with PS-2 delivers particularly strong improvements in long-term thermal stability.

2.6.5.7 Polyurethanes (PUR)

Polyurethanes are used in a wide range of applications, including flexible molded foams, flexible slabstock foams, rigid microcellular foams (RIM-PUR) and thermoplastic polyurethanes (T-PUR). Their thermo-oxidative stability is affected by chemical structure, especially the nature of the polyol segments. Polyurethane systems based on polyester polyols generally show better resistance to thermal oxidation than those built on polyether polyols.

During foam manufacture, the exothermic reaction between polyols and isocyanates can cause temperatures to rise to about 160 °C, resulting in yellow-to-brown discoloration (scorch). In extreme cases, spontaneous ignition may take place if the reaction temperatures exceed 170 °C. Appropriate antioxidant packages are therefore essential to control oxidative degradation of polyurethanes. Blends of a phenolic antioxidant such as AO-4 with an aromatic amine (AO-40) and a highly reactive radical scavenger like AO-34 are effective at simultaneously improving long-term thermal stability and suppressing scorch and discoloration after foaming, as illustrated by data in Table 2.15.

Table 2.15 Effect of Antioxidants on Discoloration of Polyurethane Flexible Foam

Stabilizer system	Discoloration (Y.I.), ΔE[a]
0.26% AO-4/0.04% AO-40	60.5
0.10% AO-4/0.15% AO-40/0.05% AO-34	3.6

[a] Microwave oven test; PU foam was obtained using polyether polyol/TDI = 100:61.6.

2.6.6 Technological Outlook

In any polymer blend, the component that is most susceptible to thermo-mechanical and thermo-oxidative degradation should receive priority protection through the use of appropriate antioxidants and processing stabilizers.

Modern plastics are often formulated to incorporate multiple additives, including non-antagonistic (preferably synergistic) application and process-tailored heat and oxidation stabilizers, light stabilizers, processing aids, fillers, reinforcing agents, and colorants. The ongoing trend toward such multicomponent additive packages is expected to continue, with industry preference shifting to low-dust and dust-free commercial preparations that are easier and safer to handle and use.

2.7 Questions

1. How is autoxidation defined in the context of polymer degradation, and what are the two main characteristic features of this process in polymers?
2. Describe the basic free radical mechanism of autoxidation in polymers, naming the key reaction steps and the main radicals involved.
3. What are the main phases of autoxidation in polymers? What are the typical reactions that characterize each phase of the autoxidation cycle?
4. How are the peroxy radicals $ROO^{\bullet}$ formed during the oxidative degradation of polymers? Why is this step crucial for the continuation of the degradation process?
5. How are hydroperoxides ROOH generated on polymers during autoxidation? What role do they play in accelerating polymer degradation?
6. What are the factors that determine the upper processing temperature limits of a polymer? How are these limits related to the onset of thermo-oxidative and thermo-mechanical degradation?

7. What is the difference between primary and secondary antioxidants in polymers? How does each class of antioxidants act within the oxidation cycle to protect the polymer?
8. What is the function of metal deactivators in plastics? Why are they particularly effective in applications where polymers are in contact with copper or other transition metals?
9. How do sterically hindered phenolic antioxidants act as hydrogen donors in plastics? How do they interrupt the radical chain mechanism of autoxidation?
10. How do thiosynergists and organic phosphites function as hydroperoxide decomposers in polymers (mechanism of action)? What makes them effective secondary antioxidants?
11. What are the main stabilization functions of sterically hindered amine stabilizers (HAS) in plastics? Wha are their most critical which applications in polymers?
12. What factors affect polymer degradation via chain-scission mechanisms?
13. What types of antioxidants are typically used to stabilize polyolefins during melt processing? What are the main requirements for these stabilizers?
14. What antioxidant classes are commonly used to achieve long-term thermal stability (LTTS) in polyolefins? How are their modes of action different from those used for melt stabilization?
15. How can mineral fillers and reinforcing agents influence the thermal and thermo-oxidative degradation of polyolefins? Provide examples that illustrate both beneficial and detrimental effects of these additives.
16. What types of antioxidants are used for limiting post-irradiation degradation in polyolefins? How do they interact with the stable free radicals formed during the radiation exposure?
17. What considerations are applied to the selection and dosing of antioxidants intended for the processing of polystyrene and its copolymers (such as HIPS, ABS, and SAN)? How do these stabilizers control yellowing and deterioration of polymer properties?
18. What are the main antioxidant systems for polyamides, polyesters, polycarbonates, and polyurethanes?

3 Light Stabilizers

Content of this chapter was adapted from:

F. Gugumus, "Light stabilizers" (Chapter 2), in: H. Zweifel, R. D. Maier, and M. Schiller; *Plastics Additives Handbook* (6th edition), Hanser, 2009, pp. 139–424.

3.1 Key Terms and Definitions

Optics	The science of the study of light's properties, behavior, and interactions with matter, including reflection, refraction, and absorption.
Photophysics	Study of the physical processes that occur when matter interacts with light, such as absorption and emission.
Photochemistry	Branch of chemistry concerned with reactions initiated or accelerated by light energy absorption.
Light stabilization	Techniques or additives used to prevent polymer degradation caused by exposure to UV or visible light.
Artificial weathering	Laboratory simulation of outdoor exposure using controlled light, heat, and moisture to predict material durability.
Natural weathering	Exposure of materials to real outdoor environments to assess long-term performance and degradation.

3.2 Introduction to Optics, Photophysics, and Photochemistry

We start by considering the differences between optics, photophysics and photochemistry, which can be summarized as follows:

- Optics is science about light and its interactions with matter. It focuses on phenomena such as color, absorption, transmission, reflection, refraction, diffraction, scattering and polarization, without considering any chemical or physical changes in the substances involved.
- Photophysics deals with the physical processes triggered by the absorption of light in molecules. These include electron transitions, emissions like fluorescence or phosphorescence, and non-radiative energy transfers. The chemical structure of the molecules in this case remains unchanged.
- Photochemistry studies chemical reactions initiated by light. The absorption of light activates molecules, leading to the breaking or formation of chemical bonds. Examples include photocatalytic decomposition processes as well as the synthesis of new compounds.

Let's start with optics (Figure 3.1):

- **Absorption**: When light hits a material, some of its energy can be absorbed by the material. This energy is often converted into heat or used for chemical processes. The absorbed light can never be again released in the form of light.

  Transmission: This is when part of the light is not absorbed, and instead passes through the material. Depending on the material, the light intensity may be weakened, or its direction may change.

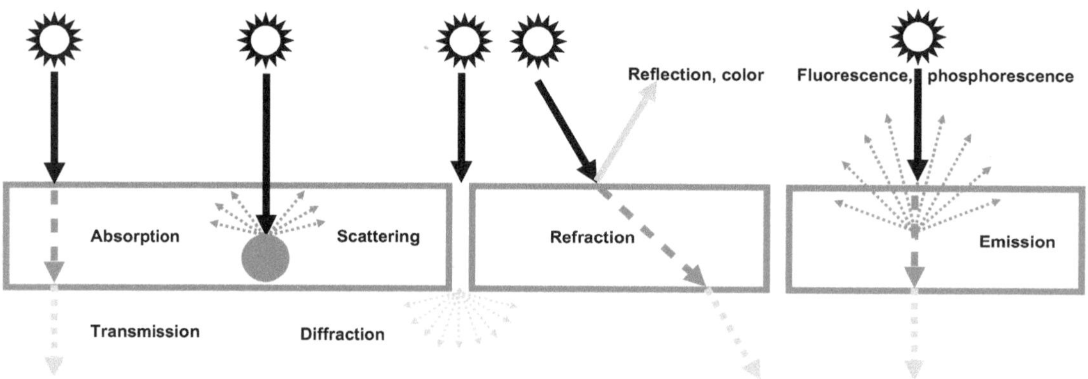

Figure 3.1 Optical phenomena

- **Light scattering** refers to the deflection of light rays when they interact with particles or molecules in a medium. This interaction causes the light to be redirected and distributed in various directions. The intensity and direction of light scattering depend on the wavelength of the light and the size of the particles it interacts with. There are different types of light scattering:
  - *Rayleigh scattering*: Occurs when particles are smaller than the wavelength of the light. It is responsible for phenomena like the blue sky, as shorter (blue) wavelengths are scattered more than longer (red) wavelengths.
  - *Mie scattering*: Occurs with particles that are roughly the same size as the wavelength of light. It leads to an even scattering of all wavelengths. That is why clouds appear white.
  - *Raman scattering*: Involves the transfer of some of the light's energy to the particles, exciting them and resulting in the scattered light having altered energy.
- **Diffraction** occurs when light spreads into shadow regions behind narrow slits or obstacles. This wave-like behavior can only be explained using the wave model of light. According to Huygens' Principle, every point on a wavefront acts as a source of new elementary waves that overlap and form a new wavefront.
- **Reflection** refers to the "bouncing back" of light waves from a surface without or only partially penetrating the material. Total reflection occurs when the light is completely reflected, such as in optical fibers.

Law of Reflection: The angle of reflection equals the angle of incidence.

- **Refraction** describes the change in a direction of light rays when transitioning between two media with different refractive indexes n (e.g., air and water), and causes light rays to appear "bent". This phenomenon is governed by Snell's Law.

Snell's Law:

$$n = \frac{c_0}{c_M}$$

where:

c_0 = speed of light in vacuum

c_M = speed of light in the medium.

- **Color** is a sensory impression created by the perception of light within the wavelength range of 400–780 nm, and which depends on the wavelengths of light reflected by an object. For example, a leaf appears green because it primarily reflects light with a wavelength of about 550 nm. Color perception is subjective and influenced by physical stimuli and the characteristics of the human eye.
- **Polarization** refers to the orientation of the oscillation plane of light waves. Unpolarized light oscillates in all directions, while polarized light oscillates only in one plane (e.g., after passing through a polarizing filter). Polarization is used in devices like cameras and sunglasses to reduce glare.

We now turn to light absoption, and start with a fundamental law:

> **Grotthus-Draper Law (First Law of Photochemistry):** Light can initiate a photochemical reaction only if it is absorbed.

Extinction E, also called optical density, specifies how much of the irradiated light of the intensity I_0 (of a defined wavelength) is absorbed by the material, this absorbed light intensity being called I_{abs}. I_0 is therefore the sum of the absorbed light I_{abs} and the transmitted light of intensity I (Figure 3.2):

$$I_0 = I_{abs} + I$$

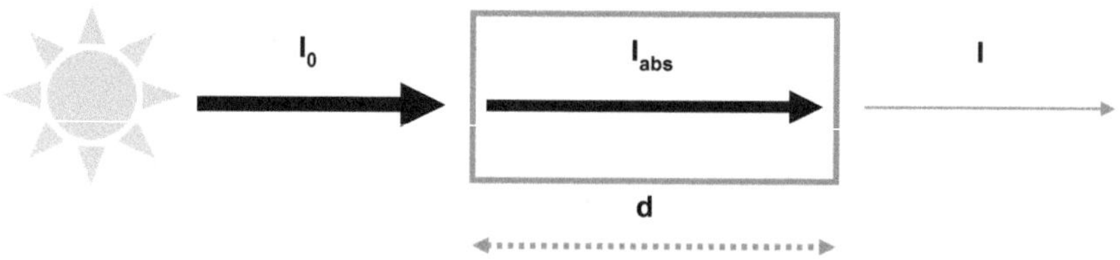

Figure 3.2 Light absorption

Extinction E is the negative logarithm of the quotient of I and I_0:

$$E = -\log\left(\frac{I}{I_0}\right) = -\log\left(1 - \frac{I_{abs}}{I_0}\right)$$

It also depends on:

- The wavelength of the irradiated light
- The thickness d of the layer through which the light has to pass through
- The concentration c of the absorbing substances
- The substance- and wavelength-specific coefficient ε.

This is expressed within another law:

Beer–Lambert Law:

$$E = cd\varepsilon$$

Total absorption is when all light of a specific wavelength is absorbed. In practice, this is often defined as an extinction value E greater than 2, i.e. when about 99% of the light is absorbed and only about 1% passes through the sample.

Total absorption is more easily achieved with an increased thickness d, higher concentration c and greater coefficient ε.

Let's dive deeper into the photophysics of molecular orbitals. When light strikes a molecule in the so-called ground state (S_0), it can boost an electron from its "home" (the HOMO, or highest occupied molecular orbital) to an "empty parking space" (the LUMO, or lowest unoccupied molecular orbital). This jump occurs only if the light's energy perfectly matches the HOMO–LUMO gap (Figure 3.3). This transition creates two singly occupied molecular orbitals (SOMOs): the original HOMO (now missing an electron) and the LUMO (now holding the excited electron). Crucially, the electron's spin remains unchanged during this process (the molecule is now in an excited, unstable state called S_1).[1)]

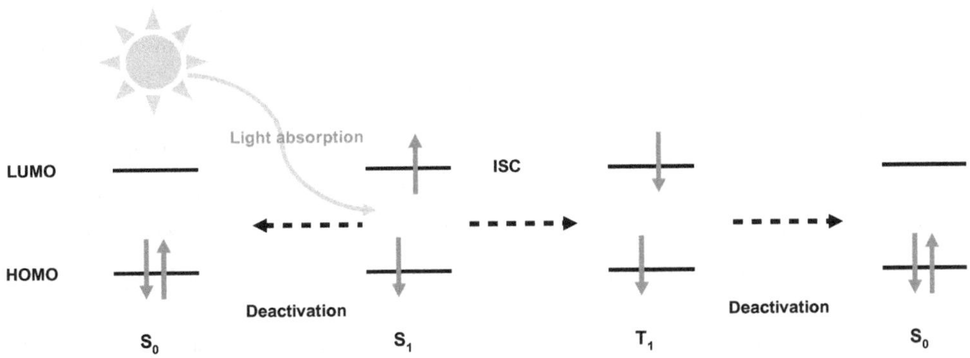

Figure 3.3 Photophysical process of molecular orbitals during and after light absorption

1 Think of spin as a tiny magnetic compass inside each electron. It can only point in two directions: up (↑) or down (↓). Although it's not a physical rotation (like a spinning top), this quantum property generates a minuscule magnetic field. According to the Pauli Principle, two electrons in the same orbital cannot be identical in every way. In part, this means that if they share the same energy level, their spins must oppose each other (↑↓). This rule explains why materials like iron exhibit magnetism and underpins technologies such as computer hard drives and MRI machines, among others.

An electron can have a spin of +½ (↑) or –½ (↓). The spin quantum number S describes the total spin of a system of electrons. The spin multiplicity S_M is calculated as follows:

$$S_M = 2S + 1$$

In the ground state S_0, there are two electrons with opposite spins:

$$S = +\frac{1}{2} - \frac{1}{2} = 0$$

This results in

$$S_M = (2 \times 0) + 1$$

(that's why it's called a singlet state).

Exercise 3.1

What if in the excited state, both electrons in the SOMOs had the same spin of +½?

The result is (2 × (½ + ½)) + 1 = 3, which explains why it is called a triplet state T.

The subscript numbers usually refer to defined energy levels (J from spin-orbit coupling) or excitation sequences. For organic molecules in photophysics, the following applies:

- S_0 = Singlet ground state.
- S_1 = First excited singlet state.[2]
- T_1 = Lowest triplet state (energetically below S_1).[3]

The subscript number (0, 1, 2) explicitly indicates the energy sequence. Now we just need to understand the meaning of "ISC" or intersystem crossing, as shown in Figure 3.3. This involves an excited electron in the S_1 state "secretly" changing its spin direction and switching to another energy state – without emitting light – but it does have consequences.

To explain these processes clearly, researchers use the Jablonski diagram – a map of energy states (Figure 3.4). Constructing an analogy with a waterfall may help here:

- Light absorption: The energy from light pumps water (electrons) from the valley (S_0) to the top of the rock (S_1). Non-radiative relaxation pathways, including vibrational energy dissipation and ISC, dominate the rapid deactivation of the S_1 state. Therefore, the lifetime of the S_1 state is in the nanosecond to picosecond range.
- Fluorescence: Some water cascades down immediately, releasing light.

[2] For many organic molecules, the S_1 lifetime is typically in the nano- to picosecond range (10^{-9} to 10^{-15} s), but it can also be in the femtosecond range for very fast processes.

[3] The lifetimes in the triplet state are significantly longer, usually in the range of microseconds to several seconds. The transition from the T_1 to the ground state is spin-forbidden, which is why relaxation proceeds slowly.

- ISC: A hidden turbine (ISC) silently redirects some water to a neighboring pool (T_1), flipping its spin (↑↑).
- Phosphorescence: Water in the T_1 pool drips slowly, releasing light over hours.
- Heat release: Some water evaporates (heat release) without any visible flow.

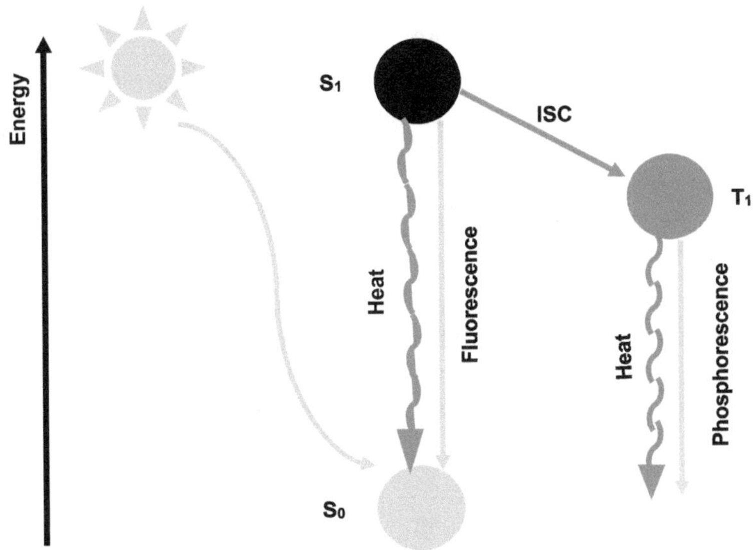

Figure 3.4 Simplified Jablonsky diagram

Intra- or intermolecular chemical reactions can occur from the excited singlet and triplet states. These reactions are then described by photochemistry. The short lifetime of the S_1 state (from nanoseconds to picoseconds) limits their applicability to fast reactions, such as isomerizations or electron transfer. Non-radiative processes (internal conversion) can reduce their quantum yield, if they occur faster than the desired reaction. The lifetime of the triplet state T_1 varies significantly depending on the molecule and its environment. Typical T_1 lifetimes range from microseconds to seconds, but in rigid environments (e.g., frozen, "glassy" matrixes), they can extend to hours.

To understand why it is that slow, we need to appreciate that the $T_1 \rightarrow S_0$ transition is spin-forbidden: the electron's spin must flip to return to the ground state, which is a slow, low-probability process. This spin restriction explains the unusually long T_1 lifetimes. Triplet sensitization is a special case, involving the indirect excitation of a substance into the triplet state through energy transfer from a light-activated sensitizer that previously transitioned to its triplet state. Important points are:

- The sensitizer remains chemically unchanged
- The energy transfer usually occurs via a mechanism that requires spatial proximity and overlap of the molecular orbitals

- The triplet energy of the sensitizer must be higher than that of the substance for the energy transfer to be energetically possible
- The substance can then undergo photochemical reactions in the triplet state that either would not occur (or would occur only very inefficiently) without sensitization. Triplet sensitizers are used in organic light-emitting diodes (OLEDs) to generate energy-efficient phosphorescence.

Let's consider two theoretical limiting cases below.

1. A "perfect light stabilizer" acts like a lightning rod, intercepting harmful energy and safely conducting it away. This is a substance that releases all excitation energy non-radiatively as heat (100% internal conversion). It protects materials from light damage by "disposing of" the energy "harmlessly". It will be discussed further in the following sections. However, in reality, even efficient stabilizers lose some energy through secondary processes.
2. An "ideal fluorescent dye" acts like a mirror and re-emits every absorbed photon as fluorescent light (100% quantum yield) that glows brightly and efficiently (see Chapter 9). In reality even the best fluorophores achieve only ~90% yield. Some of it is always lost as heat. And the wavelength of fluorescence is shorter than the absorbed light.

The fluorescence spectrum of a fluorescent dye typically exhibits mirror-image symmetry relative to the absorption spectrum of the same molecule.

This relationship arises from the Franck–Condon principle, as both absorption and emission transitions occur between similar vibrational levels of the respective electron states. However, the fluorescence spectrum is bathochromically shifted (i.e., shifted toward longer wavelengths or lower energies) compared to the absorption spectrum. This shift, known as the Stokes shift, takes place because some of the excitation energy is dissipated through rapid vibrational relaxation prior to emission.

The intersection point of the normalized absorption and fluorescence spectra approximately corresponds to the 0–0 transition energy between the electronic ground state (S_0) and the first excited singlet state (S_1). This point thus represents the energy level of the S_1 state.

Finally, photoexcited compounds can also be deactivated non-radiatively by transferring their excitation energy to other substances in the ground state. These acceptor molecules then become photoexcited or energy-rich themselves. This process is extremely important, as it enables generation of energy-rich (or excited) compounds without the need for direct light absorption in the relevant spectral range. This phenomenon is known as spectral sensitization. The acceptor is also referred to as the quencher. A distinction is made between:

- Singlet–singlet energy transfer
- Triplet–triplet energy transfer
- Electron transfer.

Electron transfer is a typical photochemical reaction, but here it proceeds with activation of the acceptor (A). It can originate from both singlet-excited and triplet-excited donors (D), resulting in the formation of a radical pair:

$$^{1,3}D^{*\cdot\cdot} + {}^{1}A_0 \rightarrow {}^{1,3}(D^{\cdot+}A^{\cdot-})$$

where:

$^{1,3}D^{*\cdot\cdot}$ = a photo-excited electron donor either in singlet or triplet state

$^{1}A_0$ = an electron acceptor in the ground state S_0

$^{1,3}(D^{\cdot+}A^{\cdot-})$ = a radical pair either in singlet or triplet state formed from a photo excited electron donor.[4)]

3.3 Light and Color

Sunlight reaching the outer layers of the Earth's atmosphere spans a continuous energy spectrum from approximately 0.7 nm up to about 3000 nm. As sunlight passes through the atmosphere, certain wavelength ranges are absorbed or scattered, significantly shaping the radiation that ultimately reaches the Earth's surface.

Short-wave UV radiation below 175 nm is absorbed by oxygen molecules in the upper atmosphere, at altitudes above 100 km. Additionally, the ozone layer in the stratosphere, located roughly 15–30 km above sea level, effectively absorbs UV radiation in the 175–290 nm range. Consequently, the UV radiation that initiates the photodegradation of plastics outdoors is primarily the 290–400 nm portion, which penetrates these atmospheric "filters" and reaches the Earth's surface.

As well as absorption, scattering sunlight by air molecules and aerosol particles (such as water droplets and dust) plays a critical role. According to Rayleigh scattering theory, the intensity of scattered light varies inversely with the fourth power of the wavelength. This means that shorter wavelengths (such as UV light) are scattered much more strongly than the longer wavelengths. This effect can lead to situations where on cloudy days, the diffuse UV radiation that reaches the surface actually increases, as

4 G. Roewer (Chapter 3), in: Autorenkollektiv, *Photochemie* [Photochemistry] (2nd edition), VEB Deutscher Verlag der Wissenschaften, 1975, p. 94.

scattered short-wavelength light preferentially reaches the ground. Studies have shown that scattering at cloud edges can enhance solar irradiance by 20% or more compared to clear-sky conditions.

The intensity and spectral distribution of global solar radiation depend primarily on the sun's position, which varies with time of day, seasons, and geographic latitude. The solar zenith angle determines the thickness of the atmospheric layer the sunlight must traverse, affecting absorption and scattering losses. Table 3.1 presents the intensity of global radiation at normal incidence on a horizontal surface, broken down by wavelength ranges. It reveals that UV radiation in the 280–400 nm range accounts only for about 6.1% of the total solar irradiance, with intensities of 5 W/m^2 in the 280–320 nm band, 27 W/m^2 in the 320–360 nm band, and 36 W/m^2 in the 360–400 nm band. In contrast, visible light (400–800 nm) comprises over 50% of the solar energy, and infrared radiation (800–2500 nm) contributes nearly 30%.

Despite its relatively small contribution to solar irradiance, UV radiation in the 290–400 nm range is the main driver of photochemical degradation in polymers.

Table 3.1 Approximate Intensity Fractions (expressed in %) of Global Radiation at Normal Incidence (AM1.5, standard spectrum, 1000 W/m^2) for Wavelength Ranges within the International Classification of Radiation Sources[5]

Wavelength (nm)	Classification	Proportion (%)
100–280	UV-C	0.0
280–315	UV-B	0.3
315–380	UV-A	4.6
380–440	Vis, violet	3.5

[5] The classification of wavelength ranges follows the conventions of DIN 5031-10 and the International Commission on Illumination (CIE). The percentage values represent approximate fractions of the global solar radiation at normal incidence (AM1.5, 1000 W/m^2) reaching the Earth's surface. The percentage values are approximate and may vary depending on atmospheric conditions, altitude, and the exact definition of spectral boundaries. The AM1.5 spectrum represents a standard reference for solar energy applications and may not correspond to all geographic locations or real-world conditions. For precise calculations, direct integration of the reference spectrum is recommended. These values were compiled with the assistance of the Perplexity.ai tool on 11 May 2025, and are based on the following references: (a) Reference Air Mass 1.5 Spectra, National Laboratory of the Rockies (NLR), *www.nrel.gov/grid/solar-resource/spectra-am1.5.html*; (b) DIN 5031: Radiometry and photometry – Part 10: Terms for the measurement of optical radiation, Deutsches Institut für Normung, 2000; (c) C. Riordan and R. Hulstrom, What is an air mass 1.5 spectrum? *Solar Energy*, 1986, 37: 491 – 496; (d) C. Honsberg and S. Bowden "Spectral distribution of solar radiation", PVCDROM (PV Education), *www.pveducation.org/pvcdrom/solar-radiation/spectral-distribution-of-solar-radiation*.

Wavelength (nm)	Classification	Proportion (%)
440–495	Vis, blue	7.1
495–580	Vis, green	15.2
580–640	Vis, yellow	14.7
640–750	Vis, red	16.6
750–1400	IR-A	32.0
1400–3000	IR-B	5.7
3000–1,000,000	IR-C	0.3

Geographical location strongly influences UV intensity. Near the equator, seasonal fluctuations are minimal, whereas at mid-latitudes (e.g., 50°N in Central Europe), UV irradiance at wavelengths of about 297.5 nm can vary by up to three orders of magnitude throughout the year. These variations are less pronounced but still significant at longer UV wavelengths such as 312.5 nm.

Table 3.2 shows the incident energy at some commonly used exposure locations. These values vary not only depending on the location, but also on the season and weather conditions. Data for other locations can also be found online.[6a)]

Table 3.2 Annual Incident Solar Energy at Some Commonly Used Exposure Locations[6)]

City	Country	Incident energy[a] (kWh/m^2)
Basel	Switzerland	1200–1300
Guangzhou	China	1300–1500
Pontecchio Marconi, Bologna	Italy	1400–1500
Melbourne	Australia	1500–1700
Singapore		1550–1650
Bogota	Colombia	1550–1700
Sao Paulo	Brazil	1550–1700
Bandol	France	1700–1800
Florida	USA	1700–2000

6 The data in this table was compiled from the following sources on 11 May 2025, supported by the Perplexity.ai tool: (a) Global Solar Atlas, World Bank Group, *https://globalsolaratlas.info/map*; (b) Photovoltaic Geographical Information System (PVGIS), European Commission, *https://re.jrc.ec.europa.eu/pvg_tools/en/tools.html*; (c) V. Quaschning, *Renewable Energy Systems: Technology – Calculation – Simulation* (10th edition), Springer Vieweg, 2023.

Table 3.2 Annual Incident Solar Energy at Some Commonly Used Exposure Locations *(continued)*

City	Country	Incident energy[a] (kWh/m^2)
Sicily	Italy	1750–1900
Darwin	Australia	1900–2100
Nairobi	Kenya	1900–2100
Spartan	South Africa	1900–2100
Almeria	Spain	1950–2100
Dhahran	Saudi Arabia	2100–2400
Arizona	USA	2100–2300
Cairo	Egypt	2200–2400
Upington	South Africa	2200–2400

[a] For an exposed plane inclined to the N or S at 45°.

Environmental factors, such as temperature and humidity, further modulate degradation processes. Higher temperatures increase effects of thermally induced oxidative degradation, which often occurs alongside of photooxidation. Water can influence the degradation through extraction, hydrolysis, or participation in photochemical reactions, especially when pigments like titanium dioxide are present.

The susceptibility of plastics to UV-induced degradation depends on their chemical structure and ability to absorb UV light. According to the Grotthus–Draper Law, only absorbed light can initiate photochemical reactions. Saturated hydrocarbons, such as PE and PP, generally absorb light only below 250 nm and thus are less prone to direct UV degradation. However, polymers containing chromophores (chemical groups with double bonds, carbonyl or other functionalities) absorb light at longer wavelengths (above 290 nm) and are more vulnerable to UV light-induced damage. Even polymers without significant absorption in the 290–400 nm range can photodegrade due to the presence of impurities or structural irregularities introduced during manufacturing, which is common for industrially produced polyolefins.

In summary, the UV radiation between 290 and 400 nm that reaches the Earth's surface, although a small fraction of the total solar energy, is the primary cause of polymer photodegradation outdoors. Its intensity is governed by atmospheric filtering, scattering, solar geometry, and environmental conditions. The chemical composition and purity of the polymer determine its sensitivity to this radiation and thus its durability under the sunlight exposure.

The mechanism of polymer degradation was discussed in Chapter 1, and the homolytic bond dissociation energies were provided in Table 1.5. To understand why chemical energy (in kJ/mol) correlates with light wavelengths, two levels must be connected:

- The "macro level": Energy is measured per mole because it works with macroscopic quantities (e.g., "193 kJ is released per mole of substance").
- The quantum level: Light consists of photons, and each photon carries a tiny portion of energy. To determine the wavelength, we need the energy of a single photon, not the entire mole.

According to a fundamental relationship in quantum mechanics, the energy of a photon is given by:

$$E_{\text{photon}} = \frac{hc}{\lambda}$$

where:

E_{photon} = energy of a single photon (in J)

h = Planck's constant (6.626×10^{-34} J·s)

c = speed of light (2.998×10^{8} m/s)

λ = wavelength (in m)

To calculate the energy per mole, the result needs to be multiplied by Avogadro's number ($N_A = 6.022 \times 10^{23}$):

$$E_{\text{mol}} = N_A E_{\text{photon}}$$

Plugging in all the constants and converting λ to nanometers (1 nm = 10^{-9} m), we get:

$$E_{\text{mol}} = \frac{119{,}627}{\lambda} \;\therefore\; \lambda = \frac{119{,}627}{E_{\text{mol}}}$$

where:

E_{mol} = homolytic bond dissociation energy (in kJ/mol)

Exercise 3.2

Calculate the wavelength of light that must be absorbed for homolytic cleavage of a H-CH_2CH_3 bond (410 kJ/mol).

Exercise 3.3

The C–F bond in CF_4 (tetrafluoromethane) has a dissociation energy of 543 ± 4 kJ/mol, one of the strongest chemical bonds known. Can it decompose in the Earth's atmosphere? Explain your logic and result.

The C–F bond in CF_4 has a very high dissociation energy, and consequently destruction of CF_4 molecules requires very high energy. The energy required to break the C–F bond homolytically corresponds to UV light with wavelengths shorter than about 220 nm. However, such high-energy UV radiation is almost completely absorbed by the Earth's ozone layer in the stratosphere, and it does not reach the lower atmosphere, where CF_4 resides. Furthermore, CF_4 is chemically inert under atmospheric conditions and does not react significantly with common atmospheric radicals like hydroxyl radicals ($OH^{\cdot}$), ozone, or other oxidants. Due to these factors (the high bond dissociation energy, the absence of reactive atmospheric species that can break the C–F bond, and the lack of sufficiently energetic UV photons reaching the troposphere), CF_4 does not decompose appreciably in the Earth's atmosphere. This explains its extremely long lifetime in the atmosphere, estimated to be about 50,000 years.

While the discussion above was focused on how sunlight, particularly its UV component, drives the photodegradation of polymers, it is important to recognize that the same fundamental principles of light absorption underline the phenomena of color and appearance in plastics.

The interaction of polymers and additives with visible light not only determines the stability of the polymers but also governs their optical properties, such as color, transparency, and gloss.

Roles of colorants in plastics are also discussed in Chapter 9. Dyes and pigments are deliberately introduced into polymer materials to achieve specific visual effects, ranging from vivid coloration to opacity and UV protection. Their effectiveness is rooted in their molecular structures that contain chromophores, chemical groups capable of absorbing light in the visible region. In particular, conjugated chemical structures, such as polyene sequences, play a central role in shifting the light absorption into the visible spectrum, thereby imparting color to the material.

Color appears when molecules absorb specific wavelengths of visible light. The amount of light which is not absorbed is reflected or transmitted, and that is what we perceive as color. For example, a molecule appears red because it "swallows" blue and green light, reflecting only red light. Absorption occurs when light energy excites electrons from their ground state to an excited state. The amount of energy required for it depends on the structure of the molecules as discussed above. Conjugated double bonds are those with alternating single and double bonds in a molecule (e.g., –C=C–C=C–C=C–C=C–). This structural arrangement allows electrons to become delocalized. They are not bound to a single carbon atom, but are "spread out" over several atoms. These delocalized electrons need less energy to be excited, and so instead of absorbing UV light (typical for isolated double bonds), they are able to absorb visible light.

The longer the conjugated molecular segment, the longer the wavelength (and the lower the energy) of the absorbed light (see Figure 3.5 and Table 3.3).

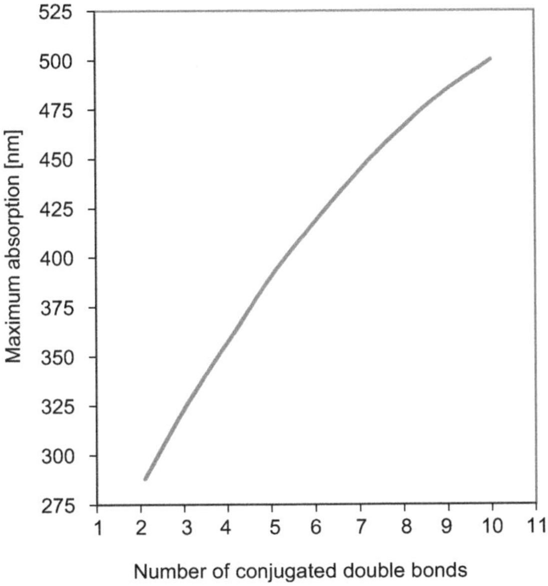

Figure 3.5 Correlation of UV–Vis peaks and the number of double bonds in polyene sequences[7)]

Table 3.3 UV–Vis Peaks of Polyene Sequences (*n* = Number of Double Bonds)[8)]

n	Isomer	Peak wavelength (nm)												
3	Triene	240	248	257	268	—	—	—	—	—	—	—	—	—
4	Tetraene	—	—	—	267	278	290	304	—	—	—	—	—	—
5	Pentaene	—	—	—	—	279	290	304	317	334	—	—	—	—
6	Hexaene	—	—	—	—	—	—	300	313	328	344	364	—	—
7	Heptaene	—	—	—	—	—	—	—	316	332	350	368	390	—
8	Octaene	—	—	—	—	—	—	—	—	332	349	367	386	410
	Average	240	248	257	268	279	290	303	315	332	348	366	388	410

[7] B. Ranby and J. F. Rabek, *Photodegradation, Photo-oxidation and Photostabilisation of Polymers*, Wiley, 1975, pp. 192 – 195.

[8] M. Schiller, *PVC Additives: Performance, chemistry, developments, and sustainability* (2nd edition), Hanser, 2022.

Examples of molecules with conjugated double bonds are:

- β-Carotene, which is present in carrots and has 11 conjugated double bonds; it absorbs blue light and appears orange (Figure 3.6)
- Chlorophyll, which absorbs red and blue light and appears green.

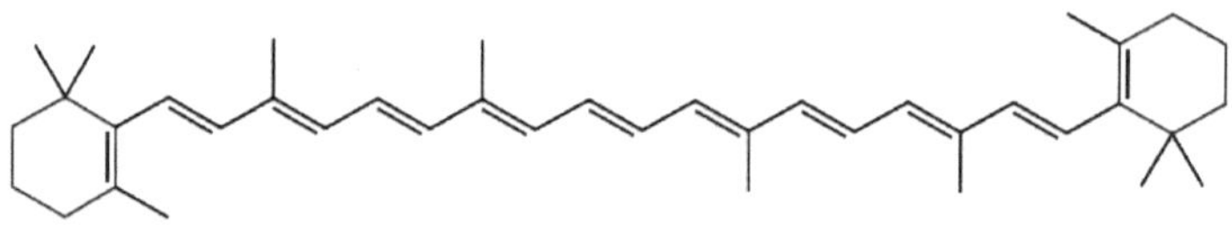

Figure 3.6 Chemical structure of β-carotene

3.4 Chemical Processes Initiated by Light Absorption

When the light falls on a polymer, it is either scattered on the surface or within the bulk of the material, or it is absorbed (Figure 3.1). According to the Grotthuss–Draper Law, only the portion of light that is effectively absorbed can induce photochemical reactions, such as polymer degradation. We now consider another key law in photochemistry:

Stark–Einstein Law (Second Law of Photochemistry): Light absorption by a molecule is a one-quantum process,[9] meaning that the sum of primary quantum yields[10] must equal one or less.

The absorption behavior of synthetic plastics is primarily determined by their chemical structure. Saturated hydrocarbons do not absorb light with wavelengths longer than about 250 nm. The presence of double bonds (chromophores) shifts absorption to longer wavelengths, especially when double bonds involved are between carbon and heteroatoms (atoms other than carbon, e.g., oxygen, nitrogen, sulfur, and phosphorus). For example, carbonyl groups (C=O) absorb light above 290 nm.

[9] Up to this point, we have attributed only wave properties to light. However, light exhibits both wave-like and particle-like properties, a phenomenon known as wave-particle duality. Depending on the experiment, light shows either interference patterns (wave behavior) or appears as discrete packets called photons (particle behavior). Both descriptions are equally valid and complementary: light is a quantum object whose nature cannot be fully described by classical concepts alone.

[10] The quantum yield indicates how efficiently a photochemical or photophysical process is. It describes the ratio between the number of events triggered by light (e.g., molecules transformed, photons emitted) and the number of absorbed light quanta (photons). A quantum yield of 1 means that each absorbed photon leads to one event.

Light with wavelengths of above 290 nm can still cause degradation in polymers like polyolefins that do not directly absorb such wavelengths due to the polymer structure. This effect is caused by trace impurities or structural irregularities commonly found in industrially produced polymers. The impurities may include catalyst residues and/or oxidation products, and these can lead to significant UV absorption and thus photochemical transformations of polymers containing them.

In semi-crystalline polymers, light scattering on crystalline segments increases the optical path length compared to amorphous segments and materials. This enhances the absorption of energy on chromophoric groups present even at low concentrations.

Polymer degradation depends on the absorbed light wavelength: different UV light wavelengths present in sunlight vary in terms of their effectiveness and can induce different degradation types. Understanding the sensitivity of plastics to different wavelengths is therefore crucial for assessing weathering factors and effects.

From the energy point of view, the sensitivity of polymers to light is not surprising. Photon energies in UV light range from 290–400 nm, corresponding to an ability to break bonds of strength 419–297 kJ/mol. This is the energy of certain bonds typically found in plastics (see also Exercise 3.2 and Table 1.3). However, this energy must first be absorbed by a specific functional group within the polymer before it can lead to bond cleavage. The ratio of the number of polymer molecules that undergo a chemical reaction to the number of photons absorbed by the material is called the quantum yield. For certain polymers, the quantum yield of chain scission reactions is in the range 10^{-2} to 10^{-5}. This means that out of 100 to 100,000 polymer molecules that absorb light, only one is actually degraded. This relatively low quantum yield, compared to the effect of small molecules in photochemistry, is attributed to the fact that the absorbed energy in the polymer is initially localized in the polymer and then distributed to adjacent regions. Furthermore, the so-called "cage effect" in solid polymers often leads to the recombination of the cleaved chains.

The chemical structures and general mechanisms of polymer degradation are discussed in Chapter 1.

3.5 Mechanisms of Light Stabilization

Studies show that even small amounts of impurities, defects in the polymer structure, or the presence of chromophores are sufficient to initiate the photooxidative degradation of plastics even if the polymer itself absorbs little UV light. To prevent or slow down this degradation, targeted UV stabilization is essential.

There are four main mechanisms for UV stabilization:

- Reducing UV light absorption by chromophores
- Deactivating the excited states of chromophoric groups
- Transforming reactive hydroperoxides into stable compounds before they can form radicals
- Scavenging free radicals immediately after their formation.

These mechanisms also form the basis for classifying light stabilizers as follows:

- UV absorbers
- Quenchers
- Hydroperoxide decomposers
- Radical scavengers.

Many light stabilizers participate in multiple mechanisms simultaneously.

3.5.1 UV Absorbers in Polymer Stabilization

3.5.1.1 Introduction and Mechanism of Action

UV absorbers are a critical class of light stabilizers used to protect polymers from the damaging effects of UV radiation. Their primary function is to absorb harmful UV light and dissipate the absorbed energy harmlessly as heat, thereby preventing UV-induced damage to the polymer chains. For effective protection of polymers, UV absorbers must themselves be highly photochemically and thermally stable; otherwise, they would be quickly consumed in side-reactions and lose their protective functionality.

A fundamental limitation of UV absorbers is that they require a certain absorption depth to provide effective protection. This means that for thin samples, such as fibers or films, either very high concentrations of UV absorbers are needed (which may cause compatibility issues with the polymer matrix), or the protection imparted by them might be limited.

3.5.1.2 Main Classes of UV Absorbers

Hydroxybenzophenones (e.g., 2,4-dihydroxybenzophenone) are among the most widely used type of UV absorbers. Their effectiveness is strongly influenced by intramolecular hydrogen bonding through *ortho*-hydroxy groups that affect both spectroscopic and photochemical properties. By introducing suitable substituents, the absorption maximum can be shifted into the 300–400 nm range, which is especially relevant for polymer protection. The protective mechanism is based on an intramolecular proton transfer and rapid non-radiative transitions that convert the absorbed energy into heat (Figure 3.7).

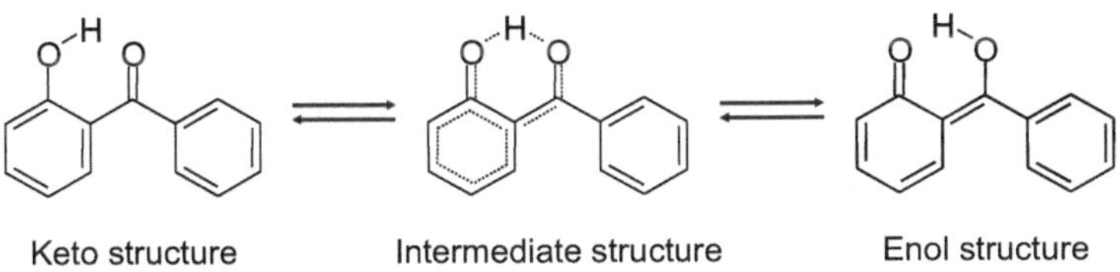

Figure 3.7 UV-light-stabilizing mechanism of 2-hydroxybenzophenones by intramolecular transfer of hydrogen

---

The strength of the hydrogen bond is crucial for the stabilizing effect. Polar media disrupt this bond, meaning that hydroxybenzophenones become less stable and may even act as photosensitizers.

2-(2′-Hydroxyphenyl)benzotriazoles are another important type of UV absorber, featuring high molar extinction coefficients and sharp absorption edges at around 400 nm. Their stabilization mechanism involves energy dissipation through excited-state proton transfer (Figure 3.8), in a fashion analogous to 2-hydroxybenzophenones (Figure 3.7). However, unlike benzophenones, benzotriazoles do not populate triplet states and almost completely revert to the ground state after the energy dissipation.

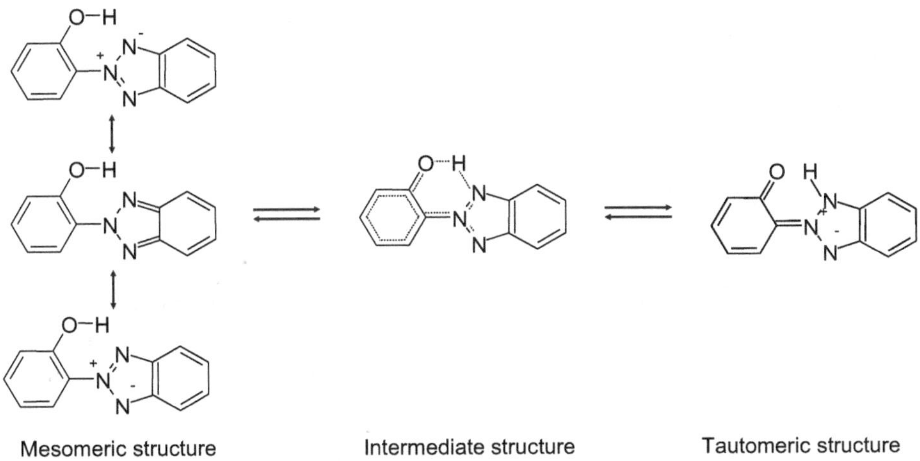

Figure 3.8 UV-light-stabilizing mechanism with 2-(2′-hydroxyphenyl)benzotriazoles, showing the involvement of mesomeric and tautomeric structures[11)]

---

11 *Resonance* (mesomerism) refers to a situation where a single molecule can be represented by two or more structures that differ only in the arrangement of electrons (as opposed to the positions of atoms). The actual molecule is a "hybrid" of these resonance forms with delocalized electrons (e.g., benzene). *Tautomerism* describes a dynamic equilibrium between two different molecules (tautomers) that have the same molecular formula but differ in the arrangement of atoms and bonds. The tautomers, such as such as keto and enol forms, can rapidly interconvert.

In recent decades, UV absorbers with novel structures have been commercialized. These newer additives belong to the class of 2-(2′-hydroxyphenyl)-1,3,5-triazines that are characterized by the significantly higher molar extinction coefficient compared to classical UV absorbers of the benzophenone and benzotriazole types (Figure 3.9).

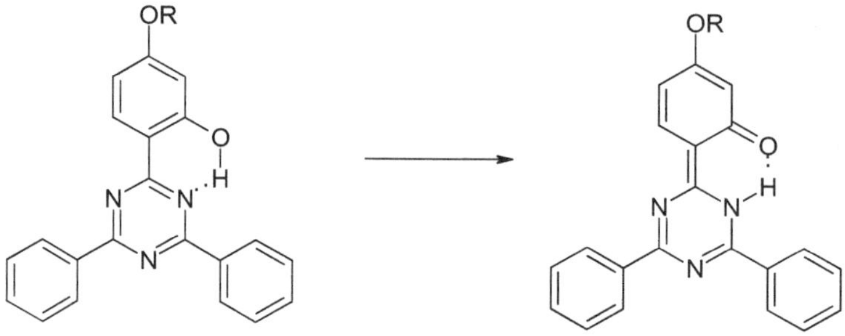

Figure 3.9 Chemical structures of 2-(2′-hydroxyphenyl)-1,3,5-triazines (R: CH_3 to $C_{12}H_{25}$)

As in the case of 2-hydroxybenzophenones and 2-(2′-hydroxyphenyl)triazenes (see Figure 3.7 and Figure 3.8), it is assumed that the energy from the singlet state is dissipated through an internal transfer of the hydrogen atom.

In contrast to the typical UV absorbers described so far, the compounds shown in Figure 3.10 cannot be considered as absorbers in the conventional sense. This is because when exposed to UV light, they rearrange to yield 2-hydroxybenzophenone, and this is likely to be the actual active species. The monobenzoate of resorcinol (top left) rearranges to form 2,4-dihydroxybenzophenone (top right). Similarly, the 2′,4′-di-*tert*-butylphenyl ester of 3,5-di-*tert*-butyl-4-hydroxybenzoic acid (bottom left) rearranges to 2-hydroxybenzophenone (bottom right). Since these precursor compounds must undergo rearrangement before effective stabilization can occur, their efficiency is rather limited. Additional challenges arise since the photo-Fries rearrangement is not quantitative, resulting in the simultaneous formation of by-products. This can lead to complications, such as discoloration. Furthermore, the compatibility of these compounds in polyolefins is quite limited, and that restricts their usage levels.

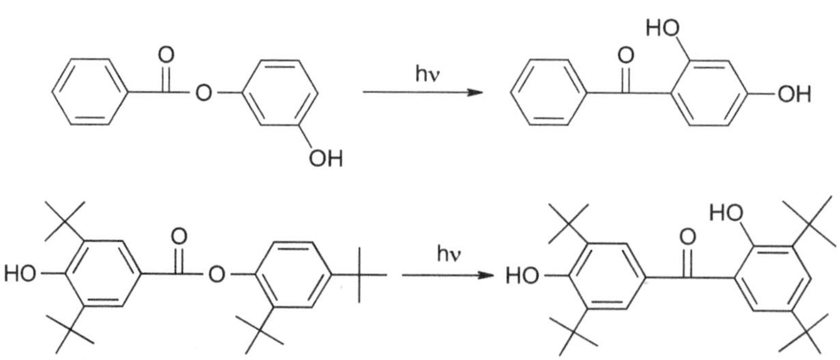

Figure 3.10 Photo-Fries rearrangement of benzoate-type light stabilizers

Cinnamates (esters of cinnamic acid; Figure 3.11) absorb UV light, with a maximum at 290–320 nm and a “shoulder” of ~370–390 nm. They are particularly efficient at converting absorbed energy into heat due to their numerous vibrational modes. Their stabilizing effect is based on both the filter effect and quenching of excited states. However, their general efficiency is limited by poor compatibility with many polymers, and that restricts their applicability.

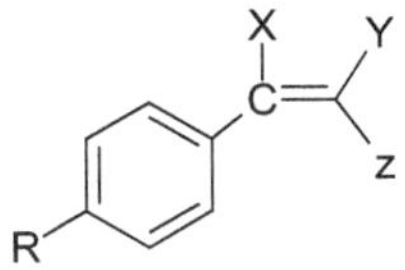

Figure 3.11
Principal structure of cinnamate-based UV light stbilizers for polymers (X: H, CH_3, C_2H_5; Y, Z: CN, O–C(O)–alkyl; R: H, OCH_3)

Oxanilides absorb even shorter wavelengths than cinnamates and are mainly used for the protection of materials in the 280–340 nm range. Their protective mechanism is almost exclusively based on the internal filter effect. Their excited singlet state exhibits a very short lifetime, enabling rapid and reversible energy dissipation. Neither triplet quenching nor radical scavenging mechanisms play a significant role for oxanilides (Figure 3.12).

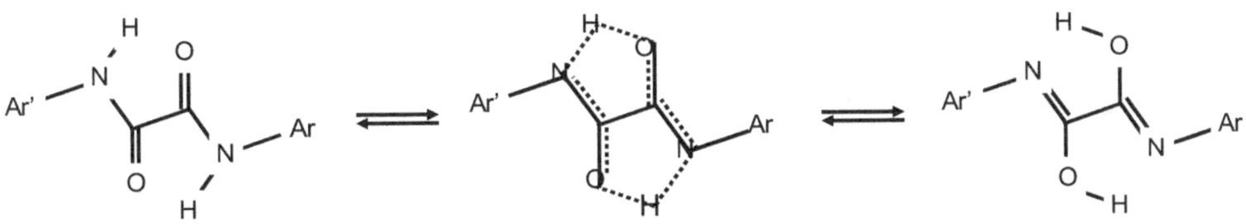

Figure 3.12 Mechanism of UV stabilization with oxanilides via intramolecular hydrogen transfer

There is also a commercially available representative of the benzoxazinone class (Figure 3.13). They absorb UV light, especially in the UVA range from 320–380 nm. Their conjugated aromatic structure efficiently dissipates energy as heat, and their high heat resistance (degradation >370 °C) enables applications in polymers processed at elevated temperatures, including PET and PC. Benzoxazinones are typically more expensive than the conventional UV absorbers.

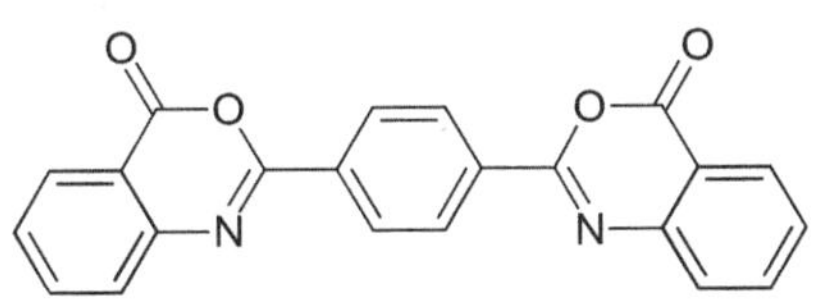

Figure 3.13
Chemical structure of benzoxazinone

Titanium dioxide demonstrates multiple functionality. It exists mainly in three structural forms: anatase, brookite, and rutile. Anatase and rutile are technically viable pigments, with the high diffraction indexes of 2.55 (anatase) and 2.75 (rutile) resulting in excellent tinting strength and hiding power. However, rutile is also an effective UV absorber (while anatase acts as a photocatalyst that accelerates the decomposition of polymers).

The process of UV absorption in titanium dioxide takes place as follows. In the ground state, the electrons of the titanium dioxide are located in the so-called valence band (vb) (Figure 3.14). When the light hits titanium dioxide and is absorbed (below 400 nm), an electron (e^-) is lifted from the valence band into the so-called conduction band (cb). This process leaves a positively charged hole (h^+) in the valence band. With a certain probability, electron falls back into the hole. When this happens, the absorbed light energy is discharged as thermal energy (heat).

The more likely the recombination of the electron and the hole, and the faster this takes place, the more pronounced the UV stabilizing effect.

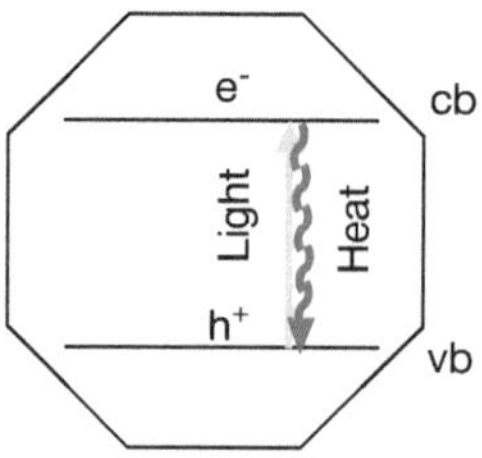

Figure 3.14
Photochemical processes in titanium dioxide during and after irradiation with UV light

Carbon black (flame black) acts as a light stabilizer in plastics, primarily by serving as a UV light absorber and blocker. It protects polymers from ultraviolet radiation by absorbing and scattering UV light, preventing it from penetrating deeply and causing photochemical degradation. Unlike chemical stabilizers, carbon black works through a physical mechanism, converting UV light energy into heat without undergoing chemical changes itself. This makes it one of the most effective inorganic light stabilizers, widely used in materials requiring high UV light resistance, such as pipes, films, and outdoor housings.

3.5.2 Quenchers

Quenchers were already mentioned in Section 3.2. They are capable of deactivating a substance from either the ground state (S_0) or excited singlet (S_1) or triplet (T_1) states, without radiation, through energy or electron transfer. In other words, they quench the excited state. Energy transfer from excited carbonyl groups to the quencher is one commonly assumed mechanism. Another mechanism involves quenching of excited molecular oxygen (singlet oxygen) formed or present within the polymer. In contrast to UV absorbers, the effect of quencher-based light stabilizers is largely independent of sample thickness. This makes them particularly effective in thin-walled articles such as films or fibers. For example, nickel-containing quenchers have been extensively used to stabilize agricultural films. In practice, only nickel chelates exhibiting the quenching mechanisms, and described above, are applied as light stabilizers. Specific representatives of this type of additive, which offer excellent light-stabilizing performance, include nickel phenolates and nickel dithiocarbamates.

However, in 1990, the International Agency for Research on Cancer (IARC) classified nickel compounds as Group 1 carcinogens (carcinogenic to humans), and since then, use of nickel chelates has steadily declined. Environmental and health concerns associated with these chelates also play a major role here, not only during the handling of heavy metal compounds but also due to the uncontrolled release of these hazardous substances into the environment at the end of the polymer article's life.

3.5.3 Hydroperoxide Decomposers, Radical Scavengers and Hindered Amine (Light) Stabilizers (HALS)

The previously used definition of light stabilizers[12)] is very broad, so that hydroperoxide decomposers and the substances described in the following sections have also been included. Hydroperoxides play an essential role in the photo-oxidative degradation of polymers (see Figure 1.19 and Figure 1.20). The ability of additives to protect polymers by decomposing peroxide impurities was postulated in 1965 and was demonstrated in plastic films nearly 10 years later. Metal complexes of sulfur-containing compounds, such as dialkyldithiocarbamates, dialkyldithiophosphates, and thiobisphenolates, are very efficient hydroperoxide decomposers, even when used only in catalytic amounts. Zinc-based compounds are of less concern compared to the nickel-containing compounds in terms of their potential health risks.

For more information about hydroperoxide decomposers, see the sections about phosphites (Section 2.5.2.1) and thiosynergists (Section 2.5.2.2). Chapter 2 is also recommended for discussion of radical scavengers (see, for example, Section 2.4) and hindered amine (light) stabilizers or HA(L)S (Section 2.5.3). HALS are basic, mostly tertiary amines, whose ability to accept protons determines their chemical reactivity and stability within the polymer. The basicity of HALS is a crucial factor for their effectiveness as light stabilizers in various polymers. Table 3.4 shows the basicity of some classes of hindered amines based on their pK_a values (the higher the value, the more basic the amine).

[12] H. Zweifel, R. D. Maier, and M. Schiller, *Plastics Additives Handbook* (6th edition), Hanser, 2009.

Table 3.4 Comparison of Basicity of Hindered Amines Based on their pK_a Values

Structure	pK_a
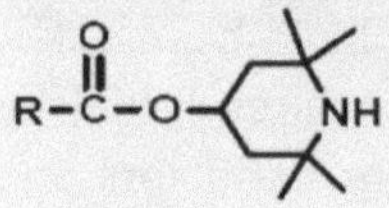	9.0
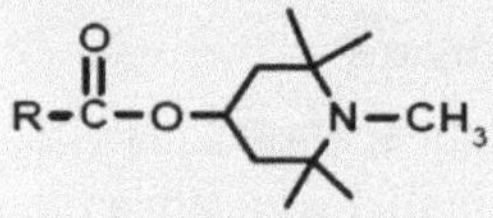	8.9
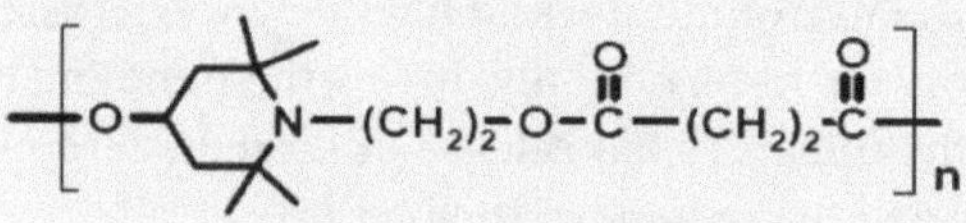	6.5
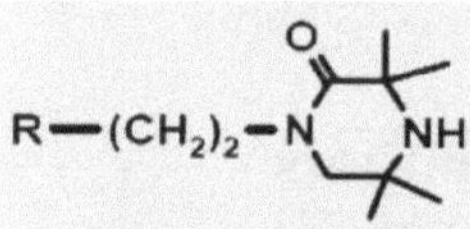	6.8
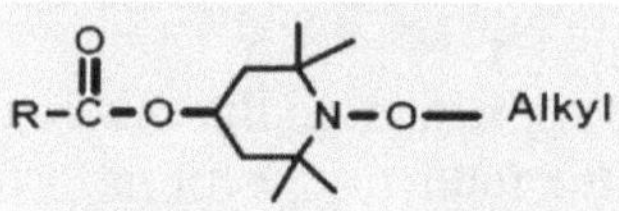	4.4

During PVC processing and aging, hydrochloric acid (HCl) is released from halogen-containing polymers like PVC (Section 3.7.6). The acid reacts with basic HALS, reduces their basicity and forms ammonium chloride. This reaction deactivates HALS and causes them to lose their polymer stabilizing effect, interrupting the Denisov cycle[13] and destroying their function as radical scavengers. Therefore, classic HALS are of limited effectiveness in many PVC compounds. However, in pigmented or modified systems, for example, in the presence of titanium dioxide or in combination with UV light absorbers, HALS-2 can significantly improve the imparted light stability. The combination of HALS-2 with the benzotriazole UV absorbers such as UVA-2 is particularly effective, whereas HALS-1 alone shows little effect in PVC.

In elastomers (Section 3.7.2) such as synthetic rubber, there is no significant release of acids like HCl during aging. Here, HALS are not deactivated, but maintain their basicity and can fully realize their light-stabilizing effect. The basicity is also benefi-

[13] The Denisov cycle works as follows: A nitroxyl radical (often represented as NOR) reacts with a harmful alkyl radicals (R$^\bullet$) in the polymer, forming a stable alkoxyamine (R–ONR). This reaction effectively stops the radical chain propagation, preventing further damage.

cial for promoting the efficiency of the Denisov cycle. Therefore, HALS are very effective light stabilizers for elastomers and are frequently used to enhance their UV resistance.

The situation is more complex in polyamides (Section 3.7.5). Carboxylic groups of polyamides can interact with HALS. This can lead to salt formation, causing migration and adverse dispersion, and resulting in reduced effectiveness of the HALS. In some cases, this may reduce the effectiveness of light stabilization or the mechanical properties of the polymers. Therefore, specially adapted, less basic HALS derivatives are often used in polyamides.

3.6 Testing of Light Stabilizers

Evaluation of the effectiveness of light stabilizers in a given substrate is conducted through indoor and/or outdoor weathering tests. Depending on the polymer's properties and its intended applications (fibers, films, profiles, etc.), various characteristics are examined. These characteristics primarily include microcracks, surface texture, gloss, chalking, color changes, and transparency, as well as mechanical properties like elongation, tensile strength, and impact resistance. For more fundamental analyses, additional methods are employed, including molecular weight distribution, viscosity in solution or melt, and spectroscopic methods like carbonyl buildup measurements using IR spectroscopy. Existing standard test methods should be used whenever possible. Experience shows that there is a rough correlation between the outdoor and accelerated weathering results.

3.6.1 Outdoor Weathering

Outdoor weathering testing of polymer specimens is carried out on racks oriented to the south (in the northern hemisphere) with the exposed surface typically inclined at 45°. The specimens should be mounted on suitable supports (preferably made of stainless steel, aluminium, or PMMA) or on frames. Outdoor weathering tests behind glass is another option. The relevant standard for this procedure is DIN 53386, corresponding to ASTM D1435 as well as ISO standards ISO 4582 and ISO 4607. Typical locations for setting up the outdoor weathering tests are listed in Table 3.2 and ref. 6b cited at this point enables determination of the incident solar energy for almost any location.

3.6.2 Artificial Weathering

An alternative to outdoor exposure test is artificial weathering testing, for which there are several test methods.

- Fluorescent lights emitting in the range 280–315 nm simulate the UV-B radiation of sunlight, of which normally very little reaches the Earth's surface due to absorption in the atmosphere, especially in the ozone layer. This wavelength range has the highest energy level of those discussed here. Using a UV-B light source, differences between test samples are often revealed after exposure for less than 1000 h. However, the tests conducted under this strong radiation allow only for limited predictability about natural weathering performance.
- Fluorescent lights emitting in the range 315–400 nm simulate the UV-A radiation of the sun, which also reaches the Earth's surface. UV-A radiation has less energy than UV-B but still has significantly more power than the visible spectral range of the sun.
- Xenon lamps emit a light spectrum that very closely resembles that of the sun. These lamps are therefore used for conducting artificial weathering. If the apparatus also incorporates alternating dry and wet/rain phases as well as light and dark cycles, then the result closely resembles natural weathering. A further advantage of this method is that compared to the natural weathering, other environmental factors are excluded from it.
- Tungsten emitters are still used for artificial weathering today. However, the UV proportion in their spectrum is rather small, while the proportion of near-infrared radiation is relatively high.
- High-pressure mercury lamps do not emit a continuous spectrum (unlike xenon lamps), and have strong emission bands at 254, 303, 313, 333, 365, 405–408, 436, 546, and 577–579 nm. They should be used only if the intent is to excite specific components in a sample.

The Bandol wheel is an apparatus for artificial weathering, and involves a medium-pressure mercury lamp with the emission spectrum similar to that mentioned above. A vertical sample holder accommodates up to 70 samples, and shutters allow light–dark cycles. Not only effect of water but also environmental contaminants, such as rust, pollen, soot, or even "acid rain", can be studied with this test design.

In Table 3.5, some typical "myths" of artificial weathering tests and their shortfalls are summarized. It lists parameters that can be changed during the artificial weathering investigations and the standards related to them.

Table 3.5 Accelerated Weathering Tests: Myths and their Interpretations[14)]

Reason for poor correlation with outdoor tests	Myth	What really happens
Short wavelength light sources (outside of the solar spectrum).	Shorter wavelengths of radiation have more energy to degrade materials faster.	High energy contained in light with shorter wavelengths causes unnatural photochemical changes.
Continuous exposure to light.	Maximizing the amount of time a material is exposed to light shortens the test time.	Some materials need a "rest period" for certain chemical reactions to take place. Since the natural outdoor exposure will always have a "dark cycle", it makes sense to simulate this as part of design of the artificial weathering tests.
High light intensities (especially with artificial light sources).	Blasting samples with high irradiance is the only way to get the weathering acceleration needed.	Some photochemical changes during exposure may be altered by high irradiance levels in a way that does not occur under normal light intensities.
Abnormally high specimen temperatures.	Rapid acceleration can be achieved at higher temperatures due to faster degradation rates.	Exposure of specimen to unrealistic temperatures often cause different types of degradation that does not correlate with outdoor exposures.
No temperature cycling, while keeping temperature high.	Keeping the test temperatures high enables results to be obtained more quickly.	Natural temperature cycling may cause additional physical changes to materials as a result of the expansion and contraction of materials.
Unnatural levels of moisture.	Soaking specimens in excessive moisture increases degradation.	The absorption/desorption cycle of water causes physical stress that may result in more degradation than a moisture-saturated environment.
Absence of pollutants or other biological agents.	Since these are "secondary" factors, their effect on the test results is insignificant.	Laboratory weathering chambers are rarely, if ever, used to replicate the effects of pollutants or other biological factors, but they are an inherent part of the natural weathering process and may cause lower than expected correlation with the natural weathering.

[14] Correlation and acceleration, *Sunspots: Material Testing Product and Technology News*, Atlas Material Testing Solutions, 2001, 31(65): 14, *www.atlas-mts.com/knowledge-center/archive/sunspots*.

3.6.3 Influence of Colorants

Initial testing of light stabilizers is usually performed in non-pigmented substrates. However, commercial plastic materials are usually colored. Colorants (Chapter 9) can influence the lightfastness of the stabilized plastics in several ways when assessing weatherability. They can act as sensitizers and thus drastically reduce the lightfastness of UV-stabilized samples. This occurs primarily with some organic yellow and organic red pigments, but also with inorganic pigments.

Some other colorants, such as phthalocyanine blue and phthalocyanine green, result in a significant increase in the UV stability of thick areas, especially in stabilized samples. Some colorants can act as UV absorbers and can prevent harmful UV radiation from reaching the deeper layers of plastic materials.

The sample temperature is strongly affected by pigmentation. The surface temperature of irradiated samples, for example, depends on the intensity of the radiation absorbed by the sample, the ambient temperature, and the air circulation. Table 3.6 shows the measurement results for PVC and other polymers. It is evident that the surface temperature increases significantly when the pigmentation changes from white to gray or black. In general, oxidation reactions are strongly dependent on surface temperature. Accelerated oxidation reactions are to be expected in dark pigmented samples.

Table 3.6 Surface Temperature of Several Pigmented Polymers during Weathering

Polymer	Color	Surface temperature (°C)	Remarks
PVC	White	33	Weather: Cloudy Outside temperature in the sun: 39 °C Black plate temperature: 51 °C Latitude: 51.6°N Time: 12:00–02:00 Date: August 1973
	Yellow	38	
	Red	40	
	Blue	41	
	Green	43	
	Gray	47	
	Brown	49	
	White	~35	Air temperature: 37 °C Black plate temperature: 50 °C
	Dark brown	~46	
PMMA	White	~44	
	Black	~49	
PE	Black	52	

However, surface temperatures are not dependent on the pigment alone; the chemistry of the plastic material also has a strong influence. Differences in surface temperatures between light- and dark-pigmented samples are less pronounced in PMMA samples than, for example, in PVC samples. This finding can be attributed to a significantly higher absorption of infrared radiation by PMMA compared to PVC.

The surface temperature values shown in Table 3.6 do not represent an upper limit. Even in temperate climates, temperatures above 60 °C can be measured on dark pigmented samples. In tropical climates, dark pigmented samples can reach temperatures of up to 80 °C due to long exposure to the sunlight and high air temperatures. In unfavorable cases, e.g. with insufficient air circulation, even 90 °C has been reached.

3.6.4 Technical Aspects of Light Stabilization

Analysis of light stabilizers is often performed using established analytical methods, such as thin-layer chromatography (TLC), gas chromatography (GC), and high-performance liquid chromatography (HPLC), similar to those used for other plastic additives. Commercial light stabilizers are usually available as powders or viscous liquids. To enable precise dosing and homogeneity, they are often added in a masterbatch during granulation along with other additives, such as antioxidants. The usefulness of light stabilizers is determined not only by their protective effects, but also by their cost, chemical, physical, and toxicological properties. These factors can severely limit their use, especially in applications with special requirements (e.g., food contact).

Stabilizers should not degrade during processing even at temperatures up to 300 °C, and most light stabilizers meet this requirement. Discoloration can be caused by the polymer itself or by the stabilizer system. In polyolefins, yellowing is usually attributed to additives. Some discolorations disappear after brief exposure to UV irradiation. Volatile stabilizers may cause processing issues by clogging equipment.

Light stabilizers are often used at higher concentrations than antioxidants. Their compatibility with a polymer matrix is critical especially for polar light stabilizers in non-polar plastics, such as polyolefins. Polymeric HALS exhibit lower migration than low molecular weight additives designed to extend long-term heat stability.

Toxicological properties are an important factor for selecting the light stabilizers properly especially for food contact applications. Migration and extraction of additives must be considered as well. Low-dust or dust-free additive forms are also commercially available and intended for improving occupational safety during their handling. UV absorbers primarily protect the top layer, while the additives functioning via other mechanisms (e.g. HALS) act throughout the entire material.

3.7 Light Stabilization of Selected Plastics

It was already pointed out that approximately three-quarters of all the light stabilizers used in plastics are added to polyolefins, and due to their economic importance, the stabilization of this group is covered comprehensively in this section. Other plastics are given due consideration and discussed less extensively.

Light stabilizers always produce minute amounts of radicals that initiate polymer degradation (Figure 1.19). Therefore, it is sensible and necessary to combine light stabilizers with primary and secondary antioxidants, hydroperoxide decomposers, etc. (see Chapter 2) in order to prevent or significantly delay the degradation of plastics and loss of their properties (Figure 1.20).

3.7.1 Polyolefins

Representatives of the following light stabilizer classes are mainly used for PP light stabilization: 2(2′-hydroxyphenyl)benzotriazoles, 2-hydroxy-4-alkoxybenzophenones, and sterically hindered amines (HALS) (Table 3.7).

Table 3.7 Light Stabilizers for Use in Polyolefins[15)]

Code	CAS no.	Remark	Typical dosages (%)				
			PP	HDPE	LDPE	LLDPE	Metallocene-PE
UVA-1	2440-22-4	2(2′-Hydroxyphenyl)benzotriazoles	—	—	0.15–0.30	—	—
UVA-2	3846-71-7		—	—	0.10–0.30	—	—
UVA-3	3896-11-5		0.05–0.40	0.10–0.30	0.30	—	0.20–0.60
UVA-4	3864-99-1		0.20–0.50	—	0.07	—	—
UVA-5	25973-55-1		—	—	—	—	0.60
UVA-19	1843-05-6	2-Hydroxybenzophenone	0.30–0.50	0.05–0.30	0.10–0.60	0.15–0.25	0.20–0.60
HA(L)S-1	52829-07-9	Monomeric	0.05–1.20	0.05–0.20	0.15–0.40	—	—
HA(L)S-2	65447-77-0	Polymeric	0.10–0.60	0.05–0.90	0.05–1.20	0.10–0.30	0.15–0.90
HA(L)S-3	71878-19-8		0.10–0.60	0.05–0.90	0.05–1.20	0.10–0.30	0.15–0.90
HA(L)S-4	106990-43-6		0.05–0.60	0.10	—	—	0.15–0.60

15 The dosage depends on the presence of additional stabilizing components added and the requirements of the finished product. This is just a guideline.

Code	CAS no.	Remark	Typical dosages (%)				
			PP	HDPE	LDPE	LLDPE	Metallocene-PE
HA(L)S-8	41556-26-7 82919-37-7	1:1 Mixture	0.10–0.60	—	—	—	—
HA(L)S-9	63843-89-0	Monomeric	0.15–0.60	—	—	—	—
HA(L)S-11	192268-64-7	Polymeric	0.05–0.20	0.10–0.20	0.60	—	—
HA(L)S-14	90751-07-8		0.10–0.20	—	—	—	—
HA(L)S-15	219920-30-6		0.20	—	0.25	—	—
HA(L)S-16	79720-19-7	Monomeric	0.20	—	—	—	—
HA(L)S-17	106917-30-0		0.20	—	—	—	—
HA(L)S-19	131290-28-3	Polymeric	0.30	—	—	—	—
HA(L)S-20	109423-00-9	Monomeric	0.20	—	—	—	—
HA(L)S-21	124172-53-8		0.10–0.20	—	—	—	—
HA(L)S-22	199237-39-3	Polymeric	0.20	—	—	—	—
HA(L)S-23	91788-83-9	Monomeric	0.20	—	—	—	—
HA(L)S-24	64022-61-3		0.05–0.20	—	—	—	—
HA(L)S-27	115055-30-6	Polymeric	0.20	—	—	—	—
HA(L)S-28	100631-44-5		0.10–0.20	—	—	—	—
HA(L)S-29	95078-42-5	Monomeric	0.05–0.20	—	—	—	—
HA(L)S-31	78276-66-1	Polymeric	0.10–0.20	—	—	—	—
HA(L)S-33	136504-96-6		0.10–0.20	—	—	—	—
HA(L)S-39	147783-69-5	Monomeric	0.10–0.20	—	—	—	—
HA(L)S-44	182635-99-0	Polymeric	0.05–0.25	—	—	—	—
HA(L)S-53	Proprietary	Tinuvin® NOR 371 FF	—	—	0.55–0.60	—	—

3.7.1.1 PP

PP is particularly sensitive to UV radiation. Non-stabilized PP exposed to daylight loses its gloss, develops hairline cracks on the surface, results in chalking of the pigmentation, and deterioration of its mechanical properties. The same phenomena occur when PP is light-stabilized, but with a more or less pronounced time delay. The light stability of an article also depends on factors other than the effectiveness of the added light stabilizers, namely the sample thickness, the degree of orientation and crystallinity, pigmentation, and the presence of fillers, reinforcing agents, and other additives, such as phenolic antioxidants and co-stabilizers.

The effectiveness of various light stabilizer classes in PP tapes, based on artificial light exposure tests, is shown in Figure 3.15. It is evident that UV stability can be increased by factors of 2 to about 20 by using different stabilizers. The performance of HALS is remarkable. Even when used only at a fraction of the concentration compared to other stabilizers, they are superior to them.

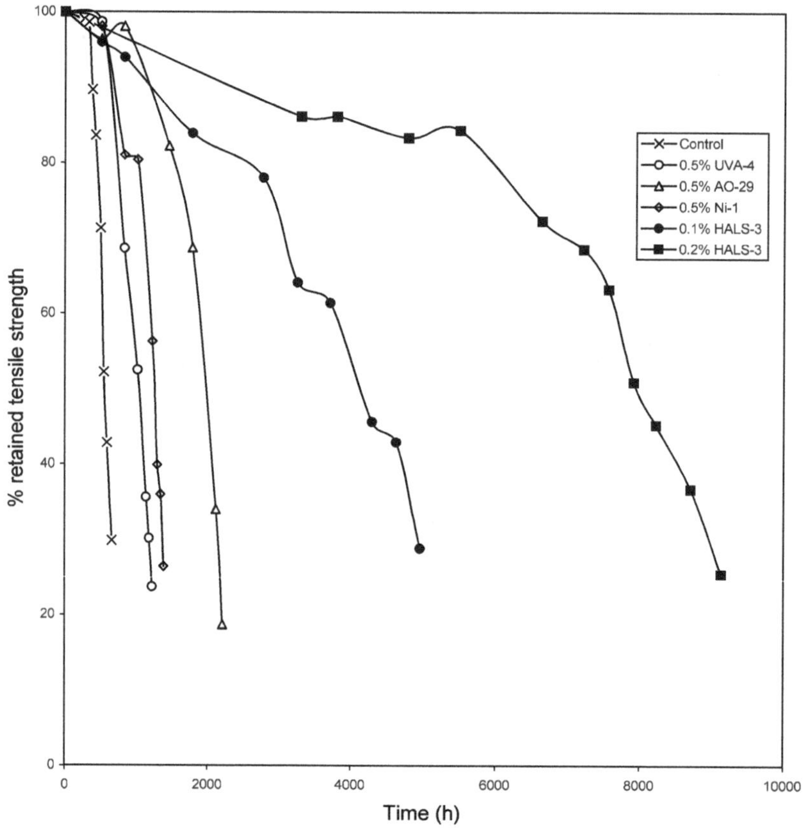

Figure 3.15 Retained tensile strength of stabilized 50 µm thick PP tapes on Xenotest® 1200 exposure. Base stabilization: 0.1% calcium stearate + 0.05% AO-18 + 0.05% PS-2

Effect of increasing HALS-1 concentrations on the imparted UV stability of PP tapes is shown in Table 3.8. Across the entire concentration range from 0.05–1.2%, the performance increases significantly with increasing concentration of HALS-1. This applies to the accelerated exposure in the Xenotest® chamber as well as to outdoor weathering in Florida (USA) and Basel (Switzerland). The performance of HALS-1 is very pronounced in the Florida weathering tests, and appears even more impressive under the natural weathering in Basel. For example, at a concentration of 0.3% HALS-1, the time to 50% tensile strength retention is approximately 3 years in Florida and approximately 9 years in Basel. These results demonstrate a clear influence of the climate zone on the service life of PP stabilized with HALS-1. The samples last longer in the temperate climate not only in terms of time but also in terms of the accumulated energy. This is mainly due to the lower portion of shortwave UV light and the lower average temperature in Basel compared to the more tropical climate in Florida.

Table 3.8 Effect of Light Stabilizer Concentration on Light Stability of PP Tapes[a]

HALS-1 concentration (%)	Time to 50% retained tensile strength (h)[b]	Energy to 50% retained tensile strength (kJ cm^{-2})[c]	
		Florida	Basel
0	495	110	165
0.05	1500	270	465
0.10	2760	690	945
0.15	4140	930	1460
0.30	6720	1930	≥3015[d]
0.60	11,300	2780	≥3015[d]
1.20	16,050	3350	≥3015[d]

[a] PP homopolymer + 0.1% calcium stearate + 0.05% AO-18 + 0.05% PS-2. Tapes 50 µm thick, stretch ratio 1:6.
[b] Exposure: Weather-Ometer WRC 600, without water spraying, black panel temperature ~53 °C.
[c] Samples all S-facing 45° aspect, direct. Experiments started in March 1978.
[d] Remaining samples were destroyed by hail after 10 years of outdoor exposure.

The higher the dosage of HALS is, the longer the mechanical performance of the final article is retained.

The effect of titanium dioxide on the UV stability of PP tapes stabilized with HALS-1 is predominantly neutral. However, quality of titanium dioxide used is crucial, since different commercial preparations can lead to significantly varying results. Other inorganic components, such as zinc oxide, can also negatively affect the UV stability of PP.

In combination with HALS-1, inorganic pigments generally show only slightly positive or negative effects. In contrast, certain organic yellow and red shades can significantly reduce the lightfastness of PP tapes stabilized with HALS.

To effectively protect thick-walled PP parts against UV-induced degradation, it is essential to preserve both their mechanical properties and appearance. In particular, loss of gloss, chalking, and the formation of surface cracks should be avoided. UV absorbers provide better protection for the deeper layers of the polymer (and thus its mechanical properties) with increasing wall thickness, while this kind of protection of the surface is less affected directly. Some improvement in surface protection can also occur through the migration of stabilizers. In addition to absorbing UV radiation, UV absorbers also contribute to the polymer protection through other mechanisms, such as quenching of the excited states.

Comparative studies show that HALS stabilizers are superior to UV absorbers, especially in maintaining impact resistance. Low molecular weight HALS are more effective than polymeric HALS, particularly in non-pigmented samples. In white-pigmented samples, the difference is less pronounced, especially with regard to chalking. Combinations of a polymeric HALS with a UV absorber can improve protection of both non-pigmented and to some extent white-pigmented materials.

3.7.1.2 PE

Since PE is susceptible to photooxidation, light protection is required for outdoor use. The light stabilizers for PE are essentially the same as for PP. The PE type, its intended application, and the desired service life of the article generally determine the amount and type of the light stabilizer that should be used.

HDPE is primarily used in blow molding, injection molding, and extrusion. The available application data were obtained with HDPE tapes and injection-molded sheets. Although HDPE films do not have a large number of outdoor applications, they are suitable for screening the performance of various light stabilizers. The exposure time required to build up a certain amount of carbonyl groups is the best test criterion, and therefore it is frequently used. The superiority of HALS over other light stabilizers is self-evident from the results. Within the HALS group, HALS-3 and HALS-11 demonstrate the best performance. UV absorbers are less effective than HALS as stand-alone light stabilizers based on both the accelerated and natural weathering tests on HDPE tapes and films. The amount of radiation absorbed by a UV absorber depends on its concentration, extinction coefficient, and sample thickness according to the Beer–Lambert law. Accordingly, the performance of UV absorbers can be expected to be superior when the specimen thickness is significantly greater than 50 µm. For more recommendations, see Table 3.8.

LDPE is primarily used in manufacturing films for coverings in construction and agricultural greenhouses. However, stabilizing LDPE against UV degradation presents specific challenges, since many light stabilizers tend to migrate or bloom to the surface due to insufficient compatibility. Early stabilization efforts relied on benzophenone- and benzotriazole-derived UV absorbers. Later, the introduction of HALS led to significant improvements, especially with the development of polymeric HALS variants that offer better compatibility and performance under outdoor weathering conditions. Effectiveness of HALS in LDPE is notably enhanced by increasing solubility of the stabilizer in the polymer. This solubility is more critical than the diffusion rate of the additive. Combinations of HALS-2 with the benzophenone-type UV absorbers like UVA-19 improve stability of LDPE, especially in thicker films (200 µm). Surprisingly, combinations of HALS with the benzotriazole UV absorbers outperform those with benzophenone-based systems, likely due to the synergistic protection mechanisms. Modern formulations with HALS-3 combined with UV absorbers can achieve a service life of over 5 years in 200 µm thick LDPE films under outdoor exposure, for example, in Florida. Concentration of stabilizers directly influences film longevity; for example, greenhouse films containing about 2% additives can last 3 years or more. The choice of stabilizer system depends strongly on film thickness and environmental conditions. For more recommendations, see Table 3.7.

LLDPE has gained significant commercial importance in recent years and is mainly used for film manufacturing by injection molding and blow molding. The light stability achieved in blown films made from LLDPE (with the butene co-monomer) is similar to that of conventional LDPE. Due to processing challenges with LLDPE, especially in film blowing, blends of LLDPE and LDPE are sometimes used. However, the composition of these blends does not significantly affect their light stability, which is comparable to that of the pure individual polymers. Tests show that several high molecular weight HALS provide effective UV stabilization in 50 µm thick blown films, and the type of polyethylene (LDPE, EVA copolymers, or LLDPE) does not result in significant differences in terms of their light stability.

The thickness of the film is a factor for LLDPE (Beer–Lambert Law): while LLDPE offers mechanical properties superior to LDPE especially in manufacturing high-quality thin films, reducing their thickness reduces their UV stability, necessitating adjustments to the stabilization system. In thick, brown-pigmented LLDPE parts, polymeric HALS outperform the benzophenone UV absorbers in terms of UV protection, though the difference is less pronounced than in the thin films. Polymeric HALS provide excellent UV stability in rotationally molded LLDPE plates, with the 1:1 HALS-2/HALS-3 mixture showing comparable performance. Synergistic effects of UV absorbers – especially the oxanilides (Figure 3.12) in combination with benzophenone-, benzotriazole-, or hydroxyphenyltriazine-based absorbers (Figure 3.9) – are similar in both PP and PE. For more recommendations, see Table 3.7.

Introduction of metallocene (“single-site” supported) catalysts has opened up new possibilities in the polymerization of various monomers. These catalysts result in polymers with narrow molecular weight distributions and precisely controlled branching, especially in polyethylene. The resulting polymers, including very low-density polyethylene (VLDPE) and ultra-low density polyethylene (ULDPE), contain lower amounts of catalyst residues, eliminating need for their removal.

Initial studies on stabilizing these new types of PE with light stabilizers show that the compatibility of UV absorbers with metallocene polyethylene (m-PE) is superior to the conventional LDPE or LLDPE. The UV stability of m-PE is comparable to that of the conventional LLDPE. Furthermore, combinations of different polymeric HALS in m-PE demonstrate pronounced synergistic effects. The effectiveness of HALS in m-PE increases proportionally to the square root of their concentration.

3.7.2 Elastomers

Rubber is often compounded with high amounts of carbon black (see Section 9.7.2), a highly effective UV stabilizer. UV stabilization of elastomeric materials with other additives is mostly limited to non-pigmented or light-colored rubbers and specialty rubbers. Even in the light-colored rubber, the effects of photooxidation are often confined to the surface, with hairline cracks and chalking being the most visible signs of photodegradation.

This topic has been extensively discussed in the literature. It has been shown that substituted salicylic acid protects natural rubber (NR) against UV radiation. Carotenoids and nickel derivatives protect *cis*-1,4-polybutadiene from photooxidation in solution. Various phenols and aromatic amines inhibit the formation of hydroperoxides during the photooxidation of polybutadiene. Organic sulfides can weaken the UV-stabilizing effect of HALS during the photooxidation of *cis*-1,4-polybutadiene.

It has also been found that HALS-1 effectively inhibits gel formation and subsequent degradation in PE-LD/SBS blends exposed to γ-radiation. Combinations of phenolic antioxidants (such as AO-18), the benzotriazole-type UV absorbers (e.g., UVA-1), and HALS (e.g., HALS-1) are used to stabilize thermoplastic rubbers (styrene block copolymers) intended for hot-melt and solvent-based sealants.

The UV stabilization of elastomers depends strongly on the specific material. The systems are classified mainly according to the monomers used and their crosslinking methods. These materials are divided into unsaturated types (BR, IR, SBR) and saturated types (EPM, EPDM). Crosslinking is mainly achieved by sulfur or peroxides. Additionally, a distinction can be made between unsaturated and saturated thermoplastic rubbers (TRs).

For unsaturated, non-crosslinked polymers, the situation is similar to that of crosslinked unsaturated polymers: UV stability is essentially achieved by adding UV absorbers. Increasing concentrations of benzotriazole-derived UV absorbers improves the retention of mechanical properties in SBS footwear, with the benzotriazole-type UV absorbers being superior to the benzophenone-type UV absorbers.

For unsaturated polymers crosslinked with sulfur, UV stability is mainly improved by UV absorbers, while HALS do not work in SBR vulcanizates. For unsaturated polymers crosslinked with peroxides, the situation is comparable.

Stabilization of saturated, non-crosslinked polymers requires different additive combinations. HALS, UV absorbers, and their combinations (especially when used with phenolic antioxidants) provide good protection against UV-induced degradation. Adding small amounts of metal deactivators further improves this protection. Combinations of polymeric HALS-3 with the UV absorber UVA-4 appear to offer the best protection.

Table 3.9 Light Stabilizers for Elastomers

Code	CAS no.	Remark	Typical dosages (%)[a]					
			EVA (14% VA)	SBR vulcanizate	Crosslinked EPDM	SBS	ASA	PUR elastomer
UVA-1	2440-22-4	(2-Hydroxyphenyl)benzotriazoles	—	—	—	0.10–0.30	—	1.00
UVA-4	3864-99-1		—	—	0.50–1.00	0.10–0.40	—	0.50–1.00
UVA-5	25973-55-1		—	—	—	—	0.15–0.30	—
UVA-10	104810-47-1		—	1.00	0.50–1.00	—		—
UVA-12	23328-53-2		—	—	1.00	—	—	—
UVA-19	1843-05-6	2-Hydroxybenzophenone	0.15–0.30	—	—	0.30	—	—
HA(L)S-1	52829-07-9	Monomeric	—	—	1.00	—	0.15–0.30	0.50
HA(L)S-2	65447-77-0	Polymeric	0.15–1.20	—	1.00	—	—	—
HA(L)S-3	71878-19-8		0.15–1.20	—	1.00	—	—	1.00
HA(L)S-8	41556-26-7 82919-37-7	1:1 Mixture	—	1.00	0.50–1.00	—	—	—

[a] The dosages depend on the presence of additional stabilizing components added and the requirements of the finished product; as a result, these numbers are just a guideline.

In saturated polymers crosslinked with sulfur, UV stability is also mainly increased by adding UV absorbers. Given the high sulfur content in these systems, it is not surprising that HALS alone do not impart good light stability. HALS/UV absorber combinations show good performance, while both the UV absorbers alone and their combinations result in significant stability improvement. Light stability of saturated, peroxide-crosslinked elastomers can be further improved by using HALS.

Similar stabilizer systems also improve light stability of PUR elastomers used as sealants as well as EVA copolymer dispersions used as binders for paints and castings. Combinations of the benzotriazole-type UV absorbers and HALS provide excellent UV stability in these cases as well (Table 3.9).

3.7.3 Hot Melt Adhesives

Adhesives are generally used to bond two materials and are therefore somewhat protected by these materials. However, UV-induced degradation can occur at the edges of the bond if the materials are more or less transparent to light. In such cases, photooxidation can cause discoloration and even loss of adhesive properties.

For hot melt adhesives based on styrene–isoprene–styrene (SIS) block copolymers (which feature two polystyrene end blocks and an elastic polyisoprene middle block), the benzotriazole-type UV absorbers are superior to the benzophenone-type UV absorbers, although there are significant performance differences among the benzotriazole-derived absorbers themselves. The most effective members of this group are UVA-11 ((2-hydroxyphenyl)benzotriazole-type, CAS 70321-86-7; see Figure 3.8) and UVA-10.

There are also notable differences in efficiency among HALS. A correlation exists between their chemical structures and the imparted performance: *N*-methylated HALS, such as HALS-8, HALS-9, and HALS-4, show the best results. Light stabilizers must not only prevent discoloration but also protect the adhesive properties. Ranking of additives that is based on adhesion retention largely corresponds to that of the protection against discoloration. However, HALS are less effective than the UV absorbers in this respect.

Adhesives based on styrene–ethylene–butylene–styrene (SEBS) block copolymers rely solely on saturated polymers, and are significantly more UV-stable than SIS-based hot melt adhesives. The difference is about a factor of three. Here too, UVA-10 and UVA-11 are among the most effective light stabilizers, and HALS provide better protection against yellowing than the UV absorbers. When adhesion retention is required, the UV absorbers offer the best protection during short-term UV exposure, while HALS better maintain adhesion characteristics for the long-term UV exposure.

The light stability of hot melt adhesives based on unsaturated styrene–butadiene–styrene (SBS) block copolymers is noticeably lower than that of the corresponding SEBS-based adhesives. Even the most effective light stabilizers cannot prevent significant adhesion loss after just a few days of artificial weathering. The choice of antioxidants has little impact on the performance of benzotriazole-derived UV absorbers. The combination of HALS-9, the UV absorber UVA-11, and the triple blend AO-18 + PS-3[16] + AO-34 is considered as the preferred system for light stabilization of SBS hot melt adhesives.

Adhesives based on SBS and SIS should generally exhibit comparable light stability. The main difference between these adhesives arises from the resins or "tackifiers" used.

The most effective stabilizers for pressure-sensitive SIS adhesives are UVA-10, UVA-11, and UVA-21.[17] HALS are almost as effective. *N*-Methyl HALS perform even better than the other HALS.

3.7.4 Styrenic Polymers

Styrenic plastics like ABS and HIPS are highly oxidation-sensitive due to their polybutadiene content, which results in rapid loss of mechanical properties. Combining HALS and the benzotriazole type UV absorbers provides good stability. For ABS, combinations of polymer-bound antioxidants with UV absorbers are recommended due to their synergistic effects. The UV absorbers protect deeper polymer layers, while HALS stabilize the surface, effectively maintaining impact strength of the materials.

These stabilizer systems are also efficient for HIPS and other impact-resistant styrene copolymers, with UVA-1 and HALS-8 particularly effective against discoloration of HIPS. UV-stabilized styrene copolymers (SAN, HIPS, ABS) often show an initial bleaching (decrease in yellowness index, YI), followed by a gradual YI increase due to photo-oxidation. This is caused by color bodies formed during the polymerization step, especially in copolymers produced with butadiene or acrylonitrile monomers. After irradiation, ABS samples may continue to yellow in the dark ("post-actinic yellowing"), but this process can be reversed by sunlight. This effect also occurs in PC, PBT, PVC, and PP, and even short-term exposure to UV light can trigger it. While HALS are essential for UV protection, they can also slightly increase dark yellowing, although antioxidants and processing stabilizers can mitigate this. In this regard, methylated HALS (like HALS-8) are more effective than the non-methylated type (like HALS-1) but comparable to the non-methylated HALS-21.

In acrylonitrile–styrene–acrylate (ASA), the light-sensitive polybutadiene phase of ABS is replaced by acrylic rubber, improving its weather resistance but slightly reducing low-temperature impact strength. Non-stabilized ASA has UV light stability comparable to well-stabilized ABS. With the proper stabilization, ASA can be made suitable for long-term outdoor applications.

---

[16] CAS no. 26741-53-7.

[17] CAS no. 147315-50-2, 2-(2′-hydroxyphenyl)-1,3,5-triazine-type (see Figure 3.9).

3.7.5 Polyamides

Various stabilizers, such as phenolic antioxidants, HALS, copper compounds, and benzotriazole-type UV absorbers can significantly slow down the photooxidation of polyamides by either scavenging radicals or absorbing UV light.

Aging of polyamides is influenced not only by light, but also by moisture and atmospheric pollutants, such as ozone and nitrogen dioxides. As a result, their stability varies greatly depending on the location and testing conditions.

Combinations of phenolic antioxidants, phosphites, HALS, and UV absorbers offer the best protection, with ternary and quaternary systems being particularly effective for PAs. While significant improvements are seen in artificial weathering tests, the advantages of stabilized samples under real outdoor conditions are often less pronounced than expected, with these differences being more pronounced in dark or black pigmented compounds. The copper/halide additive systems are commonly used despite certain drawbacks.

Similar stabilization approaches apply to various types of polyamides (PA 6, PA 6.6, PA 11, PA 12). The best results are achieved with the combinations of phenolic antioxidants, HALS (especially HALS-1, HALS-2, and HALS-8), and the benzotriazole-type UV absorbers (notably UVA-1 and UVA-11) (Table 3.10). Adding the UV absorbers to HALS further enhances light protection of these polymers. Among phenolic antioxidants, classic types like AO-3 are mainly used in combination with phosphites. Comparable stabilizing effects are seen in the presence of these additives in PA 12 and PA 11. Overall, UV stabilization of PAs is possible through selected combinations of additives, tailored to the specific polymer and its application.

In 1999, HALS-52[18] was introduced to the market. Its performance was evaluated in QUV (indoor) weathering tests. Its effective light stabilization and processing properties are attributed to the amidation reactions between the molecule and the carboxylic acid chain end groups of the polyamide.

Polyamide fibers possess inherently good UV stability, so their stabilization is usually necessary only for long-term UV exposure. Studies on PA 6 textile fibers show that the UV absorbers still impart some improvement in UV stability, but the best results are achieved with polymeric HALS stabilizers, particularly HALS-3. Increasing the concentration of HALS-3 significantly enhances light stability under both artificial and natural weathering conditions. Further experiments demonstrate that the high molecular weight HALS-4 also offers excellent protection against yellowing.

[18] CAS 42774-15-2, monomeric.

Table 3.10 Light Stabilizers for PS, SAN, ABS, PA, and PVC

Code	CAS no.	Remark	Typical dosages (%)[a]				
			PS	SAN	ABS	PA	PVC
UVA-1	2440-22-4	(2-Hydroxyphenyl)benzotri-azoles	0.10–0.20	0.25–0.30	0.50	0.25	0.10–0.20
UVA-2	3846-71-7		0.10–0.20	—	—	0.25	0.10–3.00
UVA-3	3896-11-5		—	—	—	0.30	0.15–0.30
UVA-4	3864-99-1		—	—	—	0.50	0.40
UVA-10	104810-47-1		—	—	—	0.20	—
UVA-11	70321-86-7		—	—	—	0.25–0.50	—
UVA-12	23328-53-2		—	—	—	—	0.40–3.00
UVA-16	131-57-7	2-Hydroxybenzophenone	—	—	—	—	0.10–0.20
UVA-17	131-56-6		—	—	—	—	0.10–0.20
UVA-19	1843-05-6		0.10–0.20	—	0.50–1.00	—	0.15–0.30
UVA-23	23949-66-8	Oxanilide	—	—	—	0.25	0.10–0.30
UVA-25	7443-25-6	Dimethylbenzylidene malonate	—	—	—	—	0.20
HA(L)S-1	52829-07-9	Monomeric	0.10–0.25	0.25–0.30	0.50	0.20–0.50	—
HA(L)S-2	65447-77-0	Polymeric	—	—	—	0.20–0.50	0.15–0.30
HA(L)S-3	71878-19-8		—	—	—	0.20–0.50	0.15–0.30
HA(L)S-4	106990-43-6		—	—	0.50	0.25–0.50	—

Table 3.10 Light Stabilizers for PS, SAN, ABS, PA, and PVC (*continued*)

Code	CAS no.	Remark		Typical dosages (%)[a]				
				PS	SAN	ABS	PA	PVC
HA(L)S-8	41556-26-7 82919-37-7	1:1 Mixture		0.10	0.25	0.50	—	—
HA(L)S-11	192268-64-7	Polymeric		—	—	—	0.30–0.50	—
HA(L)S-21	124172-53-8	Monomeric		—	—	0.50	—	—
HA(L)S-23	91788-83-9			—	—	—	0.50	—
HA(L)S-53		Proprietary, solid NOR		—	—	0.50–1.00	—	—

[a] The dosages depend on the presence of additional stabilizing components added and the requirements of the finished product; as a result, these numbers are just a guideline.

3.7.6 PVC

PVC has been used outdoors for decades. It is routinely exposed to severe weathering and UV radiation, which can lead to yellowing, fading, and deterioration of mechanical properties. To improve light stability, titanium dioxide is commonly used (while also serving as a white pigment) at high concentrations (up to about 12 phr). For transparent or translucent PVC compounds, the addition of the UV light absorbers is essential. (2-Hydroxyphenyl)benzotriazoles, such as UVA-1 and UVA-2, which outperform the hydroxybenzophenones (UVA-16, UVA-17), and the oxanilides (UVA-23), are particularly effective (Table 3.10). Their protective effect is increased with increasing the concentration of the UV absorber used. Both liquid and solid UV absorbers (e.g., solid UVA-2 and liquid UVA-12) impart comparable results and can be interchangeably used in many applications.

HALS, especially HALS-1, show little effect in rigid PVC, because they are deactivated by hydrochloric acid (HCl) generated during PVC processing and aging. In these processes, ammonium chloride is formed, causing the light-protective effect of HALS to be lost (see Section 3.5.3 about basicity of HALS). In pigmented compounds, such as those with added titanium dioxide, or in combination with the UV absorbers, HALS-1 can significantly enhance the protective effect. The combination of HALS-1 with a benzotriazole-type UV absorber, such as UVA-2, is particularly effective in delivering better results compared to the use of the UV absorber alone. In pigmented PVC, HALS-1 is also more effective than the UV absorbers that do not contribute to color stability.

For flexible, transparent PVC films, such as those used in greenhouses, combinations of HALS and the UV absorbers provide the best protection against yellowing and loss of mechanical properties. Overall, for maximum UV protection and outdoor color stability of PVC articles, it is recommended to use (2-hydroxyphenyl)benzotriazole-type UV absorbers (UVA-1 and UVA-2) in sufficient concentration. In pigmented or flexible compounds, HALS-1 should be added as well. HALS alone are not very effective in rigid PVC, but in combination with the UV absorbers and pigments, they can significantly improve light stability of PVC compounds.

3.7.7 PC

PC is primarily used where high impact resistance and transparency are required, such as windows in schools, transportation vehicles, and lamp covers. When exposed to UV light, PC suffers surface damage that affects its impact resistance, transparency, and color, leading to pronounced yellowing. Without UV stabilizers, PC shows some resistance to photooxidation but yellows rapidly as a result of long-term outdoor exposure to light. The degradation affects only the very thin outermost layer of PC articles, protecting the inner layers. However, this surface degradation is usually unacceptable.

To protect PC against yellowing and loss of properties, stabilization of the material is necessary. This can be achieved through either surface coatings, containing UV absorbers or the incorporation of additives. Besides antioxidants, the UV absorbers are mainly used, with attention to thermal stability, low volatility, and the initial color stability.

Various types of UV absorbers are used in practice, including the benzotriazole types such as UVA-8, as well as hydroxyphenyltriazine types such as UVA-21; they are considered adequately effective. Other UV absorbers, such as UVA-1, UVA-6, UVA-7, and UVA-11 significantly improve the lightfastness and yellowing induction time of PC at various concentrations.

The choice of UV absorber in PC depends on the application. For example, low volatility is critical in extrusion processes. Stabilization of thin PC layers is especially challenging, but the desired levels of protection can be achieved with high concentrations of UV absorbers such as UVA-21. Co-extrusion of thin UV-absorbing layers with PC is a very effective and economical method for achieving excellent UV light stability.

HALS are not used in PC because basic amines accelerate PC hydrolysis; however, tertiary amines are less problematic than secondary amines.

3.7.8 POM

The most effective combinations of the UV absorbers (UVA) and HALS for the UV stabilization of polyacetal, such as POM, involve the use of UVA-11 (a benzotriazole type) in combination with HALS-2. This pairing consistently delivers superior results in terms of gloss retention, resistance to crack formation, and color stability under a variety of artificial and natural weathering conditions. Numerous comparative studies have demonstrated that a formulation containing 0.3% UVA-11 and 0.4% HALS-2 significantly extends the time to crack formation, minimizes loss of gloss, and better preserves color compared to the individual additives.

Other combinations, such as UVA-11 with HALS-1, HALS-3, or HALS-4, also provide effective protection but generally do not match the consistent performance of the UVA-11/HALS-2 combination. Similar synergistic effects have been observed for UVA-1, UVA-5, and UVA-7 when combined with HALS-2 or HALS-8, although UVA-11 with HALS-2 remains the overall preferred system. The efficacy of these combinations is comparably high in both POM homopolymers and copolymers.

3.7.9 Polyurethanes

Polyurethanes (PUR) are produced by reacting diisocyanates with long-chain diols/polyols and short-chain diols/diamines. The light stability of polyurethanes strongly depends on their chemical structure: polyurethanes based on aliphatic isocyanates and polyester diols show the best light resistance (minimal yellowing), while polyurethanes with aromatic isocyanates and polyether polyols exhibit the lowest light stability.

Polyurethanes are used in a variety of applications, such as foams, coatings, adhesives, and fibers. UV stabilization is necessary for applications like artificial leather, textiles, and shoe soles to prevent yellowing, microcracking, and deterioration of mechanical properties.

For UV stabilization, light stabilizers such as UV absorbers (e.g., the benzotriazole types such as UVA-11), phenolic antioxidants, and HALS (e.g., HALS-2) are used. Combinations of UV absorbers and radical scavengers (like HALS) are particularly effective. The HALS-1/UVA-1 system preserves the mechanical properties of polyether polyol-based polyurethanes significantly better than the other systems, such as AO-4/UVA-1. The same applies to polyester polyol-based polyurethanes. The superiority of HALS is especially evident in the behind-glass exposure test; under direct exposure, the material discoloration occurs more rapidly than the reduction in its elongation. Aliphatic polyester polyol-based polyurethane films are superior to the aromatic ones in terms of both yellowing resistance and elongation at break.

For polyurethane sheets, the AO-18/HALS-1/UVA-1 system provides good protection against discoloration; a mixture of a phenolic antioxidant, HALS, and the UV absorber can offer comparable protection at half the concentration. In polyurethane foams, the combination of the benzotriazole-type UV absorber, phenolic antioxidant, and HALS (e.g., HALS-1/UVA-1/AO-18) is most effective. The type of phenolic antioxidants used significantly affects the yellowing protection.

Anti-scorch packages are special additive combinations used in the foaming process of polyurethanes to prevent "scorching" (burning and/or discoloration of the foam due to overheating during production). They typically consist of antioxidants and other auxiliary agents. Amine-free anti-scorch packages, such as AO-52, lower foam discoloration compared to the amine-based variants like AO-50 or AO-51.[19)]

Increasing concentrations of light stabilizers significantly improve the light resistance of polyurethane foams. Polyether polyol-based polyurethane foams are more UV-stable than the polyester polyol-based polyurethane foams. The chemical structure of the benzotriazole-type UV absorbers also influences UV stability. For white-pigmented polyurethane integral foams, liquid stabilizer systems can impart excellent stability. For polyurethane sealants, the combination of a phenolic antioxidant, a benzotriazole-type UV absorber, and HALS (e.g., HALS/UVA systems) provides very good light resistance. In non-pigmented polyether polyol-based TPU, combining HALS with a UV absorber improves light stability compared to using the UV absorbers alone. The AO-18/PS-3 combination provides a better initial color for non-pigmented polyester polyol-based TPU compared to AO-18 alone.

---

[19] AO-50: Irgastab PUR 55, proprietary composition. AO-51: Irgastab PUR 67, proprietary composition. AO-52: Irgastab PUR 68, proprietary composition.

3.7.9.1 Linear Polyesters

Linear thermoplastic polyesters include PET, PBT, copolyesters, and aromatic polyesters (polyarylates) that are produced by the condensation reaction of terephthalic acid and bisphenol. PET fibers have long been used in clothing, carpets, upholstery, and tire cords. PET and PBT, reinforced or non-reinforced, are widely applied in films, coatings, beverage bottles, electronics, sanitary products, and transportation.

Under UV exposure, PET fibers tend to lose elasticity, elongation, and tensile strength. Polyester films discolor, develop surface cracks, and eventually become brittle, with a gradual reduction in tensile strength. PET and PBT are intrinsically more light-resistant than polyurethanes or polyamides, but after prolonged weathering, they too can yellow and become brittle mainly through surface degradation, while deeper polymer layers remain mostly unaffected.

UV absorbers are effective stabilizers for PET and PBT in outdoor weathering. Their use is recommended to improve the lightfastness of dyes in polyester fibers, especially in automotive textiles. Benzotriazole-type UV absorbers are generally preferred due to their low initial color and good color stability. The effect of a hindered amine light stabilizer, HALS-1, on inhibiting PBT photooxidation has also been described.

While UV absorbers generally provide the best protection, the choice of the right UV absorber is crucial, especially for stabilizing the polyesters against yellowing. UVA-21 offers excellent protection in terms of both yellowing and mechanical properties. Tensile strength is less affected by UV irradiation than elongation.

PETG[20)] is used in outdoor-handled food containers. UVA-21, a hydroxyphenyltriazine-type UV absorber, delivers the best results in preventing loss of gloss and maintaining transmission in PETG plates. Co-extruded PETG sheets containing high amounts of UVA-21 (3.5%) in a 50 µm outer layer on a 5 mm PETG base provide excellent protection against yellowing. Similarly, three-layer PET and PETG co-extruded films with 1.25% UVA-21 in the 50 µm outer layers achieve outstanding protection.

Stabilizing PET fibers is more challenging than films, since UV absorbers are less effective in thin-wall items. The benzotriazole-type and hydroxyphenyltriazine-type UV absorbers have demonstrated their high effectiveness in fibers. HALS have not been widely tested in the polyesters, since, like other amines, they can catalyze polyester hydrolysis and cause slight browning during PET processing. However, low-molecular-weight HALS-1 performs comparably to a benzotriazole UV absorber (UVA-5) in an outdoor behind-glass weathering test. Moreover, combining HALS with UV absorbers demonstrated excellent performance.

---

[20] PETG is a glycol-modified PET characterized by its particularly high transparency, impact strength, and improved processing properties. It is produced by adding cyclohexanedimethanol (CHDM) instead of ethylene glycol during polymerization, resulting in a longer polymer block that cannot be attached to the adjacent molecular chains, as is the case with ethylene glycol. Therefore, this modification inhibits its crystallization, lowers the melting temperature, and reduces the viscosity of the melt. Its applications include film production (for example, as shrink-wrap labels for PET bottles) and 3D printing.

3.7.10 Unsaturated Polyesters

The resistance of unsaturated polyesters to light exposure is usually improved by the incorporation of UV absorbers. The benzotriazole-type UV absorbers offer better protection than the benzophenone type; however, for economic reasons, the combination of benzotriazole UVA-5 with benzophenone UVA-19 is an optimum alternative.

HALS can effectively stabilize unsaturated polyesters. Weathering of thin films and glass fiber-reinforced laminates shows that HALS reduce surface oxidation but lowers tensile strength. HALS-1 was found to be more effective than the *N*-methylated HALS-8 and HALS-9.

3.7.11 Polyester Elastomers

As already mentioned, the most valuable representatives of thermoplastic polyester elastomers are commonly referred to as segmented copolyether esters (TPE-E, also known as TPC or COPE).

In summary, light stabilization of thermoplastic polyester elastomers is primarily achieved through a combination of UV absorbers (especially the benzotriazole derivatives), pigments (carbon black, titanium dioxide), and antioxidants. For the most demanding requirements, HALS and hydrolysis stabilizers can be added to prevent both UV- and moisture-induced aging.

3.7.12 Polyacrylates

PMMA is the most widely used material in the class of polyacrylates and polymethacrylates. It is utilized in the glazing of buildings and aircraft, outdoor signage, lighting, display cases, counters, and automotive taillights. In many of these applications, PMMA is exposed to artificial UV sources and/or sunlight. While PMMA is more UV-stable than most plastics, addition of light stabilizers to it is still necessary in many cases. UV stabilizers can also impart a UV-filter effect to PMMA articles. For thick-walled PMMA used in aircraft glazing or sun-protective windows, the benzotriazole-type UV absorbers are preferred; UVA-1 is commonly used at concentrations of 0.05–0.2%.

HALS have been shown to provide excellent light stabilization for PMMA. When stabilized with HALS, PMMA demonstrates virtually no decrease in molecular weight in outdoor and accelerated weathering tests. After 5 years of weathering, PMMA plates containing UV absorbers retain their mechanical properties, such as creep and resistance to microcracking with only minor changes. However, changes are more pronounced when evaluating elongation at the onset of microcracking and resistance to organic solvents, highlighting the need to tailor stabilization to a specific application.

Resistance of PMMA to microcracking under various climatic conditions is affected by relative humidity (reversibly) and chemical degradation (irreversibly), which correlates with average daily maximum temperatures and exposure to sunlight.

Artificial weathering tests show that PMMA is highly resistant to yellowing even after long-term exposure, and the addition of UV absorbers and/or HALS does not significantly alter this behavior. However, if the transmission loss at 420 nm is considered, some HALS reduce this loss compared to the reference samples, while the UV absorbers exhibit no advantage. Regarding molecular weight changes, both the UV absorbers and especially HALS provide good protection to PMMA, with the best results imparted by using combinations of HALS (HALS-1) and UV absorbers. However, these combinations are not always suitable; for example, in tanning applications, HALS (such as HALS-1, HALS-8, and HALS-10) alone provide excellent light stability.

For thin PMMA films used in long-term outdoor applications, combining HALS and high-performance UV absorbers is very effective. Due to high extrusion temperatures (up to 280 °C), high-molecular-weight stabilizers are needed to prevent volatilization during polymer processing. Combinations of HALS-4 with UVA-8 offer excellent protection, which can be further improved by using hydroxyphenyltriazine-based UV absorbers instead of the benzotriazole types. UVA-21 combined with HALS-4 provides even better protection due to the higher inherent UV stability of the triazine chromophore.

3.7.13 Polyphenylene Ethers

Poly(2,6-dimethyl-1,4-phenylene oxide) (PPE) is a commercially significant polymer primarily in blends with styrenic polymers used for manufacturing molded parts. Due to its economic viability, various attempts have been made to stabilize PPE against photooxidation. Studies on 0.1 mm thick PPE films exposed to Xenotest showed that UV absorbers of the benzophenone and benzotriazole types effectively reduce the formation of oxidation by-products, with UVA-1 outperforming UVA-4 and UVA-19 in such applications.

HALS are also effective in PPE; for example, HALS-1 can reduce the oxygen uptake rate of it by 50%. Since PPE is mostly used in blends with styrenic polymers, and HALS are very efficient light stabilizers for PS, their use in these blends is logical. High amounts of light stabilizers are necessary to achieve the adequate protection as demonstrated by various HALS and their combinations with the benzophenone-type UV light absorbers.

3.7.14 Polysulfones

Polysulfones are used as technical materials in many fields. Similar to aromatic polyamides, only high concentrations of UV absorbers provide an adequate stabilizing effect in polysulfones. It has been proposed to use suitable quenchers to enhance photostability of polysulfones.

Additionally, 1,3-diphenyl-2-*S*-benzylisothiourea was found to be highly effective in stabilizing polysulfones against degradation under 254 nm UV light from low-pressure mercury vapor lamps. With this stabilizer added at 0.1 wt%, both the decrease in molecular weight and the elongation reduction are significantly diminished. Although the wavelength used in this study is more suitable for simulating specific conditions in the laboratory than for natural weathering, the findings can also be used for terrestrial applications.

Use of the benzotriazole-type UV absorbers like UVA-11 in cast films of polyethersulfone (PES) at relatively high concentrations leads to significant improvements compared to the non-stabilized reference samples in terms of both discoloration and mechanical properties. These results were similar in polysulfone films.

3.7.15 Epoxy Resins

Data on the stabilization of epoxy resins is rather scarce. It is known that susceptibility to photooxidation depends on the curing conditions. In fact, oxidation of the methylene group, e.g., in Novolak® resins, can lead to the formation of aromatic carbonyl groups that act as photo-initiators. The photostability of epoxy resins is increased if the thermal oxidation is inhibited either by curing under vacuum or by adding chain-terminating antioxidants, such as a sterically hindered phenol. Residual epoxy groups in incompletely cross-linked specimens can also be a source of carbonyl groups. For UV stabilization of epoxy resins, the addition of UV absorbers of the benzophenone and benzotriazole types is recommended. In combination with phenolic antioxidants, they can impart good protection even for thick articles. Protecting thin-walled articles with these agents requires their addition at very high concentration. This difficulty was overcome by incorporating an epoxidized 2-hydroxybenzophenone into the surface layer of the article. High surface concentrations of UV-absorbing groups are thus possible and are needed to achieve the significantly improved UV stability.

3.7.16 Other Polymers

It would go beyond the scope of this chapter to describe the stabilization of all polymers. In many applications, combinations of HALS with UV absorbers provide the best protection. An increasing number of plastics used today are copolymers or blends, and the degradation mechanisms of such complex systems and their adequate stabilization can be derived to a certain extent from the corresponding mechanisms discussed for the components of polymer blends. These recommendations should be made with due care, and conducting experiments with the identified stabilizer systems is always encouraged.

3.8 New Developments

Two recent developments are worthy of mention here.

CYASORB® THT 4611 (Solvay S.A.) is a HALS-based UV stabilizer. Specialty plastics used under demanding conditions require stabilization solutions that provide the target UV light resistance in all relevant applications, including agrochemical and heat-resistant greenhouse films, as well as low fogging automotive parts.

Tinuvin® XT 200 FF (BASF) is a high molecular weight hindered amine for high-performance light stabilization. It is particularly suited for agricultural film applications, such as greenhouse and mulch films made from polyolefins (PP, PE) as well as polyolefin copolymers, such as EVA and EBA (ethylene–butylacrylate copolymer). Its recommended use level is 0.2–2.0%.

3.9 Questions

1. What are the main differences between optics, photophysics, and photochemistry?
2. What happens to light when it is absorbed by a material? What is the Grotthus–Draper Law?
3. What is the difference between reflection and refraction of light? Provide an example for each.
4. What is the definition of the term "extinction" (E) in the context of light absorption? What equation connects E, I, and I_0?
5. What is the Beer–Lambert Law? What are the four factors that influence the extinction of light in a material?
6. What is the role of HALS in plastics? What are the mechanisms of action for HALS and UV absorbers?
7. How do UV absorbers protect polymers from degradation by light? Give an example of chemical classes used as a UV absorbers for plastics.

4 Heat Stabilizers for PVC, C-PVC, and Other Polymers

Content of this chapter was adapted from:

- G. Sarti, P. Frenkel, Y. Kanade, and M. Schiller, "Stabilizers and lubricants for PVC and CPVC" (Chapter 4), in: R.P. Krock, M. Schiller, P. Frenkel, J.W. Summers, and C.A. Daniels, *PVC Handbook* (2nd edition), Hanser, 2025, pp. 109–217.

4.1 Terms and Definitions

Static heat stability test	Evaluation of a material's resistance to degradation at a constant high temperature without significant mechanical stress and/or shear.
Dynamic heat stability test	Assessment of a material's heat resistance under conditions involving high temperature and mechanical shear.
Early color	Initial coloration of a material immediately after starting processing, often used to assess heat stability.
Color-hold	Ability of a material to retain its original color over time during exposure to heat.
Long-term heat stability	Capacity of a polymer to withstand extended exposure to heat without significant degradation throughout processing
In-process shear	Mechanical stress applied to a polymer during processing, affecting molecular orientation and heat stability.
Functions of PVC heat stabilizers	Prevent heat-induced PVC degradation, minimize discoloration, and maintain mechanical properties during processing.

4.2 Mechanism of PVC Degradation

The mechanism of PVC degradation (initiation, propagation and termination phases) has been debated since the 1950s. Its ionic nature,proposed and elucidated by Starnes, Fisch, and Bacaloglu, is the most comprehensive; it explains not only the catalytic effect of hydrochloric acid (HCl) and Lewis acids[1] but also the role of free radicals in PVC degradation (Chapter 1). According to this mechanism, the propagation follows the initiation in an ionic or quasi-ionic fashion. The chlorine atom adjacent to the allylic position is formed after the initial dehydrochlorination step. Step by step, the polyene sequences grow, and more HCl is liberated. Polyene sequences containing 5–20 conjugated double bonds are formed, which are responsible for the yellow-to-brown discoloration of the dehydrochlorinated PVC.

Lewis acids and HCl are known to catalyze PVC degradation, promoting the zipper chain mechanism of PVC degradation, yielding more polyenes that are eventually transformed into condensation products via crosslinking of the polyenes.

Substances capable of (a) substituting allylic chlorine defects on PVC molecules and establishing a stronger bond with that carbon, (b) scavenging HCl, Lewis acids, and free radicals formed during PVC decomposition, and (c) shortening double bonds of the polyene sequences via the Diels–Alder reaction[2] or other means, play a crucial role in stabilizing this polymer and preventing, delaying, or slowing down its thermal degradation.

Therefore, PVC decomposition proceeds through the following stages: initial release of HCl and formation of allylic chlorine defects, polyene sequence formation and their cross-linking with simultaneous formation of aromatic hydrocarbon by-products, followed by the breakdown of the cross-linked structures to yield a char residue and predominantly volatile hydrocarbons.

4.3 Types and Functions of PVC Heat Stabilizers

Since PVC starts degrading at temperatures below the processing temperatures, it must be thermally stabilized in order to be processed. The stabilization process is focused on delaying and slowing down the zipper chain decomposition reaction proceeding

---

1 The Lewis acid-base concept was introduced by Gilbert Newton Lewis in 1923. According to this concept, a Lewis acid is a chemical species that is capable of accepting an electron pair from a Lewis base.

2 The Diels–Alder reaction is a chemical reaction between a conjugated diene and a substituted alkene, commonly termed the dienophile, to form a substituted cyclohexene derivative.

through the formation of conjugated double bonds, followed by the cross-linking that leads to discoloration and substantial increase in melt viscosity during processing.

Substitution and deactivation of the allylic chlorine, a weak point of the PVC structure that can act as a starting point for the zipper chain decomposition, is the main strategy that is effective in stabilizing PVC. This is done by substituting the allylic chlorine with a group that has a stronger chemical bond, and which therefore requires more energy to be broken.

In the stabilization multicomponent package, primary stabilizers and co-stabilizers are combined in the proper ratio to maximize the heat stabilizing effect and improve the initial color, color hold, and long-term thermal stability of the PVC compounds and articles. The primary stabilizers replace the allylic chlorine and prevent formation of conjugated double bonds and specifically improve the initial color. Co-stabilizers that usually work synergistically with the primary stabilizers improve the color hold and long-term heat stability by scavenging HCl and Lewis acids, as well as shortening the polyene sequences. Therefore, primary stabilizers and co-stabilizers are the fundamental elements of PVC heat stabilizer compositions.

The following are the main types of commercial heat stabilizers:

- Solid and liquid mixed-metal carboxylate stabilizers (SMMs and LMMs) include calcium–zinc carboxylates (CZS) and barium–zinc carboxylates (BZS) with zinc soaps/salts acting as primary stabilizers
- Solid and liquid heavy-metal-free stabilizers[3] (also known as calcium organic stabilizers (COSs) or organic-based stabilizers (OBSs)), with nitrogen-containing organic molecules acting as primary stabilizers and with no zinc salts included
- Paste calcium–zinc carboxylate stabilizers (PCZs), with zinc soaps acting as primary stabilizers. These stabilizers are storage-stable suspensions of solid stabilizer components in plasticizers
- Solid lead-based stabilizers based on lead metal soaps or salts with the cooperation of essential salts containing PbO that are capable of acting as strong acid scavengers
- Organotin stabilizers based on alkyltin-derived molecules, resulting in articles with remarkable heat stability, clarity, low initial color, and color hold.

Many stabilizer combinations of these types also exist. All of these types of stabilizers usually contain a number of organic and/or inorganic co-stabilizers. Each type con-

[3] According to the original definition, "heavy metals" are those with a density of >5 g/cm^3. However, the general public understands the term in a broader sense to mean metals that are dangerous for humans and for the environment as a whole. Currently, 11 definitions of heavy metals are based on the atomic weight, five definitions specify certain chemical properties, two definitions refer to toxicological principles, and a total of 13 definitions are based on the metal density, ranging from 3.5 – 7.0 g/cm^3, and with the most commonly used defining them as >4.5 g/cm^3.

tributes a distinct set of attributes in terms of processability and performance of the final compound. Consequently, the appropriate stabilizer choice takes into consideration the target performance requirements, processing configuration, and regional regulations and practices.

4.3.1 Mixed-metal Carboxylate Stabilizers

Solid mixed-metal carboxylate stabilizers (calcium–zinc carboxylates and barium–zinc carboxylates) contain two or more of the compounds listed in Table 4.1. As can be seen, most acids used for manufacturing the metal carboxylates (soaps) come from renewably sourced fats and oils. As a result, the bio-based content of the metal laurates, metal stearates, and metal oleates (measured according to ASTM D6866) is close to 100%.

Table 4.1 Selected Metal Carboxylate Stabilizer Components and Their Characteristics

Metal soap/salts	CAS no.	M.p. (°C)	Molecular weight (g/mol)	Metal content (%)	Density (g/mL)
Barium dibenzoate	533-00-6	385	379.6	36.2	2.00
Barium di-2-ethylhexanoate	2457-01-4	>300	423.7	32.4	1.39
Barium dineodecanoate	55172-98-0	—	479.8	28.6	—
Barium dilaurate	4696-57-5	260	535.3	25.6	—
Barium dioleate	591-65-1 (95465-85-3)	— (57)	700.3 (680.3)	— (20.2)	— (1.24)
Barium distearate	6865-35-6 (91002-07-2)	>225 (141–143)	704.3 (685.3)	— (20.0)	— (1.25)
Calcium dibenzoate	2090-05-3	297–323	282.3	14.2	1.42
Calcium di-2-ethylhexanoate	136-51-6	160	326.0	12.3	1.07
Calcium dineodecanoate	27253-33-4	Semi-solid	382.6	10.5	—
Calcium dilaurate	4696-56-4	183	438.7	9.1	1.11
Calcium dioleate	142-17-6	—	602.0	6.6	—
Calcium distearate	1592-23-0 (85251-71-4)	—	607.0 (588)	6.6	— (1.08–1.12)
Calcium di-12-hydroxystearate	3159-62-4	120–166 (decomp.)	639.0	6.2	1.08

Metal soap/salts	CAS no.	M.p. (°C)	Molecular weight (g/mol)	Metal content (%)	Density (g/mL)
Calcium behenate	3578-72-1	—	719.2	5.5	1.08–1.12
Magnesium dilaurate	4040-48-6	43.8	422.0	5.7	—
Magnesium dioleate	1555-53-9	Liquid	586.0	4.1	—
Magnesium distearate	557-04-0 (91031-63-9)	— (86–88)	591.2 (572.0)	4.1	1.03
Zinc dibenzoate	553-72-0	297–323	307.6	21.3	1.5
Zinc di-2-ethylhexanoate	136-53-8	Liquid	351.8	18.6	1.18
Zinc di-n-octanoate	557-09-5	131–138	351.8	18.5	1.185
Zinc hydroxy (2-ethylhexanoate)	—	Liquid	—	22.0	—
Zinc dineodecanoate	27253-29-8	Liquid	407.9	16.0	1.08
Zinc dilaurate	2452-01-9	128	463.4	14.1	1.18
Zinc dioleate	557-07-3	Liquid	628.3	10.4	—
Zinc distearate	557-05-1 (91051-01-3)	128 (121)	632.3 (613.4)	10.3 (10.7)	— (1.11)
Zinc di-12-hydroxy-stearate	35674-68-1	150	664.3	9.8	1.15

- Correct chemical names of the metal salts are shown in the Table. However, the prefix „di" is often omitted for simplification.
- CAS numbers and properties in brackets are for commercial grades of C16-C18 fatty acid metal salts

4.3.2 Stabilizing PVC with Mixed-metal Carboxylates

The simplest mixed-metal carboxylate stabilizers contain a combination of zinc and calcium soaps/salts or zinc and barium soaps/salts (Table 4.1). The zinc compound acts as a primary stabilizer substituting for up to two labile allylic chlorines, as shown in Figure 4.1(a) according to the mechanism proposed by Frye and confirmed by others.

(a) $2\ PVC\text{–}Cl + Zn(O\text{–}CO\text{–}R)_2 \rightarrow 2\ PVC\text{–}O\text{–}CO\text{–}R + ZnCl_2$
(b) $Zn(O\text{–}CO\text{–}R)_2 + 2\ HCl \rightarrow 2\ HO\text{–}CO\text{–}R + ZnCl_2$
(c) $Ca(O\text{–}CO\text{–}R)_2 + ZnCl_2 \rightarrow 2\ Zn(O\text{–}CO\text{–}R)_2 + CaCl_2$
(d) $Ca(O\text{–}CO\text{–}R)_2 + 2\ HCl \rightarrow 2\ HO\text{–}CO\text{–}R + CaCl_2$

Figure 4.1 Mechanism of stabilization of PVC by mixed-metal carboxylates

Additionally, zinc carboxylates can also act as HCl scavengers (Figure 4.1(b)), and both reactions yield zinc chloride ($ZnCl_2$). Undesirably, $ZnCl_2$ exhibits a robust Lewis acidity, thereby facilitating the zipper chain decomposition process and promoting the cross-linking of PVC polyenes. Consequently, it becomes imperative to deactivate it. In the stabilization process, calcium and barium soaps/salts play a dual role. They scavenge the undesirable $ZnCl_2$ Lewis acid via ion-exchange and regenerate zinc salt/soap as shown in Figure 4.1(c). Furthermore, calcium and barium soap/salt also act as HCl scavengers, as shown on Figure 4.1(d).

The calcium chloride ($CaCl_2$) generated through the reactions of Figure 4.1(c) and (d) is not a Lewis acid, and, in contrast to $ZnCl_2$, it does not promote the degradation of the PVC chains. Therefore, mixtures of zinc and calcium or barium soaps or salts act as the most typical mixed-metal stabilizer systems, where the zinc compound substitutes for the allylic chlorine and controls the initial color and color retention, and the calcium and barium compounds scavenge HCl, deactivate $ZnCl_2$, and regenerate the zinc component, controlling the long-term thermal stability.

The mixed-metal carboxylates alone may not be enough to maximize the stabilizing effect required for PVC processing and the article service life. Therefore, incorporating co-stabilizers and other additives is needed to achieve that.

The higher the zinc content of a zinc salt present in a stabilizer, the lower the imparted heat stability if the stabilizers are added at the same dosage expressed in phr (parts of the components per 100 parts of PVC resin).

Hydrotalcites ($Mg_3ZnAl_2(OH)_{12}(CO_3)\cdot 3H_2O$, Section 4.3.5) are the only exceptions from both of the above rules.

4.3.3 Nitrogen-containing Primary Organic Stabilizers

The first organic substances acting as primary stabilizers were discovered in 1937; therefore, they are as old as lead-based stabilizers. Urea derivatives were introduced as primary stabilizers in the Luvitherm process in Germany during WWII. In the 1960s, the first β-aminocrotonates were used to stabilize films for packaging. The most commercially used are uracil derivatives, mainly 6-amino-1,3-dimethyluracil (DMAU), triethanolamine (TEA), triethanolamine perchlorate sodium salt (TEAP), and, to a lesser extent, β-aminocrotonates, urea derivatives, thiourea, and dihydropyridines (DHPs) (Figure 4.2). Many of those can also be used in both CZS and stabilizers that are free from heavy metals and zinc; DHP and uracil derivatives combined with stearoyl benzoyl methane or dibenzoylmethane improve early color and color hold. The organic primary stabilizers can substitute for and deactivate the allylic chlorine, as shown in Figure 4.3.

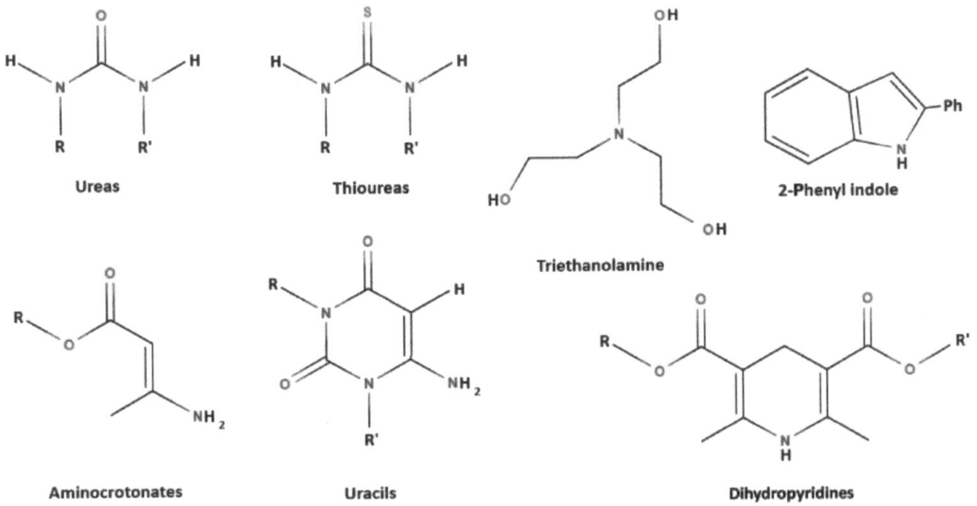

Figure 4.2 Selected nitrogen-containing primary organic stabilizers

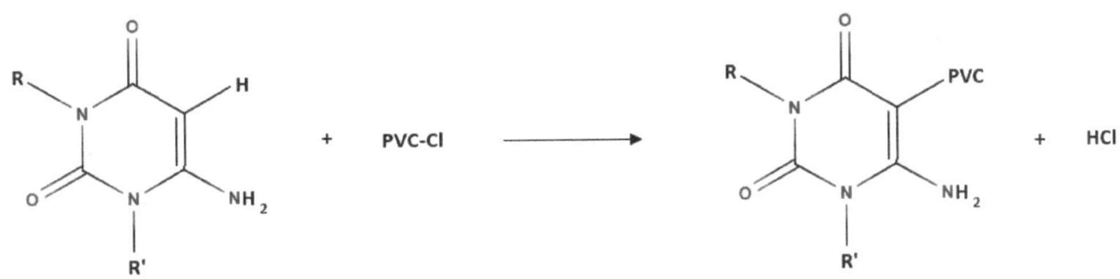

Figure 4.3 Mechanism of stabilization of PVC by uracil derivatives

Typically in solid organic stabilizers (OSs), the primary stabilizers are nitrogen-containing organic molecules, such as 6-amino-1,3-dimethyluracil. Solid OS may also contain co-stabilizers, such as sodium perchlorate, and acid scavengers, such as hydrotalcites and zeolites, for addressing long-term thermal stability.

The primary organic stabilizers control the initial color but are typically used in combination with co-stabilizers to boost the long-term heat stability of PVC. Examples of those co-stabilizers are sodium perchlorate and triethanolamine-perchlorate sodium salt (TEAP), the reaction product of TEA and sodium perchlorate. TEAP can be employed in articles intended to come in contact with food according to Regulation (EU) no. 10/2011.

4.3.4 Organic Co-stabilizers

Certain polyols,[4] such as pentaerythritol, sorbitol, and di-pentaerythritol (Figure 4.4), are often used as co-stabilizers. Their primary functions are to chelate $ZnCl_2$ and scavenge HCl through acid-catalyzed substitution of OH groups by Cl atoms. Chelation of $ZnCl_2$ and HCl scavenging effectively "quench" (form complexes with) the by-products that catalyze PVC degradation, thereby enhancing its long-term heat stability.

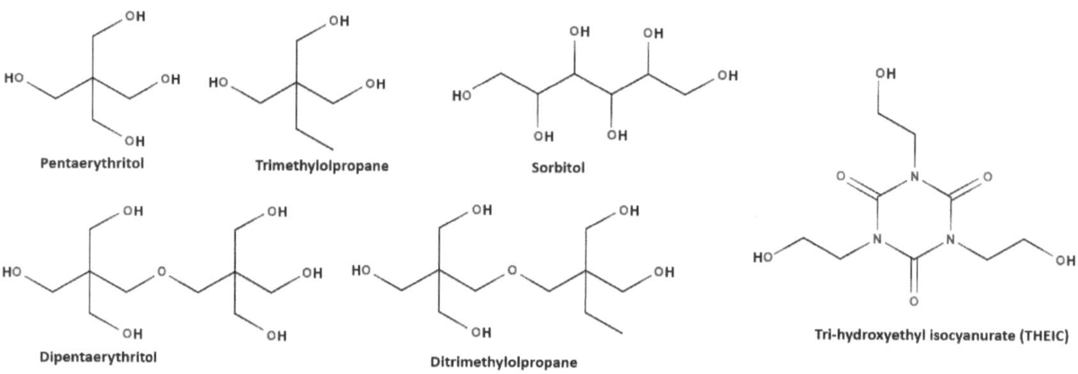

Figure 4.4 Polyol co-stabilizers

Primary OH groups are more capable of reacting with HCl, and the higher the number of OH groups per polyol molecule, the greater the thermal stability they impart as co-stabilizers.

Polyols may cause PVC plate-out[5] during processing. For example, pentaerythritol exhibits a propensity to sublimate[6] during the process and condense back on the cooler tooling and calibrators of the extrusion line, or on the molds of the injection molding machine. Due to their appreciable water solubility, polyols may not be suitable for use in applications where water pickup can be deleterious. THEIC is a nitrogen-containing polyol, and as a result, in addition to chelating of $ZnCl_2$, it acts as an acid scavenger and/or a primary stabilizer (see Section 4.3.3).

4 Polyols (also known as polyhydroxy compounds) are organic compounds with several hydroxyl groups (-OH).

5 This refers to the formation of undesirable deposits on metal equipment parts during PVC processing due to various reasons.

6 Sublimation refers to the direct transition of a substance from the solid to the gas state, and this can result in the formation of deposits on manufacturing equipment.

1,3-Dicarbonyl compounds, also called β-diketones, are a class of co-stabilizers used as a component of mixed-metal carboxylate stabilizers. β-Diketones, such as dibenzoyl methane (DBM), stearoyl benzoyl methane (SBM), and metal acetylacetonates, where the metals are calcium or zinc, are examples of those co-stabilizers (Figure 4.5).

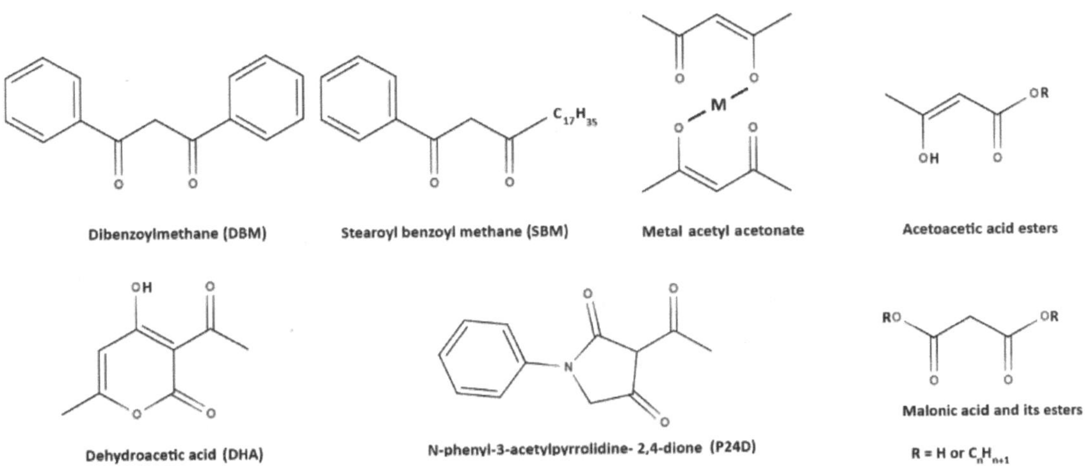

Figure 4.5 Selected β-diketone co-stabilizers

SBM, DBM, and both calcium and zinc acetylacetonates are most utilized in CZS, while both SBM and DBM are commonly used in both liquid and solid CZS as well as in liquid BZS. The pure grade of SBM is also suitable for food contact applications in both the US and EU.

1,3-Dicarbonyl compounds are capable of displacing the allylic chlorine in the presence of a zinc-containing component (Figure 4.6). Additionally, as a result of the keto–enol tautomerism, they effectively chelate $ZnCl_2$ according to Figure 4.7 and deactivate (by complexation) its Lewis acidity that is detrimental to PVC stability.

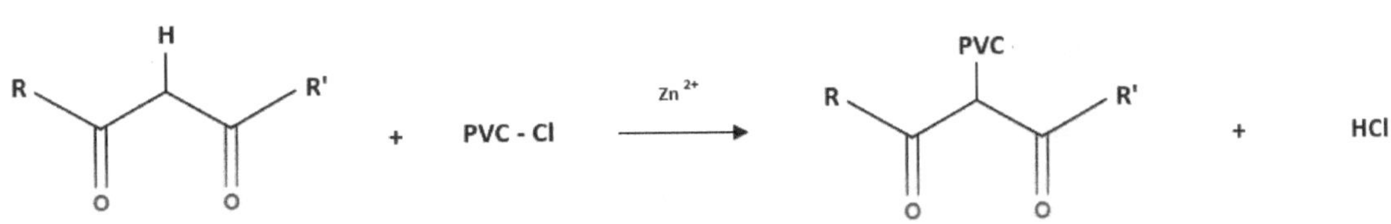

Figure 4.6 Mechanism of stabilization of PVC by 1,3-dicarbonyl compounds

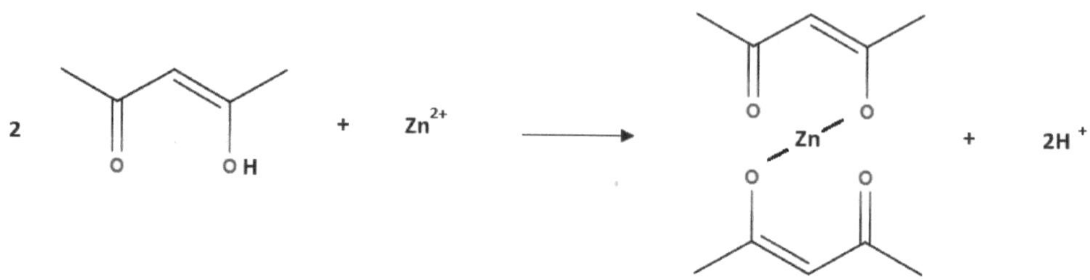

Figure 4.7 Mechanism for chelation of $ZnCl_2$ by 1,3-dicarbonyl compounds

Therefore, 1,3-dicarbonyl compounds usually play a dual role: improving the initial color and suppressing the $ZnCl_2$-catalyzed PVC degradation and cross-linking.

Calcium acetylacetonate (CaAcAc) improves both the initial color of PVC (due to the presence of the β-diketone functionallity) and the long-term thermal stability (due to the presence of calcium). Zinc acetylacetonate (ZnAcAc) enhances the initial color and shortens the long-term heat stability. Both metal acetylacetonates also affect gelling and extrusion behavior.

Rheologically, CaAcAc and ZnAcAc behave analogously to their corresponding metal stearates when present in dry blends.

Epoxidized oils (such as epoxidized soybean oil and epoxidized linseed oil) as well as epoxidized esters of fatty acids (such as epoxidized 2-ethylhexyl tallate) are widely used as co-stabilizers in combination with all types of stabilizers and co-stabilizers. The epoxy groups react with HCl (Figure 4.8), resulting in the ring-opening formation of a chlorohydrin and the extended long-term heat stability of PVC.

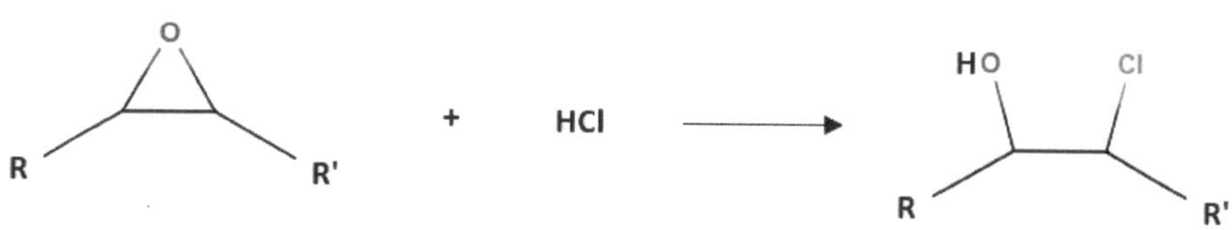

Figure 4.8 Mechanism for scavenging of HCl by epoxides

There are reports suggesting that the epoxy group can also substitute for the allylic chlorines on the PVC backbone in the presence of zinc, cadmium, or alkyltin chlorides. Some other authors claim that epoxides are also capable of deactivating Lewis acids. However, the main mode of action is likely to be HCl scavenging.

Combinations of epoxy compounds with zinc soaps and sodium perchlorate result in the remarkable heat stability of plastisols, used for manufacturing artificial leather in automotive applications.

Epoxidized soybean oil (ESBO) also serves as a well-established secondary plasticizer in conjunction with primary plasticizers. ESBO is also used as a raw material for the synthesis of a diverse range of primary bio-based plasticizers.

Organic phosphite (OP) co-stabilizers can be liquid and solid. Certain liquid ones are shown in Figure 4.9.

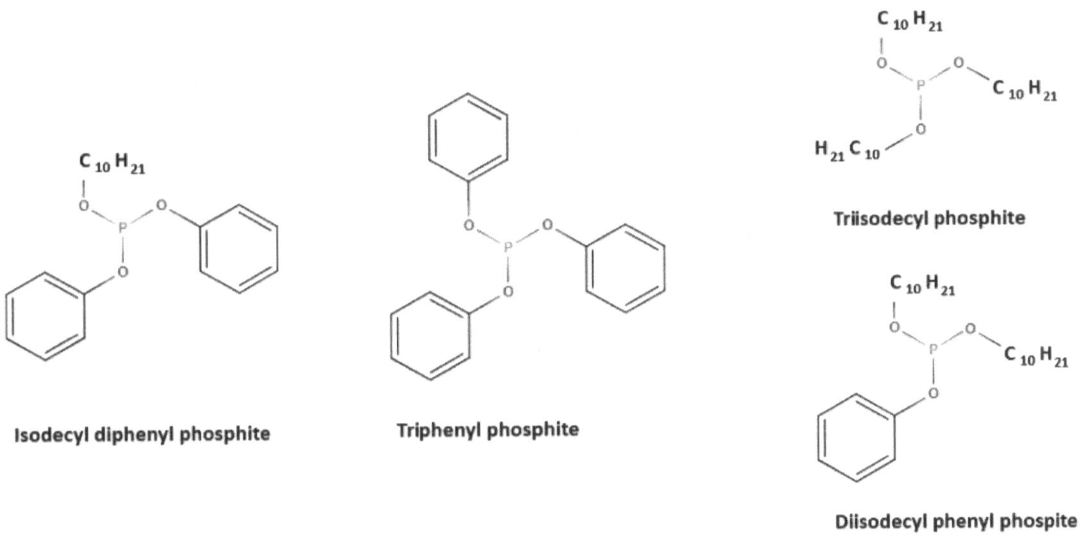

Figure 4.9 Selected liquid phosphite co-stabilizers

OPs are multifunctional additives that are capable of:

- Deactivating the allylic chlorine and improving the initial color (Figure 4.10(a))
- Extending long-term thermal stability by HCl scavenging and chelating $ZnCl_2$ (Figure 4.10(b) and (c))
- Decomposing/deactivating hydroperoxides that might be present in degrading PVC due to the exposure to UV-light and formation of free radicals (Figure 4.10(d)).

(a) PVC–Cl + $P(OR)_3 \rightarrow$ PVC–PO–$(OR)_2$ + R–Cl
(b) HCl + $P(OR)_3 \rightarrow$ H–PO$(OR)_2$ + R–Cl
(c) $M^{2+} + yP(OR)_3 \rightarrow [M(P(OR)_3)_y]^{2+}$
(d) R–O–O–H + $P(OR)_3 \rightarrow O = P(OR)_3$ + R–OH

Figure 4.10 Mechanisms of stabilization of PVC by organic phosphites

OPs can be introduced as components of LMM carboxylate stabilizers or added separately in combination with solid CZS, OS, BZS, and lead-based stabilizers. In the LMM carboxylate systems, the liquid phosphites may also serve as mutual solvents for metal soaps/salts and other components, enabling prolonged storage stability of these liquid multicomponent stabilizers.

The main drawback of these PVC co-stabilizers is that they are prone to hydrolysis. Basic and acidic components that are commonly present in the stabilizer formulations may speed up this process. Their hydrolysis can occur during PVC processing as well as during the service life of PVC articles, affecting the thermal stability, volume resistivity, and appearance of the materials, amongst other aspects. Tris(nonylphenyl) phosphite (TNPP) is suitable for food contact applications, although being on the EU SVHC list; its substitution, particularly in LMMs, is ongoing. Examples of the identified replacements for TNPP in LMM carboxylate stabilizers suitable for use in food contact applications are tris(amylphenyl) phosphites and tris(tridecyl) phosphite (TTDP).

Phenolic antioxidants are frequently used as PVC co-stabilizers. 2,6-Di-*tert*-butyl-*p*-cresol (BHT), pentaerythritol tetrakis(3-(3,5-di-*tert*-butyl-4-hydroxyphenyl)propionate) (AO 1010), octadecyl 3-(3,5-di-*tert*-butyl-4-hydroxyphenyl)propionate (AO 1076), and 4-[4,4-bis(5-*tert*-butyl-4-hydroxy-2-methylphenyl)butan-2-yl]-2-*tert*-butyl-5-methylphenol (AO CA or Topanol CA) are typical representatives of this group of co-stabilizers (Figure 4.11). Certain antioxidants, such as BHT, are used in combinations with both mixed-metal carboxylate and alkyltin stabilizers.

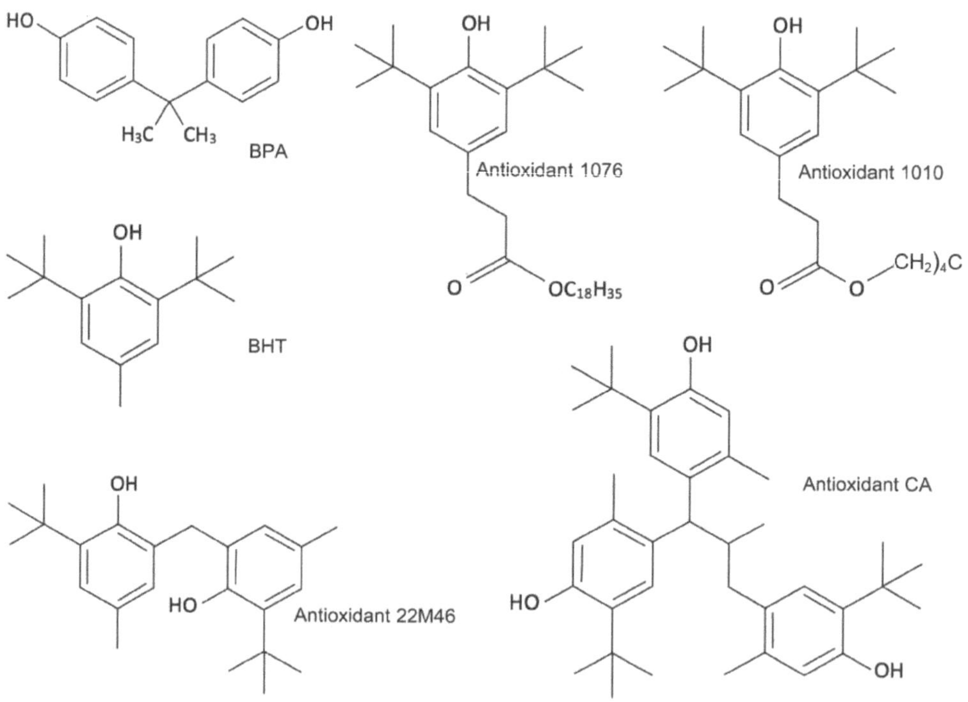

Figure 4.11 Selected antioxidant co-stabilizers

Acting as hydrogen donors (Figure 4.12), they function as free radical suppressors/scavengers, preventing the catastrophic spread of this kind of defect within the PVC matrix, preventing premature crosslinking, and extending the long-term thermal stability of both rigid and plasticized PVC compounds. For more information, see Chapter 2.

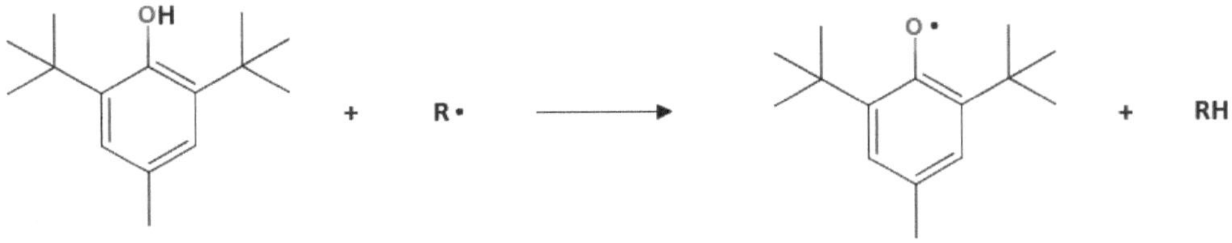

Figure 4.12 Mechanism of radical scavenging by antioxidant co-stabilizers

4.3.5 Inorganic Co-stabilizers

Metal oxides and especially metal hydroxides are commonly incorporated into low-cost PVC stabilizers. Magnesium and calcium are typical examples of those metals. It should be noted that by-products generated as a result of their reaction with HCl are water-soluble and hygroscopic. Despite this limitation, these acid scavengers are prominently used in manufacturing low-voltage PVC cable insulation and sheathing (jacketing) compounds.

Hydrotalcites (HTs) and hydrocalumites have layered doubled hydroxides (LDH) structures. Certain representatives of this group were designed for use as PVC co-stabilizers to trap HCl in between their layers. The hydrotalcites used in PVC have the chemical formula $Mg_4Al_2(OH)_{12}(CO_3)\cdot 3H_2O$. Other forms of hydrotalcites used in PVC may also contain perchlorate anions or zinc cations (HTZ), with the latter being described by the chemical formula $Mg_3ZnAl_2(OH)_{12}(CO_3)\cdot 3H_2O$.

Figure 4.13 shows the acid scavenging mechanism by hydrotalcites. The LDH structure closely resembles that of brucite, which consists of magnesium atoms that are octahedrally coordinated with six oxygen atoms in the hydroxide form. HCl reacts with the carbonate anion, liberating carbon dioxide and water and absorbing/retaining two chlorine atoms in the interlayer space.

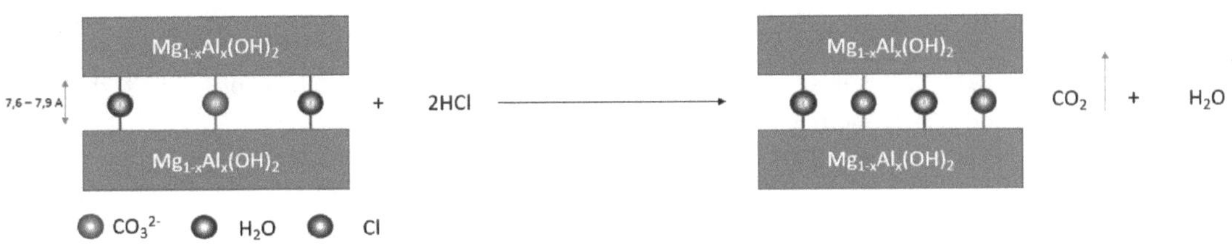

Figure 4.13 Mechanism of HCl neutralization by hydrotalcites

The LDH structure remains intact until all carbonates are depleted. Once this occurs, HCl begins to attack the hydroxide layers, ultimately leading to the collapse of the layered structure of LDH. At temperatures exceeding 190 °C, even water molecules positioned in the interlayer space may be liberated. Hydrotalcites scavenge HCl efficiently, and in doing so, they do not generate hygroscopic by-products. Consequently, they find extensive use in applications requiring extended long-term thermal stability, low water absorption, and superior electrical insulation properties. Hydrotalcites can be used in both indoor and outdoor applications, and are utilized in producing technical profiles among others.

Hydrocalumites are a subgroup within the LDH family. In terms of their compositions used in PVC applications in the past, they were mainly hydrated calcium aluminate phosphites or carbonates. Examples of those include $Ca_4Al_2(OH)_{12}HPO_3{\cdot}3H_2O$ (calcium aluminium hydroxide phosphite, CAHP) and $Ca_4Al_2(OH)_{12}CO_3{\cdot}5H_2O$ (calcium aluminium hydroxy carbonate, CAHC). The acid-scavenging mechanism of CAHP is similar to that of hydrotalcites and zinc-containing hydrotalcites. CAHP is highly alkaline, but does have higher hygroscopic stability than either hydrotalcites or zinc-containing hydrotalcites.

Zeolites are crystalline aluminosilicates. Their architecture is distinguished by the corner-sharing arrangement of AlO_4 and SiO_4 tetrahedra, forming a complex three-dimensional structure. Particularly, the composition of zeolite 4Å used for PVC stabilization possesses pore sizes that are precisely tailored for exchanging its sodium cation with proton of HCl (generated as a byproduct of PVC degradation) or with the zinc cation of $ZnCl_2$ (formed from zinc carboxylate stabilizer components) (Figure 4.14), followed by the neutralization of HCl on account of its skeletal structure.

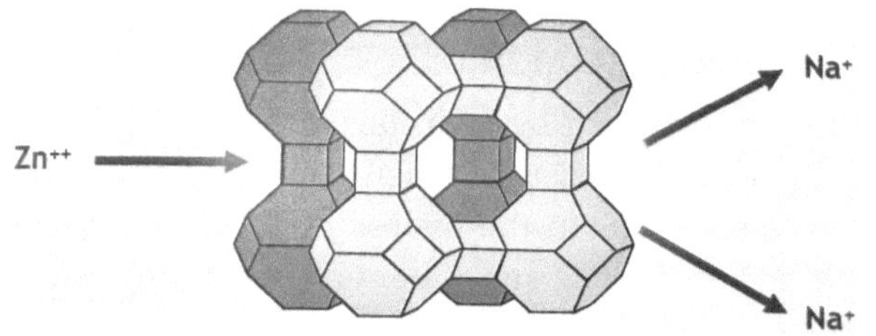

Figure 4.14 Cation-exchange mechanism within zeolite 4Å

Examples of other inorganic co-stabilizers include calcium silicate and sodium perchlorate. The first one scavanges HCl, and the second one shortens polyene sequences and consequently reduces PVC discoloration.

4.3.6 Applications of Solid Mixed-metal Carboxylate and Organic Stabilizers

In Europe, due to tightening regulations and the voluntary commitment of the entire PVC industry, the use of cadmium- and lead-containing stabilizers was phased out from 2000 to 2015. Use of lead stabilizers was also discontinued in cable compounds in North America. This process is still in progress in Asia and Australia. Over the past several decades, solid CZSs and BZSs and solid OSs have played a key role in replacing solid cadmium- and lead-containing heat stabilizers. Organotin stabilizers continue to dominate in other rigid PVC applications, such as films and fittings in Europe and most rigid compounds (including PVC pipes and profiles) in North America, since at present there is a significant performance gap between these two types of stabilizers (SMM carboxylates and organotin stabilizers), particularly in terms of the imparted transparency and whiteness.

CZSs are primarily used in the production of many rigid PVC articles, such as profiles, sheets, and pipes. They are also employed in certain plasticized PVC applications, such as hoses, tubes, gaskets, seals, membranes, soles, and profiles. CZSs are almost exclusively used in the production of cable compounds, including power, control and communications, automotive, offshore/inshore, shipboard, and rail stock. Certain CZS grades can be used in food-contact-suitable flexible PVC compounds, cap liners, and conveyor belts, and also in refrigerator components such as profiles, compact and foamed sheets, and frames.

Solid BZSs are used in interior automotive applications and high-end plasticized compounds.

Solid OSs are mainly used in rigid pipes and fittings where their performance might be superior to CZSs. Currently, the 6-amino-1,3-dimethyluracil-containing systems dominate in pipes in the EU, mainly in southern Europe (Spain and Portugal); thiourea-based OS are effective in manufacturing rigid films. Solid OSs are also used in manufacturing plasticized PVC applications, including roofing membranes and automotive components.

4.3.7 Applications of Liquid Mixed-metal Carboxylate Stabilizers

The commercial availability of liquid co-stabilizers, such as epoxy compounds and phosphite esters, enabled the development of storage-stable LMM grades based on barium alkylphenates, and cadmium and zinc soaps for use in plasticized PVC articles.

Cadmium-based liquid and solid stabilizers started to be phased out in the EU in the mid-1990s when the Council Directive 91/338/EEC, enacted in 1991, came into effect. From 2001 to 2007, the European PVC industry proactively committed to a voluntary initiative, resulting in prohibition of cadmium-based stabilizers.

Originally, certain neutral salts of calcium, barium, and zinc, such as 2-ethylhexanoates and oleates, were used for the preparation of liquid mixed-metal carboxylate stabilizers (see Table 4.1) in combination with liquid organic co-stabilizers as solvents. A step-change improvement was the development of metal-rich liquid "overbased" carbonates/carboxylates, containing >10% calcium and >30% barium. These are translucent micellar colloidal dispersed systems. The above-mentioned excess over neutrality (by reference to most barium-containing intermediates) is present as a carbonate of the micellar structure that is formed in situ by reacting metal carboxylates with CO_2.

LMM stabilizers find extensive commercial use in manufacturing various plasticized or semirigid PVC articles through calendering, extrusion, injection molding, roto-molding, and coating processes. Articles, such as roofing and geomembranes, can be produced with liquid barium-zinc carboxylate stabilizer (LBZS) through calendering and flat-die extrusion. Extrusion can also be applied to manufacture of transparent hoses and tubing, seals, and gaskets using LMM stabilizers. Liquid calcium-zinc carboxylate stabilizer (LCZS) is an excellent choice for manufacturing transparent injection-molded soles and blown films. In flooring, the layers within luxury vinyl tiles (LVTs), especially the clear topcoat, can be produced with both LBZS and LCZS through the calendering process. In automotive, LMM stabilizers are extensively used in calendering production of sun visors, artificial leather, dashboards (PVC/ABS or ASA/PVC), and decorative films.

In certain applications, solid mixed-metal carboxylate stabilizers have successfully replaced LMM stabilizers. Coatings produced via spreading, spraying, and dipping, as well as rotational molding, are the notable exceptions, where this kind of substitution has not taken place yet.

Special liquid additives called "kickers" are capable of decomposing azodicarbonamide (ADCA, ACA, ADA), a commonly used chemical foaming agent, at lower processing temperatures during the manufacture of foamed compounds, in order to achieve the desired density, thickness, and/or aesthetics in the final articles. Their typical chemical compositions feature zinc carboxylates or zinc–potassium carboxylates.

LBZS impart relatively good early color and transparency, and excellent long-term heat stability compared to LCZS, and therefore they have been actively used as replacements for liquid cadmium-based stabilizers since the 1970s.

LCZS, on the other hand, show shorter long-term heat stability but impart remarkable initial color and transparency, and are suitable for certain food contact applications.

Liquid organic stabilizers (LOS) are heavy-metal-free, and are used in niche applications where the heavy-metal-free claim is desirable or required. They impart low VOC, minimal fogging, low odor, excellent clarity, minimal initial color and good color-hold. Many of them also have inherent antistatic properties.

Paste calcium–zinc carboxylate (PCZ) stabilizers are often grouped together with LMMs because they can be pumped into the process. They are used in sensitive plasticized PVC compounds intended for manufacturing toys, packaging, and medical films and devices that are expected to be suitable for food contact.

4.4 Organotin Stabilizers for PVC

4.4.1 Chemical Groups and General Characteristics

Organotin heat stabilizers were introduced for commercial use in PVC compounds in the 1930s. Over the years they have demonstrated their outstanding efficiency, and maintained commercial viability and applicability in major PVC markets and geographic regions.

The tin atom in these stabilizers is tetravalent. It can form from one to four organometallic (tin–carbon) bonds with alkyl groups, and is capable of coordinating from one to three ligands in alkyltin compounds.

Tetra-alkyltins and trialkyltin compounds exhibit no to very little heat-stabilizing activity and therefore are not known to be used as components of PVC heat stabilizers. Additionally, trialkyltin components are known to be highly toxic. Instead, the vast majority of all the commercial heat stabilizers are based on mono-alkyltin and di-alkyltin compounds, containing three or two other ligands (L), respectively. The following rules usually apply:

Mono- and dialkyltin derivatives function synergistically in stabilizer compositions.

Alkyl groups (R) in mono- and dialkyltin compounds present in the same product are typically the same (methyl, n-butyl or n-octyl).

Alkyl groups (R) in dialkyltin compounds that are used on their own (without the presence of monoalkyltin components) as stabilizers are the same as well.

The effect of alkyl groups on the tin content, extractability, external/internal lubrication ratio, volatility, and odor of alkyltin stabilizers of similar chemical functionalities (the same ligands) reduces in the following order: methyl > n-butyl > n-octyl.

There are four main types of alkyltin heat stabilizers that represent the most significant value to the PVC industry in terms of their production volumes and performance attributes. These types are:

- Alkyltin carboxylates
- Alkyltin mercaptoacid esters

- Alkyltin mercaptoalcohol esters (also known as the alkyltin “reverse esters”)
- Alkyltin mercaptides.

Organotins are the most efficient and effective stabilizer class especially for clear and opaque rigid PVC compounds, including pipes, fittings, profiles, and films. They can effectively stabilize rigid PVC compounds if added at levels as low as 0.5%. Certain stabilizers of this type are also used in selected plasticized PVC compounds.

Organotin compounds that are currently used as stabilizers in most PVC and C-PVC compounds are mono- and/or dialkyltin compounds of the general formula $RSnL_3$ and R_2SnL_2, where the R group is methyl, n-butyl, or n-octyl, and the ligand L can be a mercaptoester, reverse mercaptoester, mercaptide, carboxylate, or sulfide. Often combinations of these ligands are used for meeting the cost-performance requirements in specific compounds or applications. n-Dodecyltin thioglycolates introduced in the 1990s as non-toxic alkyltin stabilizers did not receive broad market acceptance.

4.4.2 Alkyltin Carboxylates

In most alkyltin carboxylates, each tin atom is bonded primarily with two alkyl groups (where R is typically n-butyl or n-octyl) and two carboxylate ligands (Figure 4.15).

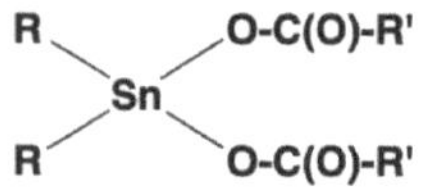

Figure 4.15 Chemical structures of dialkyltin dicarboxylate components of heat stabilizers, where R is an alkyl (for example, n-butyl or n-octyl), and R′COO is either a monocarboxylate (for example, laurate) or a half-ester of a dicarboxylic acid (for example, ethylmaleate or 2-ethylhexylmaleate)

Mono-alkyltin tricarboxylates are usually present as impurities at up to about 10%. This class of organotin stabilizers is sulfur-free, and as a result they do not cause staining of PVC surfaces when used in combination with lead stabilizers or for processing of lead-containing recyclate streams; see Section 4.4.6.

Alkyltin carboxylate stabilizers have low odor. They impart excellent clarity, good early color, UV-light stability, and overall weatherability to PVC. In general, alkyltin carboxylates exhibit limited mid- and long-term heat stability when used in rigid PVC compounds.

As a result, they are typically formulated into synergistic combinations with alkyltin mercaptoacid esters (see Section 4.4.3), known for their limited UV-light stability and superior heat stabilizing efficiency, in order to maximize both the heat-stabilizing and UV-light-stabilizing performance attributes of this kind of stabilizer compositions (Figure 4.16). These mixtures are called synergistic since they exhibit only the favorable performance attributes of each component.

To maximize heat-stabilizing and UV-light-stabilizing performance, alkyltin carboxylate stabilizers are typically formulated into synergistic combinations with alkyltin mercaptoacid esters.

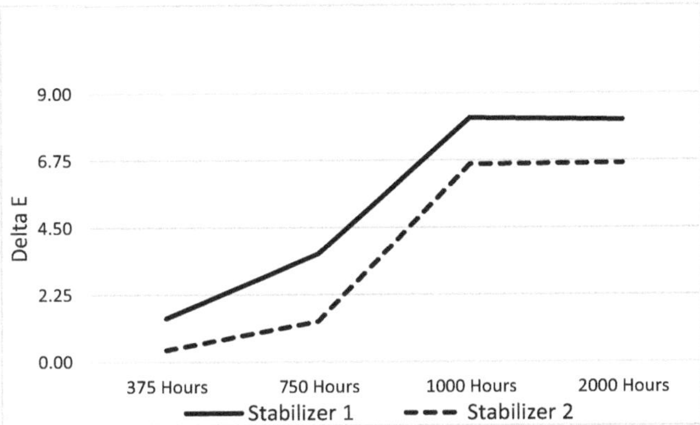

Figure 4.16 Heat stability and UV-light stability imparted by selected stabilizers, demonstrated by an accelerated weathering test. The conditions used were in accordance with ASTM G154, cycle 1. Stabilizer 1: Dibutyltin bis(2-ethylhexyl thioglycolate), torque-based decomposition time 14.0 min. Stabilizer 2: Combination of dibutyltin bis(2-ethylhexyl thioglycolate) and dibutyltin dilaurate, torque-based decomposition time 13.3 min

This synergism among the components enables broad use of these *formulated stabilizers* in manufacturing PVC articles intended for outdoor applications, including transparent panels, films, siding, fencing, and window profiles. Both butyltin carboxylate/butyltin mercaptoacid ester and octyltin carboxylate/octyltin mercaptoacid ester combinations are more weatherable than the corresponding methyltin compositions.

Larger alkyl groups sterically hinder and protect the Sn–S bonds of alkyltin thioesters against decomposition by UV light, heat, and moisture.

It should be noted that mixtures of alkyltin compounds containing different kinds of non-alkyl ligands in their chemical structures are known to equilibrate due to the ligand exchange with no change in the oxidation state of the metal center. This is a dynamic equilibrium reaction that takes place even at ambient temperature, and it is likely driven by an increase in entropy of the system (Figure 4.17).

$$R_2SnL'_2 + R_2SnL''_2 \rightleftharpoons 2R_2SnL'L''$$

Figure 4.17 Mechanism of ligand equilibration in alkyltin compounds, where R is an alkyl group, and L′ and L″are non-alkyl ligands that are not the same (L′ ≠ L″)

Therefore, in reality dialkyltin carboxylate/dialkyltin mercaptoacid ester mixtures, for example, may contain dialkyltin (carboxylate)(mercaptoacid ester) moieties.

In flexible PVC applications, dialkyltin carboxylate stabilizers impart excellent initial color, clarity, and UV-light stability.

In rigid PVC, where light stability is generally inversely proportional to the sulfur content of these stabilizers, the sulfur-free alkyltin carboxylates are employed as components of the dialkyltin carboxylate/dialkyltin mercaptoacid ester mixtures, both for imparting light stability and for reducing the odor of the stabilizer.

Alkyltin maleates, representatives of the alkyltin carboxylate stabilizer type, such as dioctyltin di(ethylmaleate), are used in manufacturing plasticized films and rigid PVC articles. Maleic acid-derived ligands were found to be more efficient in terms of the imparted heat stability compared to other carboxylate ligands of these alkyltin stabilizers. The high efficiency was attributed to the Diels–Alder reaction[2] between this ligand and polyenes of decomposing PVC molecules, in addition to the conventional mechanism of stabilization with alkyltin compounds; see Section 4.4.7.

4.4.3 Alkyltin Mercaptoacid Esters

Sulfur-containing alkyltin mercaptoacid ester heat stabilizers have a characteristic odor and usually consist of synergistic blends or reaction mixtures of a monoalkyltin tris(mercaptoacid ester) and a dialkyltin bis(mercaptoacid ester). The ratio of these mono- and dialkyltin components varies broadly (the amount of the mono-alkyltin component in these mixtures ranges from 5% to about 95%) to achieve the optimum cost-performance target characteristics suitable for the intended range of applications associated with a specific set of processing conditions. The chemical structures of typical components and compositions are shown in Figure 4.18, where thioglycolic acid and mercaptopropionic acid are esterified with iso-octanol or 2-ethylhexanol to form the thioester ligands. Lauryl alcohol and myristyl alcohol as well as iso-tridecyl alcohol are also suitable for esterifying the mercaptoacids into more externally lubricating thioester ligands. Additionally, it was demonstrated that ethylene glycol dithioglycolate can also be effectively used as a dimercaptoacid ester ligand for manufacturing alkyltin mercaptoacid ester heat stabilizers.

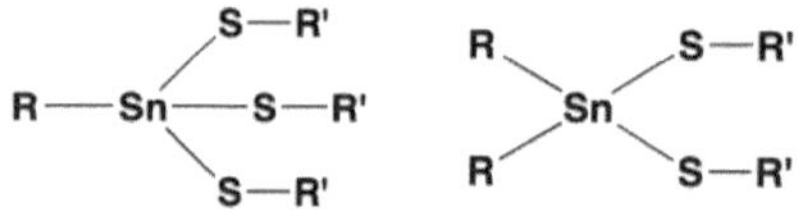

Figure 4.18 Chemical structures of alkyltin mercaptoacid ester components of heat stabilizers, where R is an alkyl group (such as methyl, n-butyl,or n-octyl), and -SR′is a mercaptoacid ester ligand (such as 2-ethylhexyl thioglycolate, iso-octyl thioglycolate, or iso-tridecyl mercaptopropionate)

The synergistic heat-stabilizing performance among the mono- and dialkyltin thioglycolate heat stabilizer components is shown in Figure 4.19. This synergistic effect has been attributed to the following:

- The monoalkyltin component is more efficient in delaying PVC dehydrochlorination through the deactivation of the allylic chlorine on the degrading PVC molecules during the initial steps of their thermal decomposition, and it manifests itself in superior initial color of the processed PVC
- The dialkyltin component is more effective in scavenging HCl (hydrochloric acid) and preventing propagation of the dehydrochlorination reaction catalyzed by the HCl, and this effect manifests itself in extending long-term heat stability of the processed PVC.

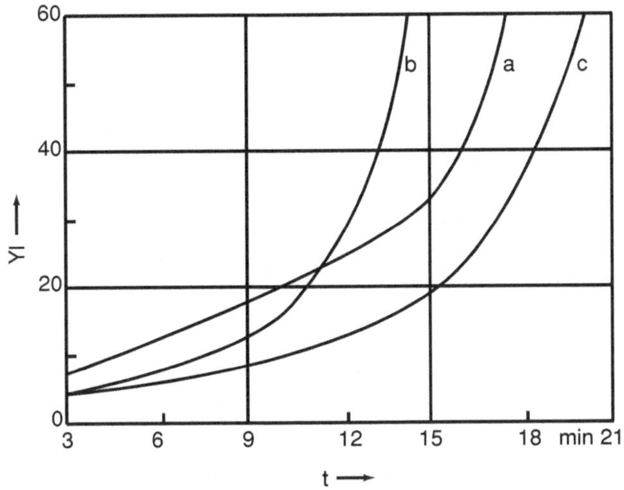

Figure 4.19 Synergism of mono- and dioctyltin isooctyl thioglycolates exemplified by measuring yellowness index (Y.I.) as a function of milling time (*t*) in the dynamic heat stability test on a two-roll mill at 200 °C: (a) dioctyltin bis(iso-octyl thioglycolate); (b) monooctyltin tris(iso-octyl thioglycolate; (c) a 20:80 combination of the mono- and dioctyltin iso-octyl thioglycolates

The versatility for achieving the optimized mono-/dialkyltin component ratios enables this kind of stabilizer to be broadly applied in a wide range of PVC compounds and applications; see also Section 4.4.7.

Alkyltin mercaptoacid esters are the most efficient heat stabilizers for processing PVC homopolymers, co-polymers, and chlorinated poly(vinyl chloride) (C-PVC); see Section 4.6. These methyl-, n-butyl and n-octyltin derivatives are used practically in all rigid PVC technical applications. These include transparent and opaque (filled and pigmented) compounds that are processed by extrusion, injection molding, calendering, and other processing options, especially under high temperature and high shear conditions, where they enable the processing window range to be extended. Additionally, mono- and dimethylitin thioglycolates as well as mono- and dioctyltin thioglycolates, are exceptionally compatible with PVC resins, and as a result were found to be suitable and qualified for use in compounds intended to come in contact with food in the US and Europe.

Alkyltin mercaptoacid ester stabilizers impart limited weatherability on PVC, including limited UV-light stability, but they can still be effectively used in outdoor applications in combinations with alkyltin carboxylates discussed in Section 4.4.2, and in compounds containing sufficient amounts of organic UV absorbers and/or titanium dioxide.

Applications, where these stabilizers are commonly used, include rigid calendered films and wraps, extruded compact and foamed sheets and profiles, injection-molded pipe fittings, molded bottles, sidings, furniture, and decking, among many others.

4.4.4 Alkyltin Mercaptoalcohol Esters

Alkyltin mercaptoalcohol esters are also known as alkyltin "reverse ester" heat stabilizers due to their particular sequencing in the ester group functionality (-COO-), which is reversed compared to the ordering (-OOC-) in the alkyltin mercaptoacid esters described in Section 4.4.3.

Typically but not always, these stabilizers have high content of monoalkyltin mercaptoalcohol ester components and a lower content of the dialkyltin components, in which the alkyl groups are either methyl or n-butyl. Most commonly, the mercaptoalcohol used for the preparation of alkyltin mercaptoalcohol esters is either 2-mercaptoethanol or 3-mercaptopropanol, which is esterified by C_{12}–C_{18} organic acids. The acids are often selected from oleic acid, tall oil fatty acids, soybean oil fatty acids, and others.

Often, alkyltin 2-mercaptoethyl esters are produced to contain also alkyltin moieties with sulfide ligands in the form of alkyltin sulfides as well as sulfur-bridged alkyltin compounds, containing Sn–S–Sn bonds in their chemical compositions, as shown in Figure 4.20. Formation of the sulfide ligands and the sulfur bridges in these chemical structures boosts the tin content and the heat-stabilizing efficiency of this kind of heat stabilizer. These stabilizers may also contain low-cost ester and/or hydrocarbon diluents. The tin content of these stabilizers ranges broadly from about 6% to about 20%. Incorporation of the sulfide ligands also allows these stabilizers to impart enhanced lubricity.

Alkyltin 2-mercaptoethyl ester/sulfides have a characteristic odor and have a low cost-in-use. They are designed to be highly externally-lubricating (an attribute that is analogous to ease of mold release). High monoalkyltin reverse ester/sulfide-containing stabilizers with 6–15% tin content, added at relatively low loadings to PVC compounds, are preferred options for use in high-output white extruded pipes of small and medium diameter.

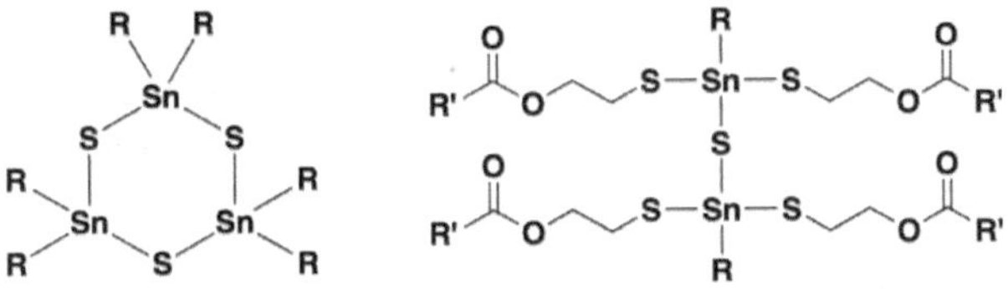

Figure 4.20
Chemical structures of selected sulfur-bridged alkyltin mercaptoethyl ester components of heat stabilizers

Monoalkyltin reverse ester/sulfides with increased concentrations of dialkyltin reverse ester/sulfide components and about 15–20% tin content are used for stabilizing thicker-walled PVC pipes of larger diameters as well as thick sheets, substrate sidings, and other profiles. As can be seen from the range of applications, these stabilizers now successfully compete against alkyltin mercaptoacid esters.

Alkyltin reverse ester stabilizers are highly compatible with PVC. As a result, many of them are qualified by the National Sanitation Foundation (NSF) for use in the PVC range composition issued by the Plastic Pipe Institute (PPI) for manufacturing potable water pipes. This qualification is based on low migration of the stabilizers out of PVC articles, and because these stabilizers impart the necessary stress resistance on pipes under pressure over an extended period of time (the stress test).

It should be noted that since fatty acids are used as raw materials for manufacturing most alkyltin reverse esters and selected alkyltin carboxylates (see Section 4.4.3), these are the only tin-containing stabilizers that contain renewable carbon in their chemical compositions; their bio-based content can reach up to 90% according to ASTM D6866.

4.4.5 Alkyltin Mercaptides

Examples of commercially available alkyltin mercaptides are monobutyltin tris(dodecylmercaptide) and dibutyltin bis(dodecylmercaptide).

Compared to the butyltins, methyltin stabilizers of this kind are not storage-stable, while the octyltins are of lower heat stabilizing efficiency due to the higher molecular weight and lower tin content. This may explain why only the butyltins of this type are commercially viable.

These premium organotin stabilizers are highly lubricating. In addition to the imparted lubricity, other performance attributes of these stabilizers include superior heat stability (especially in terms of the early color) and the so called “anti-greening effect” that refers to preventing the formation of a greenish discoloration on surfaces of extruded rigid PVC articles. These stabilizers are predominantly used in high-output extrusion applications, such as sheets and profiles.

4.4.6 Stabilizing PVC with Alkyltin (2-Ethylhexyl Thioglycolate)s

According to recently published results and consistent with a previous understanding, the most recent and complete mechanism of PVC stabilization with the use of alkyltin 2-ethylhexyl thioglycolate stabilizers can be described by the four steps shown in Figure 4.21 (see also Table 4.3 for the universal mechanism of PVC stabilization).

(a) $RSn(SR')_3 + \sim CH = CH\text{–}CH(\text{–}Cl) \sim \longrightarrow RSn(SR')_2Cl + \sim CH = CH\text{–}CH(\text{–}SR') \sim$
(b) $R_2Sn(SR')_2 + HCl \rightarrow R_2Sn(SR')Cl + R'SH$
(c) $R_2Sn(SR')_2 + R_2SnCl_2 \rightarrow 2\ R_2Sn(SR')Cl$
(d) $R_2Sn(SR')_2 + RSn(SR')_2Cl \rightarrow R_2Sn(SR')Cl + RSn(SR')_3$

Figure 4.21 Mechanism of stabilization of PVC by alkyltin 2-ethylhexyl thioglycolates: (a) Substitution and deactivation of allylic chlorine on PVC; (b) HCl scavenging; (c) Scavenging of alkyltin chlorides; (d) Regeneration of monoalkyltin tris(thioglycolate) (thioglycolate–chlorine ligand exchange)

The first step (Figure 4.21(a)) involves the substitution and deactivation of allylic chlorine defects formed on the thermally degrading PVC backbones as a result of initial dehydrochlorination. It is accomplished by the S_N2 mechanism, primarily with $RSn(SCH_2COOR')_3$ monoalkyltin tris(2-ethylhexyl thioglycolate)s and to lesser extent with $R_2Sn(SCH_2COOR')_2$ (dialkyltin bis(2-ethylhexyl thioglycolate)s). The formed HCl reacts primarily with dialkyltin bis(2-ethylhexyl thioglycolate)s (Figure 4.21(b)). Monoalkyltin tris(2-ethylhexyl thioglycolates) are also capable of effective HCl scavenging.

Both mono- and dialkyltin (2-ethylhexyl thioglycolates) were shown to react with and scavenge the corresponding alkyltin chlorides. Due to high content of the dialkyltin components in many alkyltin heat stabilizers of this type, dialkyltin bis(2-ethylhexyl thioglycolate)s are likely more active with scavenging dialkyltin dichloride Lewis acids (Figure 4.21(c)) that can possibly be formed during this step.

This step shows that the presence of unreacted alkyltin (2-ethylhexyl thioglycolate)s in the properly stabilized and compounded PVC is required to ensure that dialkyltin dichlorides and/or alkyltin trichlorides (Lewis acids that function as catalysts of PVC dehydrochlorination), even if formed, are quickly scavenged by the stabilizer excess and transformed into the corresponding alkyltin monochloro-(2-ethylhexyl thioglycolate)s.

Since it has been established that properly processed PVC does not contain alkyltin chlorides (low boiling and volatile alkyltin derivatives), it indicates that vinyl compounds stabilized with alkyltin (2-ethylhexyl thioglycolate)s are safer than previously thought with respect to potential exposure to their vapors or leachates.

It was also found that while functioning as PVC heat stabilizers, most of the dialkyltin bis(2-ethylhexyl thioglycolate) is converted into the corresponding monochloro-(2-ethylhexyl thioglycolate) derivative, and a disproportionately smaller portion of the monoalkyltin tris(2-ethylhexyl thioglycolate) is transformed into the monochloro-bis(2-ethylhexyl thioglycolate) and very small amounts of the dichloro-(2-ethylhexyl thioglycolate). This lower (comparted to dialkyltin bis(2-ethylhexyl thioglycolate) "apparent conversion" of the monoalkyltin tris(2-ethylhexyl thioglycolate) is attributed to the thioglycolate–chloride ligand exchange between the mono- and dialkyltin moieties; it results in the regeneration of the monoalkyltin tris(2-ethylhexyl thioglycolate) (Figure 4.21(d)).

As can be noted, alkyltin monochloro-(2-ethylhexyl thioglycolate)s are formed at each step of the universal mechanism of PVC stabilization. While detailed structural studies were not conducted in molten PVC, it was experimentally found that in the formed mono- and dialkyltin monochloro-(2-ethylhexyl thioglycolate)s, the remaining 2-ethylhexyl thioglycolate ligands are considerably less active as stabilizers. This change in reactivity was explained by the formation of penta- and/or hexa-coordinated tin complexes between one of the carboxyl oxygens of the thioglycolate groups and tin atoms, which in turn reduces the polarity of the Sn–S bonds compared to the initial stabilizer components.

Only one ligand per dialkytin bis(2-ethylhexyl thioglycolate) molecule and up to two ligands per monoalkyltin tris(2-ethylhexyl thioglycolate) molecule (where the amount of the formed monoalkyltin dichloro-(2-ethylhexyl thioglycolate) is very small) are actually involved in stabilizing PVC under normal processing conditions.

To obtain properly processed PVC articles with minimum degradation and lowest discoloration, a certain excess of the alkyltin 2-ethylhexyl thioglycolate stabilizer should remain present in PVC during processing to ensure the absence of alkyltin chlorides that may potentially be formed during compounding. Should increased yellowing of the processed PVC be observed, this may be attributed to the presence of the non-scavenged alkyltin chlorides (Lewis acids) in the polymer matrix, indicating that either the processing conditions need to be adjusted, or the amount of stabilizer needs to be increased.

4.4.7 Universal Mechanism of PVC Stabilization

Despite the unique attributes of each type of heat stabilizers, a discernible set of fundamental principles has emerged. This can be referred to as the "Universal Mechanism of PVC Stabilization". This conceptual framework, first described by Frenkel,[7] encapsulates the interconnected modes of action of heat stabilizers comprising at

[7] (a) P. Frenkel, "Modification of polymers via post reactor chemistries and additives" (PMSE 008A), ACS Fall Meeting "Advances in Establishing Universal Mechanism of PVC Stabilization", August 2022; (b) P. Frenkel, "Advances in stabilizing PVC with alkyltin thioesters & universal mechanism of PVC stabilization", Vinyltec, September 2022.

least two synergistic components, denoted $M'L_n$ and $M''L_m$, as shown in Table 4.2. This mechanism serves as a unifying thread that describes the diverse stabilizing actions across several classes of heat stabilizers, and provides a comprehensive understanding of the intricate dynamics that are at play during PVC stabilization.

Table 4.2 Universal Mechanism of PVC Stabilization (M = alkyl or metal, L = ligand or anion)

Stabilization mode	Mechanism	Examples
Deactivation of allylic chlorines	$M'L_n + PVC\text{–}Cl \rightarrow M'L_{n-1}Cl + PVC\text{–}L$	Figure 4.1(a), Figure 4.21(a)
Scavenging HCl	$M''L_m + HCl \rightarrow M''L_{m-1}Cl + HL$	Figure 4.1(b), Figure 4.1(d), Figure 4.21(b)
Scavenging Lewis acids and component regeneration	$M'Cl_n + M''L_m \rightarrow M'L_m + M''Cl_n$	Figure 4.1(c)
	$M'Cl_n + M'L_n \rightarrow M'LCl_{n-1} + M'L_{n-1}Cl$ $M'L_{n-1}Cl + M''L_m \rightarrow M'L_n + M''L_{m-1}Cl$	Figure 4.21(c), Figure 4.21(d)

The Universal Mechanism of PVC Stabilization includes the following three modes of action:

- Mode A: Substitution/deactivation of allylic chlorine
- Mode B: Scavenging HCl
- Mode C: Scavenging Lewis acids and component regeneration.

In the case of mixed-metal carboxylate stabilizers, Mode C is realized in one step, while in the case of mono- and dialkyltin (2-ethylhexyl thioglycolate) stabilizers, Mode C is realized in two separate steps (Table 4.2).

4.5 Lead Stabilizers

Lead stabilizers are one of the oldest types of stabilizers, with mass-production and use beginning in the 1930s. Due to the high concerns about their toxicological effects and the voluntary commitment of the PVC industry in Europe, lead stabilizers were phased out from the manufacture of rigid PVC articles, such as pipes, fittings, window frames, general-purpose profiles, and foamed and compacted sheets, in 2015. This trend continues now in Asia (China and India) and Australia.

Neutral (also called normal) lead stearate (NLS) and dibasic lead phthalate (DBLPh) are the only lead-based stabilizers capable of functioning as primary stabilizers. In contrast, dibasic lead stearate (DBLS), dibasic lead phosphite (DBLP), and tri- and tetrabasic lead sulfate (TLS, TTLS) are the main co-stabilizers that function in concert and in combination with the primary lead stabilizers when stabilizing PVC. The primary stabilizers can be added into the formulation directly or formed in situ by reacting PbO with stearic acid, which may contain other co-stabilizers and ingredients. Table 4.3 includes the most typical lead stabilizer components.

Table 4.3 Selected Lead-containing Stabilizer Components and Their Calculated Characteristics

Name	Molecular formula	Molecular mass (g/mol)	PbO (%)	Pb (%)
Lead oxide (litharge)	PbO	223.2	100.0	92.8
Dibasic lead phosphite (DBLP)	$2PbO \cdot PbHPO_3 \cdot 0.5H_2O$	742.6	90.2	83.7
Tribasic lead sulfate (TBLS)	$3PbO \cdot PbSO_4 \cdot H_2O$	990.8	90.1	83.6
Tetrabasic lead sulfate (TTLS)	$4PbO \cdot PbSO_4 \cdot H_2O$	1196.0	93.3	86.6
Neutral lead stearate (NLS)	$Pb(C_{18}H_{35}O_2)_2$	755.2	29.5	27.4
Dibasic lead stearate (DBLS)	$2PbO \cdot Pb(C_{18}H_{35}O_2)_2$	1201.6	55.7	51.7
Dibasic lead phthalate	$2PbO \cdot Pb(C_8H_4O_4) \cdot 0.5H_2O$	826.6	81.2	75.4
Tetrabasic lead fumarate	$4PbO \cdot Pb(C_4H_2O_4) \cdot 2H_2O$	1250.0	89.0	82.9
Tribasic lead maleate	$3PbO \cdot Pb(C_4H_2O_4) \cdot 0.5H_2O$	999.8	89.3	82.9

The substances in Table 4.3 function as HCl scavengers, yielding lead(II) dichloride ($PbCl_2$), which has low water-solubility, neither promoting the water uptake nor affecting the thermal stability of the PVC matrix.

This class of stabilizers imparts extended long-term thermal stability, a property that is proportional to the PbO content in the stabilizer. For example, TTLS (93.3% PbO) imparts better thermal stability than TBLS (55.7% PbO).

The use of lead stabilizers in cable compounds results in excellent electric insulation characteristics. Lead salts are also known as multifunctional additives. Some of them act as lubricants and primary stabilizers (NLS, DBLS), others as antioxidants, UV-light stabilizers, and long-term thermal stabilizers (DBLP). Therefore, compositions of the lead-based stabilizer one-packs are less complex (with fewer components) compared to Ca/Zn-based one-packs.

Lead stabilizers are used in certain parts of Asia and the Middle East in a variety of PVC applications, such as window and technical profiles, pipes and fittings, foamed and compact sheets, seals and gaskets, roofing and geomembranes, hoses and cables, and soles and uppers.

Lead stabilizers are not suitable for manufacturing transparent PVC articles.

4.6 C-PVC Stabilization

4.6.1 Mechanism of C-PVC Thermal Degradation

There are similarities and differences in the decomposition of PVC and C-PVC (see Section 4.2 for PVC). Similar to PVC, C-PVC pyrolysis occurs through two main steps. Dehydrochlorination is the first decomposition step for these non-stabilized polymers. Dehydrochlorination, with the evolution of HCl from C-PVC, starts at temperatures about 20 °C higher than it does in PVC. The weight loss of PVC during dehydrochlorination is equivalent to its amount of chlorine and equal to about 55–56%, whereas the value for C-PVC is about 40%. The second step for both non-stabilized PVC and C-PVC thermal decomposition is attributed to cyclization, aromatization, chain-scission, and oxidation of the carbonaceous char residues formed during the dehydrochlorination step.

The second decomposition stage of C-PVC takes place at 370–560 °C. The second step of the thermal degradation of PVC and C-PVC is characterized by the evolution of hydrocarbons (such as benzene) and chlorine-containing hydrocarbons, respectively. According to thermogravimetric analysis and differential scanning calorimetry, weight loss during the second step of thermolysis is about 20% for both polymers. The weight of the pyrolysis residue in the case of PVC (about 10%) is significantly smaller than that in the case of C-PVC (40%). Incorporation of heat stabilizers into PVC and C-PVC compounds substantially changes the mechanism of thermal degradation of these materials.

Although the composition of C-PVC consists of essentially of CH_2, CHCl, and CCl_2 groups, many possible arrangements of these three groupings can produce very complex molecular configurations. The ultimate structure of C-PVC depends on the degree and the method of the chlorination process used for obtaining the polymer.

The chlorination of PVC produces a change in the physical properties and makes the formed C-PVC more thermally stable compared to PVC.

The improved heat stability of C-PVC can be attributed in part to elimination of certain reactive defects and formation of long sequences of -CHCl- units. In general, the higher the chlorine content, the greater the thermal stability of C-PVC.

The higher thermal stability of C-PVC in turn enables it to be processed at higher temperatures but also requires more robust and potent heat stabilizers for this polymer compared to PVC, since the more viscous and tough the melt of chlorinated polymers, the higher the in-process shear, and the more energy is needed to process them.

4.6.2 Heat Stabilizers for C-PVC

The range of the commonly used and/or discussed C-PVC heat stabilizers is narrower than that of PVC. C-PVC heat stabilizers include some of the same (but not all) components that are used for stabilizing PVC. The commercially viable and/or patented heat stabilizer components and their combinations intended for use in manufacturing C-PVC are summarized in Table 4.4.

Table 4.4 Heat Stabilizers and Their Components for C-PVC

Name	Chemical formula	CAS no.
Dimethyltin bis(2-ethylhexyl thioglycolate)	$Me_2Sn(SCH_2COOEtHex)_2$	57583-35-4
Dibutyltin bis(2-ethylhexyl thioglycolate)	$Bu_2Sn(SCH_2COOEtHex)_2$	10584-98-2
Dioctyltin bis(2-ethylhexyl thioglycolate)	$Oct_2Sn(SCH_2COOEtHex)_2$	15571-58-1
Aluminosilicates (zeolite 4Å)	$Na_2O(Al_2O_3)(SiO_2)\cdot xH_2O$	1318-02-1
6-Amino-1,3-dimethyluracil	$C_6H_9N_3O_2$	6642-31-5
Disodium adipate	$NaOOC(CH_2)_4COONa$	7486-38-6
Disodium sebacate	$NaOOC(CH_2)_8COONa$	17265-14-4
Sodium perchlorate	$NaClO_4\cdot H_2O$	7791-07-3
Barium stearate	$Ba(OOCC_{17}H_{35})_2$	6865-35-6

Disodium hydrogen phosphate was one of the first patented heat stabilizers for C-PVC. Dibutyltin bis(2-ethylhexyl thioglycolate) and dioctyltin bis(2-ethylhexyl thioglycolate) are currently the most effective, commercially viable C-PVC heat stabilizers, and are used for this purpose in large quantities around the world. The purity of these alkyltin stabilizers is >90%. The corresponding monoalkyltin tris(2-ethylhexyl thioglycolate)s, where the alkyl groups are n-butyl and n-octyl, are present as unintended impurities and are not known to contribute advantageously to the heat stabilization of C-PVC. Methyltin (2-ethylhexyl thioglycolate)s are more volatile. Nevertheless, they are also already used for stabilizing C-PVC in India, which is one of the largest markets for C-PVC.

Use of lead stabilizers, such as tribasic lead sulfate, in C-PVC is limited due to their toxicity.

Certain nitrogen-containing substances, such as 6-amino-1,3-dimethyluracil and sodium aluminosilicates (such as zeolite 4Å) of the general formula $Na_2O(Al_2O_3)(SiO_2)\cdot xH_2O$ are broadly claimed as either C-PVC stabilizers or co-stabilizers that are used optionally in combinations with each other as well as with mixed-metal carboxylates, alkyltin mercaptides, hindered amine antioxidants, β-diketones, epoxidized oils, phosphites, and their mixtures. It should be noted that if present, zinc carboxylates are only added as minor components.

Disodium adipate has been claimed to be used as a co-stabilizer for C-PVC. Disodium sebacate has been used as a co-stabilizer in combination with zeolite 4Å and 6-amino-1,3-dimethyluracil. Sodium perchlorate functions as a C-PVC co-stabilizer when used in combination with antioxidants and nitrogen-containing substances (such as 6-amino-1,3-dimethyluracil, triethanolamine, ureas, thioureas, and dihydropyridines).

Hydrotalcites are known to be excellent HCl scavengers in PVC and are mentioned in some patent references as co-stabilizers for C-PVC, but are not known to be used commercially to any great extent. This is most likely due to their thermal instability and degradation under higher processing temperatures and in-process shear typical for processing of C-PVC.

Recently, mixtures containing high proportions of barium carboxylates were found to be especially effective in replacing alkyl tin mercaptides for the stabilization of C-PVC against thermal decomposition, and also for increasing the impact resistance of those compounds. C-PVC is broadly used in manufacturing pipes and fittings intended for handling hot potable water.

4.6.3 Mechanism of C-PVC Stabilization

Historically, it was thought that the defect structures on C-PVC molecular chains were similar to those of polymer molecules found in PVC, which includes allylic chlorine, and therefore the mechanisms of thermal stabilization of C-PVC compounds were expected to be similar to those of PVC.

Taking into consideration that (a) the recently obtained C-PVC thermolysis data demonstrated its greater thermal stability, and (b) the composition of the most efficient C-PVC heat stabilizers contain monoalkyltin tris(thioglycolate)s and/or zinc carboxylates only as minor components, it can be concluded that the mechanism of C-PVC thermal stabilization is different from that of PVC, at least in part.

The main difference manifests itself in the practically insignificant formation of the allylic chlorine defects as an initial step of the C-PVC dehydrochlorination under elevated temperatures and in-process shear, and it in turn eliminates the need for deactivating those defects via the S_N2 mechanism with either monoalkyltin tris(2-ethylhexyl thioglycolate)s or zinc carboxylates. Not including those components in the composition of heat stabilizers also eliminates the possibility of forming Lewis acids derived from those components, when their synergists are completely used up during polymer processing.

Therefore, the main mode of C-PVC stabilization against degradation under high heat and shear conditions is likely *scavenging of HCl*, which is known to catalyze propagation of the polymer thermal degradation processes. Most C-PVC co-stabilizers, shown in Table 4.4 (dialkyltin bis(2-ethylhexyl thioglycolate)s, disodium sebacate, aluminosilicates, disodium phosphate, tribasic lead sulfate, and barium stearate), are known for their capacity to scavenge HCl when used as components of PVC heat stabilizers, and are likely to function in the same manner in C-PVC.

Similar to that of PVC, the HCl scavenging mode of C-PVC stabilization with organotins and mixed-metal carboxylate stabilizers can be expressed by Figure 4.21(b) and Figure 4.1(d), respectively.

Among other components of C-PVC stabilizers, the antioxidants are known to function effectively in delaying and slowing down the free-radical crosslinking of the partially dehydrochlorinated polymer, and sodium perchlorate reduces the length of polyene sequences, maintaining low color and also delaying the crosslinking step. Therefore, the second mode of C-PVC stabilization can be called *anti-crosslinking*.

6-Amino-1,3-dimethyluracil is a PVC heat co-stabilizer known to react with and deactivate allylic chlorine defects (Figure 4.3) as opposed to scavenging HCl, while stabilizing the polymer. Since there are no allylic chlorine defects in substantial quantities on C-PVC molecules formed as a result of thermal decomposition during their compounding, and therefore, there is no need for deactivating them, the role of 6-amino-1,3-dimethyluracil in stabilizing C-PVC is not well-defined.

It is interesting to note that the time to full degradation and crosslinking in the case of C-PVC is shorter than that of PVC under their respective process conditions. Additionally, the color development from the beginning of the dynamic heat stability testing until complete polymer degradation is not as pronounced for C-PVC as it is for PVC, indicating that the average length of polyene sequences formed as a result of C-PVC dehydrochlorination appears to remain more constant than that of PVC under their respective process conditions. This mechanism of C-PVC stabilization was proposed by Frenkel in 2024.

4.7 Acid Scavengers in Halogen-free Polymers

Acid scavengers are not only used to stabilize halogen-containing polymers, such as PVC and CPVC. They are also used for the stabilization of polyolefins alongside antioxidants (Chapter 2) and other additives (Chapter 6 and Chapter 7), which are essential for polymer processing. Acid scavengers, also known as acid binders, antacids, or, less specifically, co-stabilizers, contribute significantly to the overall performance of polymeric materials. Addition of acid scavengers to polyolefins is essential for suppressing decomposition reactions triggered by catalyst residues remaining in the polymers. These residues are by-products of Ziegler–Natta-type or metallocene catalysts that are used in polyolefin production. Compositions of these catalyst residues vary depending on the polymerization process and consist of compounds containing transition metal chlorides. These residues may adversely affect the performance characteristics of polymers and cause corrosion of metal surfaces that come into contact with the polymer melt. Similar corrosion effects are sometimes caused by other additives, such as brominated or chlorinated flame retardants, with which the acid scavengers may also react.

In addition to their ability to react with/absorb acids, the acid scavengers must also meet typical requirements for use in polymers. These characteristics include commmercial availability, standardized quality, low-to-no hygroscopicity, good compatibility with polymer matrixes, melting points below the plastic's processing temperature, and particle sizes that enable adequate dispersibility.

In a classic neutralization reaction, a Brønsted base reacts with a free Brønsted acid (such as HCl), as shown in Figure 4.22.

$$M(OH)_2 + 2HCl \rightarrow MCl_2 + H_2O$$

Figure 4.22 Example of Brønsted base/acid neutralization reaction, where M = Mg, Ca

To be considered for use in polymers, all oxides and hydroxides must:

- Be free of undesirable contamination (e.g., iron and other heavy metals) that could adversely affect the polymer's chemical stability
- Contain no discoloring particles that may interfere with the appearance of the end articles
- Have a particle size of <20 µm to avoid causing surface imperfections in the final articles.

Oxides and hydroxides of calcium and magnesium, as well as zinc oxide, are suitable for use as additives in many polymers. In addition, more complex compounds, such as hydrotalcites and hydrocalumites (already described in Section 4.3.5), can also be considered. These compounds neutralize hydrogen chloride due to their basic properties, as shown in Figure 4.13.

Another class of compounds capable of neutralizing acids are the fatty acid salts of divalent metals, e.g., stearates of calcium, magnesium, or zinc. These react according to Figure 4.1(b) and Figure 4.1(d), respectively, to form free stearic acid and the corresponding metal chlorides. In PVC, the reaction products do not cause any problem.

However, when used in polyolefins, a loading of calcium stearate even below 1500 ppm can be problematic in some applications due to the hygroscopic behavior and water solubility of the formed calcium chloride. In terms of their reactivity toward acids, the stearates are weaker bases and acid scavangers compared to the hydroxides.

In addition to neutralizing free acids, such as HCl, certain metal soaps are also capable of neutralizing Lewis acids. This involves ligand exchange reactions. For example, calcium stearate chemically reacts with the residues of the single-site catalysts.

For polymers used in processes such as the production of fibers or biaxially oriented films, where unmelted impurities in the polymer melt can cause quality defects in the finished articles, filtration tests are often conducted for quality assurance. Acid scavengers can lead to filter clogging during processing. Inorganic neutralizers, such as hydrotalcites and zinc oxide, do not melt at the temperatures at which the polymers are processed. Recommended dosages of acid scavangers used in polyolefins are summarized in Table 4.5.

Table 4.5 Recommended Loadings of Acid Scavengers in Polyolefins

Polymer	Substance	Loading (ppm)
PP	Calcium stearate	500–1200
	Sodium stearate	500–800
	Hydrotalcite	100–500
LLDPE	Zinc stearate	500–1200
	Calcium stearate	500–1500
	ZnO	200–800
	Hydrotalcite	200–500
HDPE (Ziegler-Natta[a])	Zinc stearate	300–1000
	ZnO	200–800
	Hydrotalcite	200–500
HDPE (Phillips[a])	Zinc stearate	200–500
	Calcium stearate	200–500

[a] See Section 1.7.

In polyacetals (POM), the fatty acid salts are used as co-stabilizers to neutralize the residual formic acid produced during formaldehyde oxidation. Concentrations of calcium stearate and calcium 12-hydroxystearate acid scavengers in POM are typically in the range 0.1–0.3%. Metal stearates are added to polyacetals not only as acid scavengers but also as pigment-dispersing aids. Combinations of carboxylic acid salts with nitrogen-containing compounds like melamine, dicyanodiamide, and PA 12 are also used. In recent years, hydrotalcite has also taken market share from metal stearates in POMs, just as has already been the case with polyolefins.

Undesirable interference between polymer additives has also been reported. Calcium stearate is known to react with and reduce the effectiveness of sodium benzoate, a nucleating agent, forming sodium stearate and calcium benzoate (Figure 4.23). Organic clarifiers that also act as nucleating agents do not react with calcium stearate.

$$CaSt_2 + 2\,NaBz \rightleftharpoons CaBz_2 + 2\,NaSt$$

Figure 4.23 Interfering reaction of polymer additives, where St = stearate; Bz = benzoate

Hydrotalcite-type acid scavengers were demonstrated to significantly limit the viscosity-reducing properties of fluorocarbon-based polymer processing aids. Both calcium and zinc stearate slightly influence the effectiveness of these fluorine-containing polymers at low shear rates. Zinc oxide, in contrast, shows no interactions with these processing aids.

Similar to the trends observed with other additives, dust-free metal stearates are gaining market share. Additive one-packs are also used more frequntly now to address filling problems associated with fine organic and inorganic components.

The effect of using acid scavengers was investigated in the processing of PP produced with metallocene catalysts (see Chapter 1). Both the corrosion and the color retention tests favored the use of calcium stearate in metallocene-catalyzed syndiotactic PP. These results lead to the conclusion that acid scavengers will continue to be required for polyolefin production.

4.8 Questions

1. What is the mechanism of PVC thermal degradation?
2. What are the main types and functions of PVC stabilizers?
3. What is the definition of heavy-metal-free heat stabilizers? Is zinc a “heavy metal”?
4. What is the composition of nitrogen-containing primary organic stabilizers for PVC?
5. What is the mechanism of action of nitrogen-containing primary organic stabilizers for PVC?

6. What are the components of mixed-metal carboxylate stabilizers for PVC?
7. What is the mechanism of action of mixed-metal carboxylate stabilizers for PVC?
8. What are the main types and the mechanism of action of organic co-stabilizers?
9. What are the main types and the mechanism of action of inorganic co-stabilizers?
10. What are the main applications of solid mixed-metal carboxylate and organic stabilizers in PVC?
11. What are the main applications of liquid mixed-metal carboxylate stabilizers in PVC?
12. What are the main types of organotin stabilizers used in PVC?
13. What are the main applications of each organotin stabilizer type in PVC?
14. What is the mechanism of stabilizing PVC with alkyltin (2-ethytlhexyl thioglycolate)s? What is the purpose of having a certain excess of unreacted organotin stabilizer components present in the processed PVC?
15. Describe the main points and significance of the Universal Mechanism of PVC Stabilization.
16. What are the most common components of lead stabilizers for PVC? Are they suitable for manufacturing transparent PVC articles?
17. What is the mechanism of C-PVC thermal degradation?
18. What are the main stabilizers used in C-PVC?
19. What is the mechanism of C-PVC stabilization?
20. What are the acid scavenger types used for stabilizing PP?
21. What are the acid scavenger types used for stabilizing POM? What is the function of fatty acid salts as additives of polyacetals?
22. What are the advantages of using hydrotalcites instead of metal stearates in non-halogenated polymers?

5 Plasticizers

Content of this chapter was adapted from:

- P. Frenkel, M. Schiller, and P. Harmon, "Plasticizers" (Chapter 5), in: *PVC Handbook* (2nd edition), ed. R. P. Krock, M. Schiller, P. Frenkel, J. W. Summers, and C. A. Daniels, Hanser, 2025, pp. 219–283.

5.1 Terms and Definitions

Hardness	Measurement of a soft material's resistance to indentation, commonly used for elastomers and soft plastics.
Permanence	The duration an additive (such as a plasticizer) maintains its intended effect within a material without significant physical loss.
Efficiency	The degree to which a specific amount of additive (such as a plasticizer) achieves the desired effect in a polymer formulation.
Substitution factor	The ratio used to replace one additive with another, while achieving equivalent performance.
Diffusion	Movement of molecules from an area of higher concentration to an area of lower concentration within a medium.
Volatility	Tendency of a substance to vaporize under given conditions, impacting processing and performance.

Low-temperature flexibility	Ability of a material to bend or deform without cracking when subjected to low temperatures.
Flame retardancy	Capability of a material to resist ignition, reduce flame spread, or self-extinguish when exposed to fire.
Heat stability	Resistance of a material to property loss or decomposition when exposed to heat.
Structural isomers	Compounds with the same molecular formula but different structural arrangements of atoms.
Homologs	Members of a chemical series differing by a repeating structural unit, such as $-CH_2-$.
Life cycle assessment	Systematic evaluation of environmental impacts of a product from raw material extraction to disposal.
Bio-based content	Proportion of a material carbon derived from renewable biological resources to the total carbon content, expressed as a percentage.

5.2 Market Trends

In 1951, the International Union of Pure and Applied Chemistry (IUPAC) developed a universally accepted definition for a plasticizer as a substance or composition incorporated in a material (usually a plastic or an elastomer) to increase its flexibility, workability, or distensibility.

The worldwide market for plasticizers has grown from more than 4.6 million metric tonnes in 2003 to 10.9 million metric tonnes in 2022, with approximately 90% used in PVC applications. In North America in 2022, plasticizer consumption was about 1 million metric tonnes, and the plasticizer market in Europe was about 1.2 million metric tonnes. The region of the world with the largest plasticizer production is Asia Pacific, with nearly 7 million metric tonnes produced annually.

The use of dioctyl phthalate (DOP, or more correctly DEHP, di-2-ethylhexyl phthalate) prevailed as the preferred general-purpose plasticizer for PVC until the late 1970s. Changes in costs and availability of the raw materials that serve as plasticizer feedstock have caused a significant reduction in the number of plasticizer suppliers and plasticizer products in use commercially. There are still over 50 commercially available plasticizers, including a number of new ones produced from petroleum and bio-based renewable feedstocks. Approximately 70% of the worldwide consumption is comprised of five plasticizers: di-2-ethylhexyl phthalate (DEHP), di-iso-nonyl phthalate (DINP), di-iso-decyl phthalate (DIDP), dipropylheptyl phthalate (DPHP), and di-2-ethylhexyl terephthalate (DOTP). DEHP has been substituted over the past 20 years by higher-molecular-weight phthalates such as DOTP, DINP and other plasticizers. There has been signficant growth in DOTP demand globally, and it now has the highest consumption among all general-purpose plasticizers.

5.3 Mechanisms of Plasticization

For a plasticizer to be effective, it must be thoroughly mixed and incorporated into the PVC polymer matrix. This is typically achieved by heating and mixing until either the resin dissolves in the plasticizer or the plasticizer dissolves in the PVC resin. The plasticized material is then molded or shaped into the useful products and cooled. Plasticizers, as an additive category, exhibit a range of characteristics in terms of both the ease with which they form the plasticized material and in the resulting mechanical and physical properties of the formed plasticized compound.

Several theories have been developed to account for the observed characteristics of the plasticization process. Plasticization is described by three primary theories with some modifications.

- The *Lubricating Theory* states that, as the system is heated, the plasticizer molecules diffuse into the polymer and weaken the polymer–polymer interactions (van der Waals' forces). Here, the plasticizer molecules act as shields to reduce polymer–polymer interactive forces and prevent the formation of a rigid network. This lowers the PVC glass transition temperature (T_g) and allows the polymer chains to move rapidly, resulting in increased flexibility, softness, and elongation. A technical term for these movements is "crankshaft motion" or crankshaft-like motion. There is one "crank" per about 100 carbon atoms in the PVC chain.
- The *Gel Theory* considers the plasticized polymer to be neither solid nor liquid but an intermediate state, loosely held together by a three-dimensional network of weak secondary bonding forces. These bonding forces, acting between plasticizer and polymer, are easily overcome by applied external stresses, allowing the plasticized polymer to flex, elongate, or compress.
- The *Free Volume* theory considers the internal space available within a polymer. As this "free volume" is increased, more space becomes available for molecular or polymer chain movement. A polymer in the glassy state has its molecules packed closely, but this packing order is not perfect. When the free volume is low, the molecules cannot move past each other very easily. This makes the polymer rigid and hard. When the polymer is heated to above T_g, the thermal energy and molecular vibrations create additional free volume which allows the polymer molecules to move past each other rapidly. This has the effect of making the polymer system more flexible and rubbery. Free volume can be in-

creased through modifying the polymer backbone, such as by adding more side chains or end groups. When small molecules such as plasticizers are added, this also lowers the T_g by separating the PVC molecules, adding free volume and making the PVC soft and rubbery.

When a plasticizer is incorporated into a PVC formulation, molecules of PVC can rapidly move past each other. If the plasticizer was *uniformly* incorporated into the PVC, it would behave similarly to an uncured rubber, with lots of creep and high compression set. However, a thermoplastic elastomer, such as PVC, includes meltable physical crosslinks (in this case PVC crystallites) in order to make the material thermoplastic and give it a physical cure.

The plasticizer must *not* be a powerful solvent for all the parts of the PVC structure, but instead it must selectively enter only the amorphous PVC segments. It must not enter and destroy the crystalline parts of the structure (which are present to about 15%).

The mechanistic explanation of the plasticization phenomenon considers interactions of the plasticizer with the PVC resin macromolecules. It implies that the plasticizer molecules are not permanently bound to the PVC resin molecules, but are free to self-associate and to associate with the polymer molecules at certain sites, such as amorphous sites. As these interactions are weak, there is a dynamic exchange process. Different plasticizers yield different plasticization effects because of the differences in the strengths of the plasticizer–polymer and plasticizer–plasticizer interactions.

At low plasticizer levels, the plasticizer–PVC interactions are the dominant interactions, while at high plasticizer concentrations plasticizer–plasticizer interactions can become more significant. This can explain the observed *"anti-plasticization" effect*, whereby plasticizers added at low levels (<15 phr in the absence of any functional filler) increase the rigidity of the plasticized PVC compounds as measured by hardness, modulus, tensile strength, elongation, and low-temperature properties.

For a plasticizer to be effective and useful in PVC, it must contain two types of structural segments, polar and non-polar. The polar portion of the plasticizer molecule must be able to coordinate reversibly with the PVC polymer, thus softening the PVC, while the non-polar portion allows the plasticizer–PVC interaction to be controlled, ensuring that it is not so powerful that it destroys the PVC's crystallinity. It also adds free volume, contributes shielding effects, and provides lubricity.

Examples of polar segments of plasticizer structures are the carbonyl group of the carboxylic ester functionality and to a lesser extent an aromatic ring. An example of a non-polar segment is the aliphatic side-chain of an ester.

The balance between the polar and non-polar portions of the plasticizer molecule is critical for controlling its solubilization effect. If a plasticizer is too polar, it can destroy PVC crystallites. If it is too non-polar, incompatibility issues between PVC and the plasticizer can arise. Figure 5.1 provides quantitative information about compatibility, volatility, and solubility temperature of *o*-phthalates as a function of the number of carbon atoms in the alcohols used for their synthesis. Useful tools for estimating plasticizer compatibility are: (a) the non-polar/polar ratio method developed by Van Veersen and Meulenberg (as discussed above) and (b) the solubility parameter methods (see Section 5.4.4).

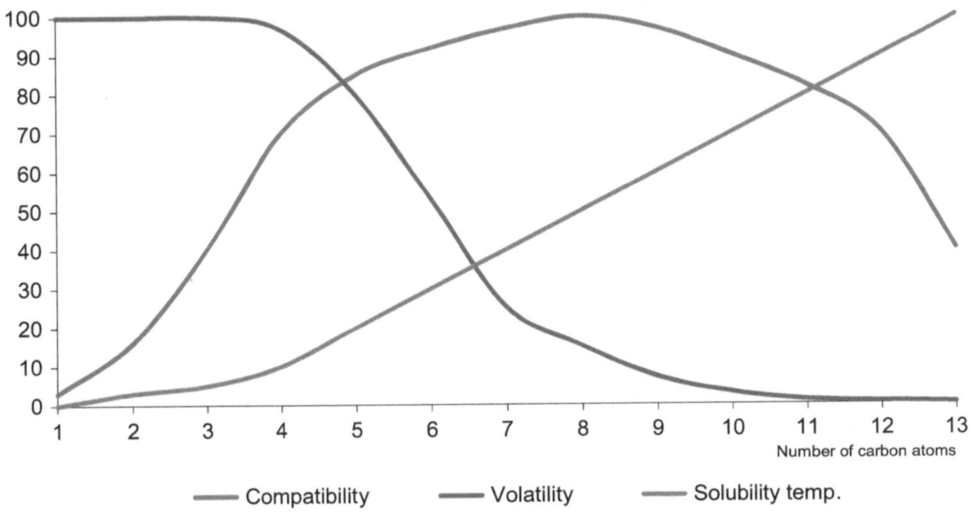

Figure 5.1 Compatibility, volatility, and solubility temperature as a function of the number of carbon atoms in the alcohol-derived component of *o*-phthalate plasticizers

5.4 Characterization of Plasticizers

5.4.1 Efficiency

Plasticizer "efficiency" can be quantified by using an A-type durometer, which measures the Shore A hardness of a PVC compound. Similar comparisons can be made by measuring other mechanical properties. However, the reliability of the Shore A hardness method and the common practice of using the test support its widespread applicability for quantifying and comparing plasticizing efficiency. Figure 5.2 depicts the quantitative determination of plasticizer efficiency, expressed as the "substitution factor" (SF). In this example, the hardness values resulting from use of DINP as a plasticizer for PVC were compared to that of DEHP.

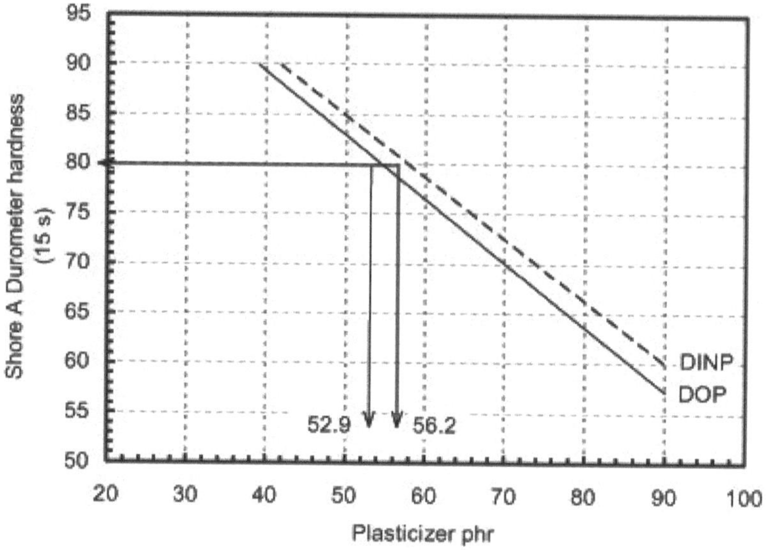

Figure 5.2 Shore A durometer hardness of PVC compounds plasticized with DINP vs DEHP (shown as DOP here). Image credit: John Wiley (reprinted with permission)[1)]

It shows that a Shore A durometer hardness of 80 ("80A") is achieved with 52.9 phr DEHP and 56.2 phr DINP. Thus, the substitution factor (SF) for DINP vs DEHP is 1.06, as calculated using Equation 5.1:

$$\text{Substitution factor (SF)} = \frac{\text{phr}_{\text{plasticizer,80A}}}{\text{phr}_{\text{DEHP,80A}}} = \frac{56.2 \cdot \text{phr}_{\text{DINP}}}{52.9} \quad (5.1)$$

The SF indicates that DINP is 6% less efficient than DEHP. In other words, DINP needs to be added at a level that is 6% higher than that of DEHP in order to achieve the same hardness (or softness). It has been found that SF remains constant over plasticizer levels ranging from about 20–90 phr. If other attributes are not of significance, then the question about this kind of substitution rests on comparative formulation costs (also known as cost-in-use). In reality, it is known that when compared to DEHP at the equivalent hardness, DINP-plasticized PVC exhibits preferable low-temperature properties as well as significantly reduced plasticizer loss due to volatilization and diffusion. Most commercial and developmental plasticizer grades have been evaluated in PVC over a wide range of use levels (phr). Table 5.1 summarizes substitution factors (relative plasticizing efficiency values) determined experimentally for commercial plasticizers, along with their acronyms, chemical compositions, density, molecular weights, and solubility parameters.

---

1 L. G. Krauskopf, in: *Handbook of PVC Formulating*, ed. E. J. Wickson, 1993, Wiley, p. 171.

Table 5.1 Selected Plasticizers and Their Characteristics

Substance	Abbreviation	Molecular weight (g/mol)	SF	Density (g/mL)	HSP δ ($MPa^{1/2}$)
Branched *o*-phthalates					
Butyl benzyl phthalate	BBP	311	0.94	1.119	19.15
Di(2-ethylhexyl) phthalate	DEHP	390	1.00	0.990	18.18
Di-iso-nonyl phthalate	DINP	418	1.06	0.970	17.92
Di-iso-decyl phthalate	DIDP	446	1.10	0.966	17.94
Di-iso-tridecyl phthalate	DTDP	530	1.27	0.952	17.89
Linear *o*-phthalates					
Di(linear nonyl) phthalate	L9P	418	0.99	—	—
Di(linear C_7,C_9,C_{11}) phthalate	7911P	~418	1.00	—	—
Di(linear C_9,C_{11}) phthalate	911P	~446	1.07	—	—
Di(linear C_{11}) phthalate	DUP	474	1.16	—	—
Trimellitates					
Tri(2-ethylhexyl) trimellitate	TEHTM	546	1.14	0.988	17.94
Trioctyl trimellitate	TEHTM (TOTM)	546	—	0.980	17.79
Tri-iso-nonyl trimellitate	TINTM	588	1.27	0.978	17.89
Aliphatic diacid diesters					
Di(2-ethylhexyl) adipate	DEHA	370	0.95	0.925	17.26
Di-iso-nonyl adipate	DINA	398	1.00	0.920	17.25
Di-iso-decyl adipate	DIDA	426	—	0.919	17.31
Di(2-ethylhexyl) azelate	DOZ	412	—	0.915	17.19
Di(2-ethylhexyl) sebacate	DOS	426	—	914	17.21[a]

Table 5.1 Selected Plasticizers and Their Characteristics (*continued*)

Substance	Abbreviation	Molecular weight (g/mol)	SF	Density (g/mL)	HSP δ (MPa$^{1/2}$)
Phosphate esters					
Tri(2-ethylhexyl) phosphate	TOF	435	1.00	—	—
Tricresyl phosphate	TCP	368	1.31	—	—
Epoxides					
Epoxidized 2-ethylhexyl tallate	OET	424	0.96	0.928	17.61
Epoxidized soyabean oil	ESBO	940	1.10	0.952	16.90
Epoxidized linseed oil	ELSO	988	—	1.035	17.56
Dibenzoates					
Diethylene glycol dibenzoate		314	—	1.200	20.78
Dipropylene glycol dibenzoate		342	—	1.120	19.64
Citrates					
Tributyl-*O*-acetyl citrate	ATBC	402	—	1.052	17.81
Tris(2-ethylhexyl)-*O*-acetyl citrate		570	—	0.984	17.55
Miscellaneous					
Di(2-ethylhexyl) terephthalate	DOTP (DEHTP, DEHT)	390	1.08	0.980	18.00
Di-iso-nonyl cyclohexane 1,2-dicarboxylate	DINCH	396	1.08	0.950	17.53
Acrylonitrile–butadiene rubber (30–40% acrylonitrile)	NBR	—	2.40	0.963–0.989	20.2–21.5[a]
Ethylene–vinyl acetate copolymer (18–28% vinyl acetate)	EVA	—	2.60	0.935–0.949	17.9 ± 0.1[a]
Chlorinated polyethylene, 37% Cl	CPE	—	—	1.160	18.85[a]

[a] M. Schiller, unpublished results.

5.4.2 Low-temperature Performance Characteristics

The low-temperature flexibility (T_f) and low-temperature brittleness point (T_B) of plasticized PVC compounds are measured and compared at equivalent room-temperature hardness (using an “80A” durometer) as well as at equal plasticizer use levels (50 phr).

For a given alkyl group configuration, the trimellitate family of plasticizers impart comparable-to-slightly superior low-temperature properties compared to the *o*-phthalate counterparts. This result is attributed to the lower plasticizing efficiency of trimellitates and the need to add them at higher levels compared to the *o*-phthalates. Adipate plasticizers, on the other hand, impart significantly superior low-temperature properties (lower by 25–35 °C) compared to their *o*-phthalate counterparts, in spite of the fact that they are more efficient (substitution factors of <1.00) in providing the target room-temperature hardness compared to the phthalates. More linear alkyl groups in the plasticizer structures contribute superior low-temperature properties (lower by about 5–7 °C) compared to the corresponding branched isomers. They also have lower substitution factors (or higher plasticizing efficiency) to meet the target room-temperature hardness. It should be noted that performance test results measured on finished PVC articles may be influenced by non-formulation-dependent factors. For example, incompletely fused, improperly conditioned, or otherwise damaged specimens may show unexpected failures in mechanical property tests.

5.4.3 Permanence (Transience)

Plasticizers have strong affinity to the PVC polymer chain, but they are not chemically bonded or grafted to the polymer backbone (although epoxy plasticizers may be an exception to that). Therefore, non-grafted and non-copolymerized plasticizers can be separated from the PVC matrix by extraction with solvents, oils, or water, as well as by surface rubbing, volatilization, or migration into adjacent media.

It was found that two controlling mechanisms (other than effects of degradation) are at play under conditions that contribute to plasticizer losses. These are: (a) the rate of loss that occurs at the surface of the specimen and (b) the rate at which the plasticizer diffuses to the surface; see also Chapter 11. The slowest rate is the controlling factor. For example, most plasticizers have extremely low solubility in water, and therefore, exhibit surface-controlled loss rates under aqueous environments.

Plasticizer losses due to extraction by oil media (in which many plasticizers are highly soluble) are controlled by diffusion rates.

Plasticizer losses associated with volatilization are influenced by their boiling temperature, vapor pressure, solvation, and oxidative stability. Polymeric plasticizers of high molecular weight (≈1000 to >500,000 Da)[2] and bulky molecular structures show low volatility and excellent permanence attributable to their low diffusivity.

Di(n-alkyl) phthalates impart lower volatility and superior oxidative stability compared to their branched counterparts. All phthalates, trimellitates, and aliphatic dicarboxylic acid diesters show excellent hydrolytic stability under aqueous environments. Diffusion-controlled transience is poor for the aliphatic dicarboxylic acid diesters, but is good for branched *o*-phthalates. Linear di-n-alkyl phthalates are measurably less extraction-resistant than the branched phthalates, but significantly superior to the aliphatic dicarboxylic acid diesters. The oil-extraction resistance of trialkyl trimellitates is inferior to that of *o*-phthalates. This is apparently attributable to their lower plasticizing efficiency, as well as the increased proportion of alkyl moieties in their molecular structures.

5.4.4 Miscibility (Compatibility)

The terms miscibility and compatibility are essentially interchangeable with respect to the behavior of components added to PVC. PVC systems behave as solid solutions and abide by the three-dimensional solubility parameter concept developed by Hansen. PVC is inherently a polymer with notable polarity, and correspondingly, most plasticizers also exhibit a degree of polarity. The polarity of plasticizers reduces as the number of carbon atoms they contain (derived from the alcohol raw materials) increases. Typically, this polarity reduction is gauged qualitatively. The δ value, derived from the Hildebrand solubility parameter (HSP), provides a quantitative assessment of intermolecular interactions, and has proved especially valuable in assessing material compatibility. Materials with similar δ values generally tend to be more compatible. The δ value of PVC is 19.5–19.8 $MPa^{1/2}$. Schiller has calculated and verified δ values for most commercially available plasticizers (Table 5.1).

Figure 5.3 illustrates the correlation between δ and the number of carbon atoms in *o*-phthalate plasticizers.

[2] Da (dalton) is an alternative term for the atomic mass unit (amu or u), and is commonly used in biochemistry to express molecular masses. 1 Da is equal to 1/12 the mass of a carbon-12 atom. In polymer science, the units kDa (kilodalton = 1000 Da) and MDa (megadalton = 1,000,000 Da) are often used. The numerical value of the molecular mass in Da is equal to the molar mass in g/mol, which simplifies its use in polymer chemistry.

As the alcohol chain length increases in *o*-phthalate plasticizers, δ decreases. Similar tendencies have also been observed within other plasticizer types.

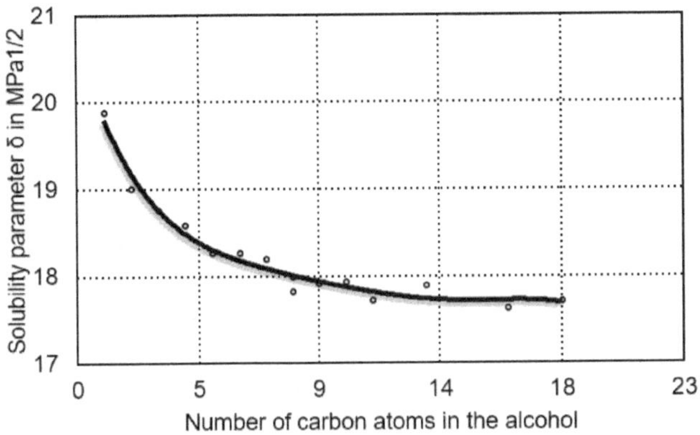

Figure 5.3
Correlation between the δ solubility parameter and the number of carbon atoms introduced into *o*-phthalates with alcohols

Substances with solubility parameters in the range 17.2–20.2 $MPa^{1/2}$ should be compatible with PVC and can be considered as good solvents or plasticizers for it. Extensive practical experience spanning decades supports the notion that phthalates featuring 4–13 carbon atoms in their alcohol-contributed segments are compatible with PVC. This corresponds to the δ range 17.89–19.15 $MPa^{1/2}$. Trimellitate plasticizers are also aligned with this range, but conversely, adipates, sebacates, and azelates have slightly lower δ values (17.19–17.62 $MPa^{1/2}$), indicating somewhat reduced compatibility. DOTP, having a δ value of 18.00 $MPa^{1/2}$ is also outside the established range for PVC plasticizers. This value is consistent with the lower compatibility of DOTP with PVC compared to its isomeric counterpart DEHP.

Naert[3)] proposed to group plasticizer performance attributes (Figure 5.4) as follows:

- "Economics" – Price, efficiency, and density
- "Compatibility" – Exudation, migration, extraction resistance, and solubility temperature
- "Processing" – Gelation and rheology
- "Performance" – Weathering, volatility, and effectiveness at low temperature.

A variety of isomeric structures and homologs that vary in terms of their plasticizing performance can be encompassed within each plasticizer family.

[3] D. Naert, "General-purpose phthalate plasticizers and their alternatives", PVC Formulation Technical Conference, Cologne, Germany, March 2009.

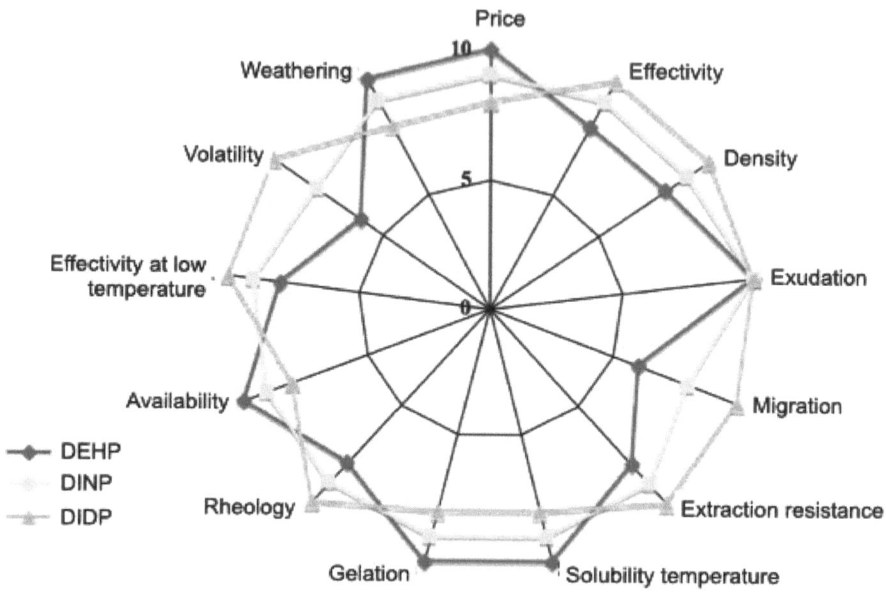

Figure 5.4 Performance characteristics of general-purpose phthalates[2)]

5.5 Plasticizer Structural Isomers and Homologs

Within each plasticizer type, there can be several structural isomers. Those have to do with either different alkyl group branching (known as chain isomers; for example, n-octyl-, i-octyl- and 2-ethylhexyl *o*-phthalates) or the positioning of ester functional groups (known as positional isomers or regio-isomers; for example, DEHP and DOTP).

Homologs are a series of plasticizers of one chemical type whose successive members differ only in the addition of one methylene group (for example, DINP and DIDP).

Property variations imparted by isomeric structures are primarily attributable to the composition and placement of alkyl moieties. Plasticizers made with alcohols of more linear (less branched) molecular structures are more efficient; they impart low temperature flexibility, reduced volatility and increased oxidative stability. Krauskopf differentiated plasticizer performance effects based on five degrees of alkyl group branching in commercial di-C_8, di-C_9, and di-C_{10} *o*-phthalates. These were grouped as follows:

- *Normal:* 100% non-branched alcohols; applicable to a limited number of commercial products based on mixtures of normal C_8 and C_{10} alcohols
- *Linear:* A mixture of normal and monomethyl-branched alcohols. The resultant alcohol is approximately a 70:30 molar ratio of normal and 2-methyl-branched isomers

- *Slightly branched (SLBR):* A mixture of monomethyl and dimethyl-branched alcohols. These alcohols are a random mixture of monomethyloctanols and dimethylheptanols
- *Moderately branched (MODBR):* Primarily dimethyl or monoethyl (i.e., 2-ethylhexyl) branched alcohols. These alcohols are used for the "general-purpose" (GP) plasticizer category
- *Highly branched (HIBR):* Trimethyl-branched alcohols, the main example being 3,5,5-trimethylhexanol. These alcohols are susceptible to oxidation and exhibit increased resistance to diffusion.

These variations in the degree of branching demonstrate measurable effects on properties of both plasticizers and PVC compounds containing them with respect to plasticizing efficiency, low-temperature properties, diffusion, volatility, heat stability, and resistance to oxidative degradation. The normal C_9 phthalate has a substitution factor (SF) of 0.94, while the plasticizing efficiency of the "linear" dinonyl phthalate is equivalent to that of DEHP (SF = 1.00). When compared at equivalent hardness in PVC, the "linear" dinonyl phthalate is only slightly deficient to the normal dinonyl phthalate with respect to low-temperature properties and volatility.

The "moderately branched" (MODBR) DINP shows lower plasticizing efficiency (SF = 1.06) and deficiencies in terms of low-temperature flexibility and volatility compared to the normal dinonyl phthalate. Compared to DEHP performance at specified room-temperature hardness, DINP (MODBR) provides equivalent low-temperature properties with significantly lower volatility. The performance of the "slightly branched" (SLBR) DINP is essentially equivalent to that of a 50:50 mixture of di(n-nonyl) phthalate and moderately branched DINP (MODBR).

The *o*-phthalates are an example of an isomer family that is suitable for studying correlations among the isomeric structures and their performance imparted on plasticized PVC compounds (Figure 5.4). Key physical properties correlate well with plasticizer performance in PVC only when the comparisons are limited to homologs within a given chemical family. More extensive attempts to correlate the physical properties of neat plasticizers with their performance in PVC are usually unsuccessful.

Mild physical interactions between plasticizers and PVC polymers allow calculation of the specific gravity of the plasticized PVC compositions. Yet this interaction precludes the accurate quantification of plasticizer losses due to volatilization (i.e., the calculations are based on vapor pressure) or diffusion-controlled migration. Oxidative stability or hydrolytic stability/degradation, on the other hand, have the opposite effect when attempting to predict plasticizer transience from PVC.

In general, the color of plasticized PVC compositions is typically not altered by the plasticizers, because most commercial plasticizers are colorless (Pt–Co color <25). However, highly discolored (amber-brown) plasticizers may adversely affect the color of flexible PVC compositions.

5.6 Effect of Plasticizer Level on Compound Characteristics and Performance

Effects of varying plasticizer levels on the mechanical properties of PVC compounds are listed in Table 5.2.

Table 5.2 Typical Properties of Selected PVC Compounds Plasticized with DINP. Data reproduced with permission[4]

	Rigid	Semirigid	Flexible	Very flexible	Extremely flexible
DINP (phr)	0	34	50	80	600
Composition (wt%)	0	25	33	44	86
Specific gravity at 20 °C (g/cm^3)	1.40	1.26	1.22	1.17	1.02
Hardness (Shore A durometer, 15 s)	—	94	84	66	<10
Flexural stiffness at 23 °C (MPa)[a]	>900	69	12	3.4	—
Tensile strength (MPa)[b]	>41	31	21	14	—
Elongation, (%)[b]	<15	225	295	400	—
Brittleness (°C)[c]	>23	-16	-32	-47	—
Example applica- tions	Bottles, pipe, siding, records	Shades, shoe heels, thin films, produce wrap	Wall-cover- ing, book-binders, upholstery, garden hose	Boots, gloves, water beds	Fishing lures

[a] ASTM D 747.
[b] ASTM D 882.
[c] ASTM D 746.

Table 5.2 shows typical properties of PVC compounds intended for general-purpose applications and plasticized with DINP at levels that range from zero (a rigid PVC compound) to about 600 phr. Typically, useful commercial articles contain about 20–100 phr plasticizers; fishing lures are an exception, containing about 600 phr.

4 L. G. Krauskopf, in: *Encyclopedia of PVC* (2nd edition), ed. N. L. Nass and C. A. Heiberger, Marcel Dekker, 1988, p. 149.

The physical properties of plasticized PVC compounds, such as hardness, tensile strength, elongation, and modulus, are strongly influenced by the plasticizer use level and its chemical type/structure.

5.7 Plasticizer Classification, Composition, and Performance Attributes

Plasticization is achieved by incorporating a plasticizer into the PVC matrix through mixing and heat. Plasticizers may be classified as either *monomeric* or *polymeric*, depending on their chemical structure and molecular weight.

The IUPAC definition of a plasticizer is entirely focused on performance characteristics when combined with a polymer. There is no description of chemical structures and/or physical properties of the plasticizer *per se*. The key performance attributes of plasticized compounds are influenced by plasticizer level (usually expressed in **p**arts per **h**undred parts of **r**esin (phr)) as well as its chemical type. In addition, variations in isomeric[5)] and homologous structures within any given chemical type contribute to performance variations that have been measured in plasticized PVC compositions. The major chemical types (families) of PVC plasticizers are: branched *o*-phthalates, linear *o*-phthalates, terephthalates, trimellitates, cyclohexanoates, aliphatic dibasic acid esters, polyesters, epoxides, phosphates, and hydrocarbons.

Plasticizers can also be characterized according to the origin of their raw materials:

- Conventional petroleum-based (derived from crude oil)
- Bio-based and/or bio-attributed (derived from renewable feedstocks – see Section 5.8 and Section 5.8.2).

Furthermore, a systematic comparison of plasticizers is facilitated by segmenting all their types into three subgroups depending on their performance attributes in PVC:

- **General-purpose (GP) plasticizers** impart the desired flexibility to PVC along with an overall balance of optimum properties at the lowest cost. These plasticizers include branched dialkyl *o*-phthalates and terephthalates, more recently introduced cyclohexanoates, and low-cost hydrocarbon oils, called "extenders", which have a very limited compatibility with PVC.
- **Performance plasticizers (PP)** contribute additional performance characteristics compared to the GP type, and which are desirable in flexible PVC, albeit at somewhat higher costs. These are strong solvation abilities, low-temperature performance, and low volatility. Strong solvating plasticizers, such as dibenzoates, have

[5] Isomerism is the occurrence of two or more chemical compounds with the same molecular formula and molecular mass, but which differ in the chemical bond or spatial arrangement of the atoms.

higher polarity that can be attributed in part to aromaticity. Low-temperature flexibility is usually imparted by linear dialkyl *o*-phthalates and aliphatic dibasic acid esters, which are less solvating and have higher diffusivity. Low volatility requires high molecular weight plasticizers, such as trimellitates and polyesters.

- **Specialty plasticizers (SP)** provide properties beyond those associated with the GP and PP classes. Examples of these special attributes are low diffusivity, heat stability, and flame resistancy. Polyester plasticizers impart low volatility and low diffusivity, along with low smoke (in the absence of aromaticity) under the flame test conditions. Epoxy plasticizers also function as thermal co-stabilizer for PVC. Phosphates and halogenated hydrocarbons provide fire retardancy to PVC compounds. However, specialty plasticizers may be even more expensive than the PP grades.

The optimization and selection of plasticizer systems for performance and cost are achieved taking into consideration the target applications and associated regulations, as well as performance attributes and raw material costs, by selecting the relevant isomers and homologs of a specific chemical type and (if necessary) a combination thereof. *o*-Phthalates are still the most widely used class of plasticizers in PVC, and contribute the most complete array of required performance characteristics in flexible PVC. In addition, their cost and commercial availability support their applicability on the market.

Examples of GP plasticizers include the phthalates DEHP, DINP, DPHP, DIDP, and DOTP, as well as cyclohexanoates and others. GP plasticizers are used in a broad range of applications, including consumer products, cable insulation, certain food-contact-suitable materials, PVC flooring, wall covering, and others. They typically offer the best combination of price and performance, and are available globally.

Owing to regulatory action affecting DEHP, alternatives, such as DOTP and cyclohexanoates, have been developed and have been used as replacements for it over the last 10 years. The alternatives perform similarly to the incumbent *o*-phthalates with minor adjustments to PVC formulations. For example, compared to DEHP, DOTP is less efficient (SF = 1.08 compared to 1.00 for DEHP), but it imparts:

- Slightly enhanced low-temperature performance
- 8–10 °C higher plastisol gelation temperature
- Lower volatility (e.g., >50% lower weight loss at 100 °C from 20 mil film samples).

DOTP with the carboxylic ester groups in the 1- and 4-positions is compositionally the same, while it differs structurally from DEHP, which has the carboxylic ester groups in the 1- and 2-positions.

Among the cyclohexanoates, di-iso-nonyl cyclohexane 1,2-dicarboxylate (DINCH) is produced by hydrogenation of the corresponding di-C_9 phthalate (DINP). Its performance characteristics in PVC are similar to its phthalate counterpart except for having higher volatility and lower solvency in PVC compounds. Hexamoll® DINCH was introduced by BASF as a plasticizer intended for use in applications that are health- and the environment-sensitive. Another example of the cyclohexanoates is ELATUR®, di-iso-nonyl-1,4-cyclohexane dicarboxylate (DINCD) from Evonik. This is a hydrogenated terephthalate made with iso-nonyl alcohol. Both DINCH and DINCD, non-phthalate alternatives to the *o*-phthalates, can be produced based on 1,2- and 1,4-cyclohexane dicarboxylic acid, respectively. Hanhwa, a Korian manufacturer, also introduced di-2-ethylhexyl cyclohexanoate (Eco-DEHCH), which is a hydrogenated version of DOTP.

5.7.1 Performance Plasticizers

The PP group consists of three subgroups:

- "PP-SS": strong solvation
- "PP-LT": low temperature
- "PP-LV": low volatility.

The PP-SS group, also known as "fast fusers", are plasticizers that exhibit high solvating characteristics attributed to their increased polarity and aromaticity. Thus, lower molecular weight *o*-phthalates, such as diisoheptyl, dihexyl, butyl octyl, and butyl benzyl phthalate fall into this category. These plasticizers also result in excessive fuming during processing and volatilization losses in end-use applications, and have largely been replaced in PVC applications owing to regulatory action in Europe and North America. Non-*o*-phthalate plasticizers, such as dibenzoates and alkane sulfonic acid esters of phenol (e.g., Mesamoll®) have replaced them. Tri(cresyl) phosphate also is an example of an PP-SS plasticizer.

Sulfonates also exhibit a strong solvating effect in PVC. Mesamoll® is a product of Lanxess, and is described as the phenyl cresyl ester of pentadecyl sulfonic acid. Reportedly, it is resistant to hydrolysis and diffusion-controlled plasticizer losses.

Benzoates are the esterification products of benzoic acid and selected alcohols and glycols, with preferred glycols being diethylene glycol, dipropylene glycol, and butanediol. An example of a commonly used benzoate is dipropylene glycol dibenzoate (DPGDP). Owing to its solvating strength, this is used in PVC flooring, where the benzoate segment expedites fusion and processing of the compounds, while their low molecular weight contributes to a hardened, stain-resistant surface.

The PP-LT group are low-temperature *o*-phthalates produced with the use of normal alcohols. These contribute superior low-temperature flexibility to PVC compounds. The entire family of *aliphatic dibasic acid esters* also imparts exceptional low-temperature properties.

For aliphatic dibasic acid esters, lower-molecular-weight alcohols are used with higher molecular weight acids, and *vice versa*, such that the total number of carbon atoms per molecule ranges from C_{18} to C_{26}. This allows the required non-polar/polar ratio to be maintained, ensuring good compatibility with PVC along with low-temperature properties.

DEHA is the standard and most widely used plasticizer of this class. DOZ, DOS, and DINA are used for achieving low-temperature flexibility.

The PP-LV group are low-volatility plasticizers, and this property is primarily attributed to their higher molecular weight, which also manifests itself in their lower vapor pressure. High-molecular-weight *o*-phthalates that serve as LV plasticizers include those that have molecular weights greater than DIDP (446 g/mol).

Increasing the molecular weight of *o*-phthalates increases their non-polar/polar ratio, with loss of compatibility with PVC occurring at molecular weights greater than that of DTDP (530 g/mol; i.e., alcohol chain length >C_{13}).

Two other chemical families also known for their use as low-volatility plasticizers are trimellitates and polyesters (also referred to as "polymerics"). The trimellitates are plasticizers produced from three moles of monohydric alcohols and one mole of trimellitic anhydride (TMA).

Compared to the phthalate diesters, the third ester group in the trimellitates results in a plasticizer of higher molecular weight, lower volatility, increased polarity for better compatibility, and increased entanglement with PVC matrix.

Used in combination with polyesters, trimellitates result in reduced plasticizer migration from PVC refrigerator gaskets and prevent crazing of ABS-based door liners. However, the trimellitates contribute an increased plasticizer diffusivity, if compounds containing them are immersed in oils.

5.7.2 Specialty Plasticizers

Specialty plasticizers are also divided into three subgroups:

- "SP-LD": low diffusion
- "SP-Stab": heat stabilization
- "SP-FR": flame resistance.

The SP-LD group offers low diffusion, a property that is attributed to the high molecular weight and highly branched isomeric structures of the plasticizers. DIDP and DTDP impart superior resistance to diffusion-controlled plasticizer losses, and sometimes are used in combination with more costly diffusion-resistant plasticizers.

The polyester plasticizer family is known for its outstanding performance in this performance attribute category. Liquid polymeric plasticizers are typically polyesters with a molecular weight ranging from 1000 to 8000 Da. Solid polymeric plasticizers are polyethylene copolymers, such as ethylene vinyl acetate (EVA), chlorinated polyethylene (CPE), nitrile butadiene rubber (NBR), and terpolymers like ABS (acrylonitrile–butadiene–styrene) and ELVALOY™ (co- and terpolymers of ethylene, butyl acrylate, and carbon monoxide), with molecular weights that can extend >500,000 Da. These polyesters are prepared by the esterification of polyols, such as propylene glycol or butylene glycol with aliphatic dibasic acids.

The greater the plasticizer viscosity, or molecular weight, the lower its diffusion (higher the permanence).

Polymeric plasticizers composed of branched chemical structures are more resistant to diffusivity losses than those based on linear isomeric structures; on the other hand, they are less oxidatively stable. Polarity, or the oxygen-to-carbon ratio, also increases the extraction-resistance of polymeric plasticizers. Plasticizers of low polarity exhibit increased extraction resistance towards polar extraction fluids, such as soapy water.

The SP-Stab group are epoxidized oleochemical plasticizers, which also function as thermal and UV-light stability enhancers of PVC. This plasticizer family consists of essentially two main chemical types:

- Epoxidized natural oils, such as epoxidized soybean oil (ESBO; Figure 5.5) and epoxidized linseed oil (ELSO) that have molecular weights of approx. 1000 Da and impart low volatility.
- Epoxidized alkyl esters of fatty acids, such as epoxidized 2-ethyhexyl tallate, that have PVC compatibility similar to GP plasticizers. Their molecular weight and viscosity are substantially lower than that of the epoxidized natural oils.

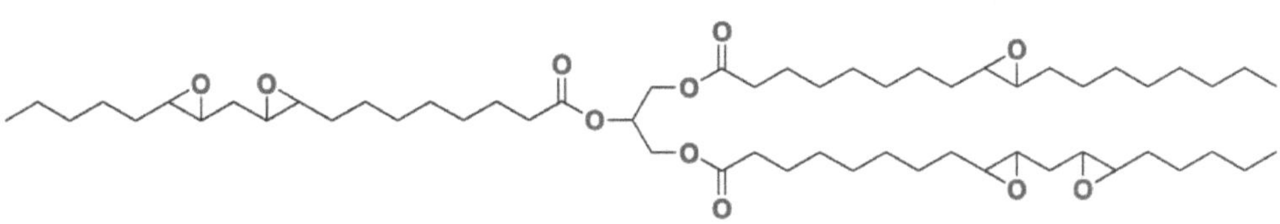

Figure 5.5 Epoxidized natural oil (an example of an epoxidized triglyceride)

Conventionally, both types of epoxy-derivatives have limited compatibility with PVC and are used as secondary plasticizers. Secondary plasticizers are used at lower concentrations compared to primary plasticizers, and they typically contribute one or more additional performance attributes. They reduce cost at 3–12 phr for achieving optimum cost–performance characteristics, including cost-in-use, heat stability, rheology, hardness, and (if necessary) the low-temperature flexibility of PVC compounds.

The SP-FR group are flame-resistant plasticizers that primarily feature halogenated phthalates and phosphate esters. Phosphate plasticizers, which may be considered "inorganic esters", are prepared by the slow addition of phosphorus oxychloride to an aliphatic alcohol or phenol. The highly aromatic tricresyl phosphate (TCP) is the most effective flame retarding plasticizer, but it generates a lot of smoke under fire conditions. Trialkyl phosphates (such as tri(2-ethylhexyl) phosphate (TOF)) are less efficient in terms of fire retardancy compared to TCP. Commercially available phosphate plasticizers contain combinations of aryl and C_8–C_{10} alkyl groups in their structures to offer a balance of flame retardancy, low smoke, volatility, and acceptable plasticizing efficiency. Incorporation of a combination of phosphate plasticizers, antimony trioxide (Section 10.6.2), and zinc borate yields excellent low-flame-generating flexible PVC compounds suitable for use in demanding applications, such as plenum, cable jacketing and insulation. Phosphate plasticizers may also be combined with phthalates to reduce the raw material cost of the compounds.

5.8 Primary Bio-based Plasticizers

5.8.1 Overview

Bio-based plasticizers are manufactured using only renewable raw materials. Bio-attributed plasticizers are produced using combinations of renewable and petroleum-based feedstocks. The attribution of bio-attributed properties is managed through the "mass balance approach", which tracks in part the inputs of biomass materials and process outputs. It's important to distinguish between verifiably renewable raw materials and bio-attributed raw materials Both bio-based plasticizers and bio-attributed plasticizers can be expected to reduce the carbon footprint of plastic materials containing them as these plasticizers at least partially made from renewable feedstocks.

Examples of the most well-known bio-attributed plasticizer types are:

- Citric acid esters
- Dicarboxylic acid esters
- Acetylated fatty acid esters
- Isosorbide diesters

- Epoxidized fatty acid esters
- Valeric acid esters.

Traditionally, for their specific performance attributes, such as low-temperature flexibility and heat stabilization, plasticizers produced from renewable feedstocks were used as secondary specialty plasticizers designed for niche applications. However, over the last decade, the pool of renewable feedstocks and processing approaches suitable for manufacturing bio-based plasticizers has broadened, along with their application range. As a result, a number of recently developed esters function now as GP primary plasticizers, meeting and often exceeding cost/performance requirements set by their conventional petroleum-based functional equivalents.

Due to the nature of the feedstocks, bio-attributed plasticizers are an excellent source of renewable carbon in plasticized PVC compounds, and further advance sustainability goals by reducing the carbon footprint (as illustrated with the use of the life cycle assessment methodology). Bio-renewable raw materials are by definition sustainable natural resources.

More often than not, bio-based chemicals and materials are of lower toxicity and less hazardous compared with the conventional plasticizers used for the same applications.

Raw materials are considered to be bio-renewable if the carbon acquisition (as a result of carbon dioxide sequestration via photosynthesis) through those resources takes place within a 10-year period and enables measurable carbon footprint reduction.

In 2011, the US Department of Agriculture established the "USDA BioPreferred® Product" label, which can be affixed on certified products and intermediates to indicate their bio-based content. The minimum bio-based content required for having a product certified was set at 25% in accordance with ASTM D6866.[6] This label was designed to incentivize manufacturing and the use of bio-based chemicals and materials. Similar certification programs for bio-based chemicals and materials have also been established in Belgium and Germany.

PVC has considerably lower total carbon content (38%) than other conventional thermoplastics like polyethylene and polypropylene (both 86%) and polystyrene (92%). Therefore, the carbon footprint of PVC is said to be lower as well.

[6] ASTM D6866 applies to any material containing organic carbon-containing components that can be combusted in the presence of oxygen to produce carbon dioxide gas (therefore, minerals and inorganic fillers, such as metal carbonates, are excluded). The method measures the percentage of new (renewable) radiocarbon ^{14}C to the total carbon. The bio-based content of a material or an assembly depends on both the bio-based content of its components and their concentration in the material or the assembly.

A variety of processes have been developed for converting bio-renewable resources into chemicals and materials. Natural oils (such as soybean oil, linseed oil, and castor oil) as well as cellulose sugars produced in plants, are suitable raw materials for the production of bio-based chemicals in general and bio-attributed plasticizers in particular.

5.8.2 Selected Types of Primary Bio-based Plasticizers

5.8.2.1 Citrates

Citrates are specialty primary plasticizers that offer good biodegradability and low toxicity compared to some conventional plasticizers, and thus are applied in PVC blood bags, controlled-release pharmaceuticals, gums, and products that come into contact with food, such as food packaging films. A wide variety of citrate plasticizers is now commercially available.

Initially, due to their relatively high production cost but favorable toxicological and eco-toxicological profiles, citrates were used primarily in the "sensitive" applications described at the start of Section 5.7. Currently, as a result of process improvements, the manufacturing cost of citrates plasticizers has reduced so that it is now comparable to that of conventional plasticizers. Market demands for citrates and their blends have grown, and led to a broadening of their use as GP plasticizers in technical applications, such as textile coatings, artificial leather, hoses, wall coverings, flooring, and others.

The preferred method of manufacturing the citric acid used in citrate plasticizers is fermentation of corn sugars. Tri-n-butyl *O*-acetylcitrate (ATBC; Figure 5.6) is one of the high-volume citrate plasticizers.

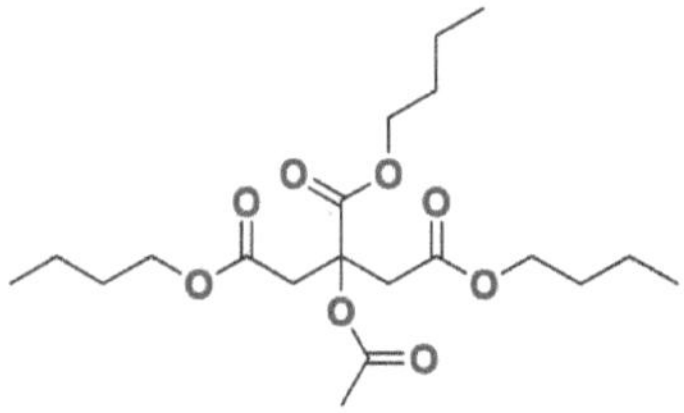

Figure 5.6
Tri-n-butyl *O*-acetylcitrate (ATBC)

It has been demonstrated that the performance of ATBC and DEHP is similar in terms of plasticizing efficiency (as determined by Shore A hardness), gelation temperature, migration, storage stability, water extraction, and volatility. As a result, DEHP can be replaced with ATBC in many applications. Citric acid esters are readily metabolized within the human body into non-toxic constituents, have no adverse effects on health during their handling and use, and are known for their safe ecological profiles; citrate plasticizers are registered on major national chemical inventories.

5.8.2.2 Dicarboxylic Acid Diesters

In addition to the citrates, a range of dicarboxylic acid diesters (in which the acids are produced from renewable raw materials) are also used as plasticizers. These include azelates, sebacates, adipates, succinates, and furoates. The first two types have traditionally been derived from renewable feedstocks. Significant advances were made in the 2010s to enable the manufacture of bio-based adipates, succinates and furoates.

The azelates are derived from azelaic acid, which in turn is manufactured by ozonolysis of oleic acid. An example of this class of bio-based plasticizers is DOZ, which is produced by reacting azelaic acid with 2-ethylhexanol. The sebacates, such as DOS, are produced by esterifying sebacic acid, which is typically derived from castor oil. Adipic acid used for the manufacture of the adipates, such as di-2-ethylhexyl adipate, was conventionally produced from petroleum-based raw materials and currently can also be derived from renewable raw materials.

Bio-based succinic acid has also become a commercially viable raw material, which in turn could be used to make plasticizers, such as di-2-ethylhexyl succinate (DEHS).

Another bio-based dicarboxylic acid that is projected to be commercially available and viable over the coming years is 2,5-furandicarboxylic acid (FDCA). FDCA is suitable for making furandicarboxylic acid diester plasticizers, such as di(2-ethylhexyl) 2,5-furandicarboxylate (FDAE) (Figure 5.7).

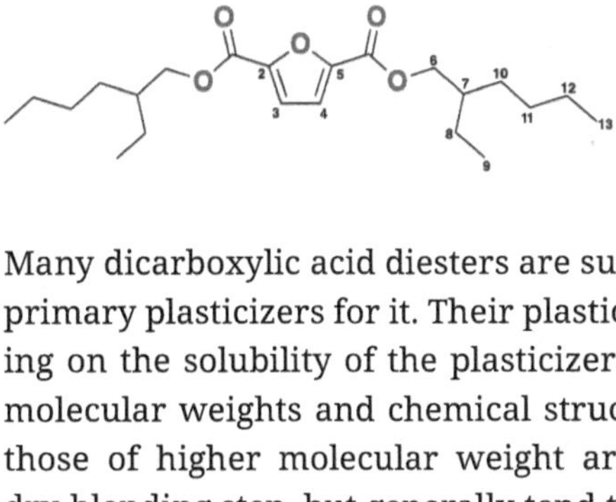

Figure 5.7
Di(2-ethylhexyl) 2,5-furandicarboxylate (FDAE)

Many dicarboxylic acid diesters are sufficiently compatible with PVC and are used as primary plasticizers for it. Their plasticization efficiencies vary considerably depending on the solubility of the plasticizers in PVC, which is in turn influenced by their molecular weights and chemical structures. Within the same family of plasticizers, those of higher molecular weight are often slower to absorb in PVC during the dry-blending step, but generally tend to yield superior retention of mechanical properties after aging at elevated temperatures.

FDAE has a relatively high molecular weight (approaching that of DEHP) and thus is considerably less volatile than some of the other dibasic acid diesters. The superior performance observed with FDAE as a primary plasticizer is attributed to its higher molecular weight and the aromatic moiety of its structure.

5.8.2.3 Acetylated Fatty Acid Esters

Grindsted® Soft-N-Safe™ (Figure 5.8) or "SNS" is an acetylated monoglyceride of hydrogenated castor oil, and is an example of the acetylated fatty acid ester class of bio-based plasticizers. It is produced in a multistep process from hydrogenated castor oil, and its two main components (a fully acetylated glycerol mono-12-hydroxystearate and fully acetylated glycerol mono-stearate) account for 95% of its total composition. The main performance attributes of SNS include processability, efficiency, and a gelling profile that is similar to that of DEHP; it also imparts considerably reduced amounts of volatile organic compounds (VOCs).

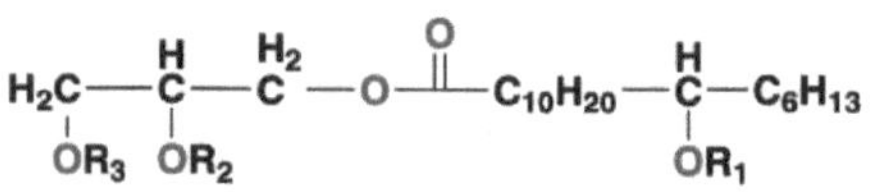

Figure 5.8 Acetylated monoglyceride of hydrogenated castor oil, where $R_1 = R_2 = R_3 = COCH_3$

Applications of SNS include toys, bottle-cap liners, screw-cap liners (for e.g. jam jars), teething rings, cling film, tubes, and conveyor belts in the food industry and medical equipment. It has also been used in a number of technical applications, such as footwear, safety gloves, flooring, wallcovering, tarpaulins, weather stripping, rainwear, wire and cables, office furnishings, and as a carrier for liquid pigments and stabilizers.

5.8.2.4 Isosorbide Diesters

Isosorbide (Figure 5.9) is a bio-based diol derived from corn starch. Polysorb® ID 37, manufactured by Roquette, is an example of a commercial isosorbide diester plasticizer. The performance imparted by Polysorb ID 37 is in many ways similar to that of DINP in terms of processability, efficiency, heat stability, low-temperature flexibility, and air release. Compared to DINP, Polysorb ID 37 has been demonstrated to improve viscosity stability and increase Congo Red decomposition time. Reportedly, the volatility (in terms of weight-loss percentage) of PVC compounds plasticized with Polysorb ID 37 was found to be higher than that of DINP. The main applications of Polysorb ID 37 include spread coatings and flooring compounds.

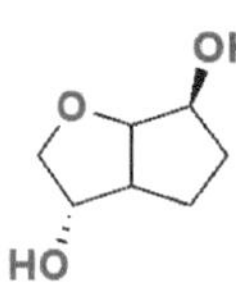

Figure 5.9 Isosorbide

5.8.2.5 Epoxidized Fatty Acid Esters

Epoxidized fatty acid monoesters, such as epoxidized octyl soyate and epoxidized 2-ethylhexyl tallate, have been known as specialty secondary plasticizers that also function as heat co-stabilizers.

Over the last decade, a number of plasticizer products based on epoxidized monoesters of fatty acids have been developed and introduced to the market as primary plasticizers. In reality, esters of this type were reported to be suitable for replacing phthalates as early as the mid-1970s. They are produced using one of the following three reaction sequences:

- Esterification of fatty acids with monohydric alcohols followed by epoxidation
- Transesterification of oils with monohydric alcohols followed by epoxidation
- Transesterification of epoxidized oils.

Certain epoxidized fatty acid ester plasticizers are fast to absorb and fuse into PVC. In particular, epoxidized methyl soyate (also described in the literature as a type of epoxidized fatty acid methyl ester, e-FAME) exhibits properties of fast-fusing plasticizers. Examples of commercial epoxidized methyl soyate plasticizers are Nexo E1 Vikoflex® 7010 and reFlex® 100, which have a bio-based content of about 95%. It was also shown that PVC plastisols plasticized with this epoxidized ester exhibit good viscosity stability, air release, and long-term heat stability. However, undesirable performance characteristics of compounds plasticized with epoxidized methyl soyate include high volatility, considerable weight loss as a result of water extraction, and in-process color development.[7)]

Recently introduced by Valtris as Santicizer® Platinum G-3000, epoxidized benzyl soyate (or benzyl epoxy soyate) demonstrates fast-fusing and heat co-stabilizing characteristics that are similar to that of epoxidized methyl soyate. However, unlike epoxidized methyl soyate, epoxidized benzyl soyate overcomes the shortcomings of the corresponding methyl derivative and imparts desirably low volatility.

In context of the manufacturing of epoxidized fatty acid alkyl esters (such as epoxidized methyl soyate described above) for use as PVC plasticizers, the hydrophilic impurity content has been shown to decrease with increasing the iodine number.[8)]

Interestingly enough, a study of single-chain epoxidized FAME derivatives revealed that epoxidation of methyl oleate resulted in oxirane oxygen values that were close to the theoretical one, with little or no apparent degradation of the epoxy ring having occurred during the reaction. In contrast, epoxidation of methyl linolenate yielded significant oxirane ring opening and formation of secondary hydroxyl groups.

Commercially available epoxidized fatty acid octyl esters are epoxidized 2-ethylhexyl tallate (Drapex® 4.4 by Galata Chemicals), epoxidized 2-ethylhexyl soyate (Proviplast PLS Green 8 by Proviron), epoxidized octyl soyate (Vikoflex™ 4050 by Cargill) and ep-

---

[7] The term "in-process color development" was first proposed by Frenkel and Chaudhary to describe a discoloration that is caused by plasticizers when the plasticized compounds are subjected to a static heat stability test. This type of discoloration usually manifests itself in the initial and mid-term heat stability region. Contrary to the degradation of PVC, this type of discoloration is associated with bio-based plasticizers and does not usually translate into the inferior mechanical properties of the plasticized compounds.

[8] The term "hydrophilic impurities" refers to epoxidized fatty acid esters containing hydroxyl groups formed as a result of opening of epoxy rings on the fatty acid chains; the concentrations of these impurities are determined by high-performance liquid chromatography (HPLC).

oxidized octyl stearate (Lankroflex™ ED6 by Valtris). PVC compounds containing these plasticizers reportedly showed enhanced low-temperature flexibility, high UV-light and heat stability, and acceptable volatility and extraction resistance. These plasticizers are also known as synergists for thermal stabilization of PVC when used in combinations with the mixed metal carboxylate heat stabilizers.

Due to its low viscosity and high efficiency, the epoxidized octyl esters of fatty acids enable the preparation of low-viscosity plastisols even if added at low concentrations. Both plastisols and organosols prepared with the use of the epoxidized octyl fatty acid monoesters esters are characterized by excellent viscosity stability. The low solvating action of these plasticizers also allows for easier control of the gelation rate in rotational and slush-molding applications. Applications of epoxidized fatty acid octyl esters include plastisols, organosols, and flexible, semi-rigid and rigid PVC formulations used for manufacturing films, sheets, shoe soles, hoses, toys, wires and cables, artificial leather, laminates, and coated materials.

There are several commercially available plasticizers (including the epoxidized fatty acid monoesters discussed above) that are produced as a result of partial (or incomplete) transesterification by reacting epoxidized vegetable oils with esters or alcohols of low molecular weight. Alternatively, similar chemical compositions can be prepared by reacting vegetable oils with the esters or alcohols first, followed by epoxidation of the prepared reaction mixture.

Reaction mixtures prepared by reacting epoxidized natural oils with esters or alcohols of low molecular weight consist of an equilibrium mixture of the epoxidized oils, epoxidized di-glycerides, epoxidized mono-glycerides and epoxidized monoesters of fatty acids.

There are publications that describe bio-based plasticizers consisting of blends of epoxidized natural oils with epoxidized fatty acid monoesters. Usually both types of components of this sort are produced separately, and ratios at which those components are blended are optimized for best performance, processability, and economics. Therefore, it is appropriate for plasticizers of this type to be called "formulated", a term that was first proposed by Frenkel and Chaudhary. In formulating flexible PVC with epoxidized fatty acid ester plasticizers, one of the key properties to keep in mind is the low-temperature flexibility of the compound. As long ago as 1953, it was recognized that different types of epoxidized fatty acid esters (when used as plasticizers for PVC) yield varying degrees of low-temperature flexibility and optimum heat-aging performance.

5.8.2.6 Pentaerythritol Tetravalerate

Pentaerythritol tetravalerate (PETPV), Pevalen™, is a bio-attributed plasticizer manufactured by Perstorp, using renewably sourced valeric acid as one of the raw materials. Pevalen was designed for use in consumer articles, such as coated fabrics, artificial leather, flooring materials, wall coverings, cables, and films. In terms of high plasticizing efficiency, exceptional UV light resistance, low volatility, and low smoke generation, its performance is comparable or superior to conventional primary plasticizers.

5.8.2.7 Bio-based Plasticized PVC Compounds and Carbon Footprint

According to completed "cradle-to-gate" life cycle assessment (LCA), the global warming potential associated with manufacturing Drapex® Alpha 200 by Galata Chemicals is 0.53 kg/kg CO_2 eq. This is about 70% lower than the average value of 1.96 for the conventional *o*-phthalate plasticizers DEHP, DINP, and DIDP, as assessed using ISO 14040:2006 and ISO 14044:2006.

Typically, incorporating primary bio-attributed plasticizers into PVC notably reduces the carbon footprint of the corresponding plasticized PVC compounds (where conventional petroleum-based PVC resin is used in combinations with bio-attributed plasticizers instead of conventional phthalates).

In many applications, plasticized PVC materials can outperform other plastics not just based on their lower cost and a broad range of valuable performance attributes, but also based on their low carbon footprint, the main sustainability criterion, which is lowered even further if the plasticizer is bio-attributed.

5.9 The Future of Plasticizers

The future of plasticizers is closely associated with the future of PVC. For many years, environmental activist groups have criticized PVC because it contains chlorine, in the mistaken belief that the production, use, and disposal of PVC products causes serious damage to the environment. Although the initial criticism was relatively unsuccessful, it has raised awareness about alternative materials that may replace flexible PVC in certain applications. In more recent years, the focus of regulators has shifted towards vinyl chloride monomer and additives used in PVC, particularly to certain metal-containing stabilizers and *o*-phthalate plasticizers. The plasticizers are targeted because under certain conditions small amounts of these can migrate from flexible PVC compounds, and because some legacy *o*-phthalates have been shown to cause or potentially cause adverse health effects. Furthermore, the future of any plasticizer or any plasticizer type is strongly dependent on the uninterrupted supply of cost-effective product grades that are sustainable and do not present unacceptable health and/or environmental risks. Until now, the demand for the most part has been met with the three primary plasticizers DOTP, DINP, and DIDP.

Interest in other classes of plasticizers that have balanced cost-performance properties as well as reduced carbon footprint and lower toxicological profile has been increasing. Citrates, polymerics, epoxidized esters of fatty acids, and benzoates are being effectively promoted not just on the basis of their cost and performance but also due to their "non-phthalate" nature.

Research on commercially viable plasticizers that are chemically bonded to PVC continues. One of the consequences of criticism of flexible PVC is that product designers have been further exploring possibilities for supplying soft, plasticizer-free PVC co-polymeric materials or PVC-based polymer mixtures. It is also expected that future technological developments will continue to intensify the inter-material competition based on the cost, performance, and sustainability of commercially viable materials.

5.10 Health, Safety, and Environmental Aspects

5.10.1 Hazard Classification

Cancer was the focus of phthalate research and regulatory concern in the 1980s and early 1990s; however, for the last 25 years the discussion has shifted to potential reproductive and developmental effects. These effects were found to be isolated to phthalate esters with the esterified alcohol components having straight carbon chain lengths of 3–6 and with the total carbon number of the esterified alcohol being C_8 or less. More specifically, the effects are seen with DIBP (2-methylpropyl; straight chain of 3 carbons), DBP (n-butyl; 4 carbons), DNPeP (n-pentyl; 5 carbons), and DEHP (ethylhexyl; 8 carbons). These effects were not observed with two C_{10} phthalates, DIDP (iso-decyl; straight chain of 7–9 carbons) and DPHP (propylheptyl; 7 carbons), and subtle effects were observed at high doses with DINP (iso-nonyl; 6–8 or 6–9 carbons depending on the source of the alcohol). These effects are now often associated with concerns about endocrine disruption.

No carcinogenic concerns have been reported for ATBC, DEHA, ESBO, DOTP, and DINCH, but the US EPA has classified DEHP as a Class C carcinogen. Neither reproductive, developmental, nor endocrine-disrupting effects have been reported for non-*o*-phthalate plasticizers.

Similar to *o*-phthalates, other common plasticizers used in PVC applications have quite low water solubilities and do not exhibit toxicity up to their water solubility limits. In terms of their environmental fate, ATBC, DOTP, ESBO, DEHA, and DINA are all readily biodegradable. The most common commercial non-*o*-phthalate plasticizers have been tested for key toxicological and environmental endpoints, and no adverse effects have been reported so far. The environmental fate, specifically biodegradation, bioaccumulation, and chronic and acute aquatic toxicity, are important criteria for products such as plasticizers. The most common *o*-phthalates currently used as plasticizers, which are based on alcohols with 8–13 carbons (e.g., DEHP through DTDP), are readily biodegradable, are not bioaccumulative, and are not toxic to aquatic organisms up to their respective water solubility limits.

5.10.2 Current Topics in Regulatory Compliance

5.10.2.1 US EPA and TSCA

The Lautenberg Chemical Safety Act was passed by the US Congress in 2016; this act provided updates to the existing Toxic Substances Control Act (TSCA), which is the primary chemicals regulation in the US. A critical part of the new regulations is a process to review existing chemicals in order to evaluate potential risk. Five *o*-phthalates, DBP, DIBP, BBP, DCHP (dicyclohexyl phthalate), and DEHP, along with phthalic anhydride, were designated as high-priority chemicals.

5.10.2.2 California Proposition 65 and *o*-Phthalates

California Proposition 65 (officially the "*Safe Drinking Water and Toxic Enforcement Act of 1986*") requires businesses to provide warnings to Californians about significant exposures to chemicals that cause cancer, birth defects, or other reproductive harm. Six *o*-phthalates have been added to the list after it was determined that they present a hazard for carcinogenicity or reproductive and development effects: DBP, BBP, DnHexP (di-n-hexyl phthalate), DEHP, DINP, and DIDP.

5.10.2.3 REACH Classification and Labelling

The following *o*-phthalates have been classified in the EU as reproductive toxins, and added or recommended for addition to the list of "Substances of Very High Concern" (SVHCs): BBP, DEHP, DBP, DCHP, DIBP, DIHepP (di-iso-heptyl phthalate), D711P and D610P (>0.3% dihexyl).

The commercially viable general-purpose high-molecular-weight phthalates (DINP, DIDP, and DPHP) are not subject to any hazard classifications. In addition, DINP was evaluated by the ECHA Risk Assessment Committee for possible classification, and the committee concluded no classification was warranted.

5.10.3 Regulated Applications of Plasticizers

5.10.3.1 Food Contact Applications

o-Phthalates have been used in several food contact applications, such as beverage tubing, cap liners and gaskets, and conveyor belts. Owing to the regulatory actions with DEHP, the consumption of non-*o*-phthalate plasticizers, such as DOTP, DINCH, ESBO, and ATBC, in these applications has grown.

In the US, several plasticizers were included in the 21 Code of Federal Regulations (CFR) that govern indirect food additives, according to the US FDA. DEHP, DINP, DIDP, DCHP, DEHA, DINA, ATBC, other citric acid esters, ESBO, and DOTP are currently listed in various 21 CFR sections.

The Food and Drug Administration Modernization Act of 1997 amended the Food, Drug, and Cosmetic Act to streamline some of the FDA's approval processes. One of the new procedures was to establish "a notification process for food-contact substances defined as any substance intended for use as a component of materials used in manufacturing, packing, packaging, transporting, or holding food if such use is not intended to have a technical effect in such food". Additionally, several Food Contact Notifications (FCNs) that are specific to the company filing the notification have been issued for the following plasticizers: PETPV, DBT, DOTP, TEHTM (tri(2-ethylhexyl) trimellitate), ESBO2EH ("2EH" refers to 2-ethylhexyl), SNS, ASE (alkylsulfonic phenyl ester) and TXIB (trimethyl pentanyl di-iso-butyrate).

In the EU, six *o*-phthalates are included in the list of food contact materials (FCMs), as listed by the European Food Safety Authority. All plasticizers and plastic additives are subject to a migration limit of 60 mg/kg (in food) with establishing specific migration limits and group migration limits for each of them. The *o*-phthalate plasticizers are DBP, BBP, DEHP, DIBP, DINP, and DIDP. Non-*o*-phthalates that have EU food contact clearance are ATBC, DEHA, ESBO, DINCH, SNS, DOTP, ASE, TEHTM, and DEHCH.

5.10.3.2 Medical Applications

Some of the most essential uses of plasticized PVC compounds are in various medical articles, such as films to make containers for storage and collection of blood products, IV solution bags and tubing for transfusions, IV infusions, enteral nutrition, dialysis, and drainage. Plasticized PVC has a high biocompatibility, it is easily sterilized, and offers a range of special performance attributes relevant to these applications, including kink resistance, when used in medical tubing. The most commonly used plasticizer in these applications remains DEHP. However, owing to potential health concerns, there has been a shift to other plasticizers or to alternatives to PVC.

In addition to DEHP, four additional plasticizers are currently listed in the European Pharmacopoeia (from the European Directorate for the Quality of Medicines) for use in a range of medical applications. Recently added plasticizers are DINCH, BTHC, TEHTM, and DEHT.

Over the years, DEHP has been demonstrated to have some unique performance characteristics when used in PVC storage containers intended for the three major blood products: red blood cells, platelets and plasma. However, it has been found that other plasticizers can provide the same or superior performance. For example, platelets are stored at room temperature typically for up to 5 days, and BTHC and TEHTM are used in these applications owing to their excellent gas permeability compared to

DEHP. DINCH-plasticized PVC has also been used for several years for pediatric platelet storage in the Netherlands. Frozen plasma may be stored in this way for as long as 1 year at –18 °C or for 7 years at –65 °C. DINCH and DOTP have been shown to be suitable for plasma storage as well. With low-temperature storage requirements, resistance to breaking and cracking has been a critical parameter. It is expected that DOTP and DINCH will be effective in this application, since their low temperature flexibility properties are superior to DEHP.

Trace amounts of the DEHP plasticizer that migrate into the blood helps stabilize red blood cells and reduce the amount of hemolysis. BTHC, a citrate plasticizer, was introduced over 25 years ago as a replacement for DEHP. Recently, several studies with DINCH and DOTP suggested that these are suitable replacements for it, as regards blood stabilization.

5.11 Questions

1. What are the main functions of plasticizers?
2. What is the mechanism of plasticization? What is anti-plasticization?
3. What is the difference between plasticizer efficiency and their substitution factor?
4. What are the main considerations for defining the optimum polarity of PVC plasticizers?
5. What are the chemical classes/types of PVC plasticizers and their performance attributes?
6. What is the "economics" group of plasticizer attributes?
7. What is the "compatibility" group of plasticizer attributes?
8. What is the "processing" group of plasticizer attributes?
9. What is the "performance" group of plasticizer attributes?
10. What are the two types of structural isomers of plasticizers?
11. What is the effect of plasticizer use level on specific gravity, hardness, and elongation of plasticized compounds?
12. What are the main chemical families/types of plasticizers?
13. What are the compositions/chemical types of general-purpose plasticizers?
14. What are the compositions/chemical types of performance plasticizers?
15. What are compositions/chemical types of specialty plasticizers?

16. What are the main performance alternatives to DEHP?
17. What is the effect of adding the third ester group in the trimellitate chemical structure on its volatility and compatibility compared to *o*-phthalates?
18. What is the secondary function of epoxy plasticizers in PVC compounds?
19. What are the chemical types of commercial bio-attributed plasticizers?
20. What are the main positive environmental impacts associated with manufacturing and use of bio-attributed plasticizers?

6 Processing Aids and Impact Modifiers

6.1 Terms and Definitions

Processing aids	
Processing aid	Additive that improves the processing characteristics of a polymer, such as flow.
Die	Tool used in extrusion or molding processes to shape molten material into a desired article.
Die swell	Expansion of a polymer extrudate after leaving the die due to elastic recovery.
Melt strength	Ability of a molten polymer to resist stretching under tension, crucial in film blowing.
Melt fracture	Surface irregularity or distortion in extruded materials due to unstable melt flow.
Additive dispersion	Uniform distribution of additives throughout the polymer matrix to ensure consistent properties.
Plate-out	Unwanted deposit of additives on processing equipment surfaces.
Process throughput	Amount of material processed per unit of time in manufacturing.
Gels (in polymers)	Solid, often crosslinked lumps within a polymer matrix that disrupt surface and mechanical properties.

Impact modifiers	
Polymer fracture	Breaking of a polymer under stress, influenced by structure, defects, and environmental conditions.
Impact resistance	Ability of a plastic to resist breaking when subjected to sudden force or shock.
Toughening	Enhancing a polymer's resistance to fracture by adding elastomers or modifying its structure.
Transparency	The degree to which light passes through plastic without scattering.
Weatherability	A plastic's ability to withstand outdoor exposure (sun, rain, temperature) without degrading.
Melt strength	Resistance of molten polymer to stretching; key for processes like blow-molding and foaming.
Glass transition temperature	Temperature range where a polymer changes from hard/brittle to soft/rubbery state.
Homopolymer	Polymer made from only one type of monomer unit.
Copolymer	Polymer made from two different monomers.
Terpolymer	Polymer made from three different monomers.
Rubber	Elastic polymer material that can stretch and return to shape.
Yield stress	The stress at which a plastic begins permanent deformation.
Ductile break	When a material stretches before breaking.
Brittle break	When a material breaks suddenly with little deformation.

6.2 Processing Aids

A processing aid, also known as a flow modifier, is a general term that refers to several classes of primarily polymeric additives that are used to modify the melt-processing of thermoplastics. This section covers processing aids for both PVC and polyolefins.

Early approaches to improve the processability of thermoplastic polymers focused on (a) reducing the molecular weight of the resin, (b) designing copolymer resins, and (c) adding small amounts of plasticizers or lubricants. Modern processing aid technology allows manufacturers to reach the desired processing performance while using standard resin grades, and it helps to avoid incompatibility issues and plate-out that can occur with low-molecular-weight lubricants or plasticizers.

6.2.1 Classification of Processing Aids

6.2.1.1 MMA-based Processing Aids

Table 6.1 Categories of PVC Processing Aids, Their Characteristics, and Applications

Parameter	Molecular weight			
	Low	Medium	High	Ultra-high
Lower molecular weight limit	50,000	1,000,000	4,000,000	8,000,000
Upper molecular weight limit	500,000	4,000,000	8,000,000	—
Effect on increasing melt elasticity	Little	Medium	Strong	Very strong
Effect on increasing melt viscosity	Reduce	Increase slightly	Increase moderately	Increase significantly
Effect on Increasing die swell	Little	Medium	Strong	Very strong
Effects on fusion	Delay	Expedite	Expedite	Expedite
Processing type, main applications and dosages	Injection molding (0.5–3.0 phr) Calendering (0.5–1.0 phr) Thermoforming (0.5–2.5 phr)	Profile extrusion (0.2–1.5 phr) Foam core pipe (foam layer 1.0–3.0 phr) Sewer, drainage and electrical pipes (up to 2.0 phr) Compact sheets (1.0–2.0 phr) Foam coextrusion (outer skin up to 3.0 phr) Calendered film (0.5–1.0 phr) Blown film (1.0–2.0 phr) Furniture foils (1.0–2.5 phr) Thermoforming (0.5–2.5 phr) Injection molding (0.5–3.0 phr) Foam extrusion (6.0–12.0 phr)		Foam extrusion (6.0–12.0 phr)

The main commercial families of processing aids are based on copolymers that typically contain methyl methacrylate (MMA), alkyl acrylates, and in some cases additional monomers, such as acrylonitrile or styrene. The molecular weight of these additives is a key parameter that strongly influences their behavior during processing.

In general, four categories of processing aids can be distinguished, based on their molecular weight[1] or chain length and their intended function in the process (Table 6.1):

- Low-molecular-weight processing aids with short chain lengths
- Medium-molecular-weight processing aids with medium chain lengths
- High-molecular-weight processing aids with long chain lengths.
- Ultra-high-molecular-weight processing aids with very long chain lengths.

The compositions of most processing aids for PVC are MMA-based.

6.2.1.2 Styrene–Acrylonitrile (SAN) Copolymer Processing Aids

SAN-based processing aids behave in many respects like MMA-based all-acrylic modifiers. They:

- Are compatible with PVC and several engineering thermoplastics, such as PC, PC/ABS, PBT, and TPU
- Follow the usual structure–property rules for processing aids of higher molecular weight and longer chains
- Increase melt strength (whereas lower-molecular-weight SAN grades promote higher melt flow)
- Can accelerate fusion in PVC in some cases more efficiently than all-acrylic processing aids of comparable molecular weight
- Can serve as carrier resins for pigments and other additives, supporting good dispersion in the polymer matrix
- Result in very little die swell when used as PVC processing aids, which makes them particularly attractive for extrusion applications, where low die swell and dimensional control are required.

6.2.1.3 Lubricating Processing Aids

There is a class of MMA–styrene copolymer processing aids that combines externally lubricating (metal-release) behavior with pronounced effects on melt rheology. These modifiers tend to delay PVC fusion when compounds are run at elevated melt temperatures and high output rates.

---

[1] The term "molecular weight" reveals very little about a polymer's chain length because of the differences in repeating units (including chain substituents). For example, the monomer methyl methacrylate (MMA) has a molecular weight of 100 g/mol, while the molecular weight of its homolog octyl methacrylate (OMA) is about 198 g/mol, nearly twice as much. Polymerizing 1000 MMA monomers into a linear chain produces PMMA with a molecular weight of 100,000 g/mol. To achieve the same molecular weight with OMA, only about 500 monomer units are needed, resulting in a polymer chain half as long as that of PMMA. If 1000 OMA monomers were polymerized, the chain length would match that of PMMA, but the molecular weight would be roughly twice the value.

In contrast to low-molecular-weight external lubricants, such as paraffin or polyethylene waxes, polymeric lubricating processing aids become partially entangled with PVC chains and therefore show much less tendency to plate-out on metal surfaces. This reduces cleaning-related downtime and helps to achieve smooth surfaces with high gloss, while strongly diminishing sticking of the PVC melt to hot tooling.

Processing aids of this type typically have predominantly linear chain architectures with only limited branching. Their main function is to modify the gelation and fusion behavior of PVC, and this depends on their chemical composition, molecular weight, and dosage in the formulation (Figure 6.1).

> The higher the processing aid dosage, the greater the gelation acceleration of PVC compounds.

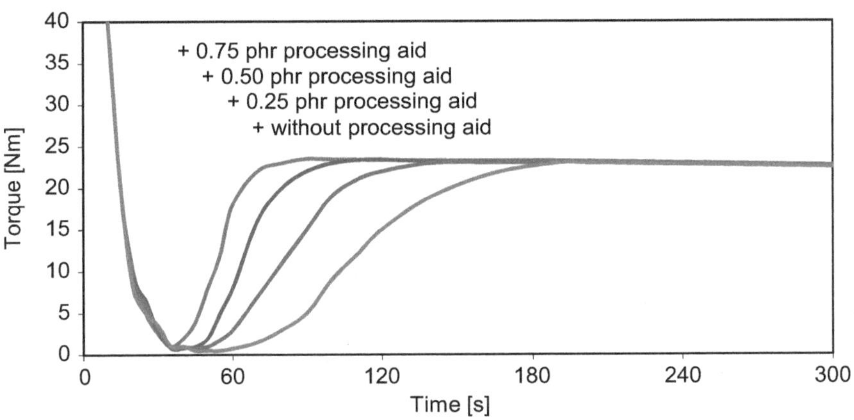

Figure 6.1 Effect of dosages of medium-molecular-weight processing aids on the gelation behavior of a PVC window profile compound

6.2.2 Mode of Action of Processing Aids in PVC

Entanglement between PVC and processing aids is a key factor in improving the processability and melt characteristics of PVC compounds. As PVC melts, the longer chains of the processing aid become tangled with the shorter PVC chains, forming a physically entangled network that can be viewed as a type of "physical crosslinking". The entangled long polymer chains of the processing aids impart rubber-elasticity on the melt, and with increasing entanglement also impart die swell (elastic rebound) during and after processing.

Processing aids are typically used in rigid PVC at about 1–3 phr, while significantly higher levels (often above 8 phr) are employed in applications that demand very high melt strength, such as cellular (foamed) PVC.

6.2.3 Performance Attributes of Processing Aids in PVC

The most widely recognized benefits of using processing aids in the melt-processing of PVC include the ability to:

- Accelerate or slow down fusion of the PVC powder formulation
- Form a more uniform and homogeneous melt
- Impart higher melt strength, preventing melt fracture of the PVC matrix and so enabling control of foam processability, cell structure, and quality
- Increase melt elasticity, enabling smooth surfaces of final articles to be achieved.
- Enable better control over die swell.

6.2.3.1 Promoting Melt Homogeneity

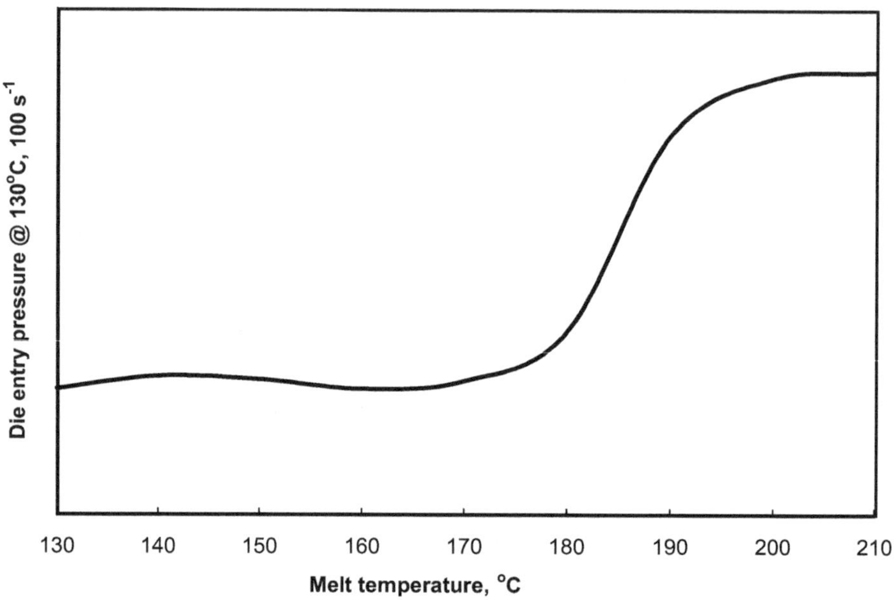

Figure 6.2 Die entry pressure as a function of melt temperature

In the processing of a PVC compound, it is important to have a homogeneous non-fractured melt to produce articles with a high-quality surface. Figure 6.2 illustrates how the morphology of the primary particles changes at higher processing temperatures after fusion has occurred – from discrete particles without mutual entanglement to a structure in which neighboring particles are interconnected. The shift from particle-dominated flow to molecular (melt) flow is significantly promoted when a processing aid is included in the formulation.

A higher entanglement force results in a more homogeneous PVC melt at normal processing temperatures.

The die entry pressure method can also be applied to demonstrate how processing aids influence the temperature at which this transition takes place. When a fusion-promoting processing aid is included in the formulation, the transition occurs at a lower temperature, leading to a more uniform melt, and a lower thermal load on the compound.

Although the detailed mechanism by which processing aids act in PVC is debated, current understanding suggests that during processing, these additives attach to the PVC grains. This promotes faster and more controlled breakup of the grains into the primary particle flow, and leads to a more uniform melt.

The fusion-expediting or fusion-delaying behavior of a given processing aid can be evaluated with a torque rheometer, in which the torque is recorded as a function of temperature.

6.2.3.2 Increasing Melt Strength and Elasticity

Melt strength describes how a polymer melt resists deformation under the load, and reflects both its elastic response and its elongation viscosity. In simple terms, elasticity characterizes the tendency of the polymer melt to return to its original state when the stress is removed. A combination of melt strength, elongation, and elasticity defines the "toughness" of a melt.

Acrylic processing aids used in PVC are generally compatible with the matrix and readily form entanglements with PVC chains, which increases melt strength. This reinforcement of the melt helps to prevent the expanding gases from rupturing cell walls or breaking through the surface skin in the production of foamed PVC articles, thereby improving foam structure and surface quality. In rheological measurements, the presence of such a processing aid is typically associated with higher die entry pressures accompanied by an increase in melt elasticity.

6.2.3.3 Controlling Die Swell

While the melt flows through the tooling, the polymer chains of PVC, as well as the processing aid, are stretched and oriented in the direction of the flow. As the melt exits the die, more space becomes available, and the stretched and oriented polymer chains partially revert to their entangled forms. The stretching tension is released, and the melt swells (Figure 6.3). This melt stress release or elastic recovery of the melt after exiting the die is called "die swell". Die swell is defined as the ratio of the extrudate diameter to the die diameter, and it exhibits a significant correlation to the molecular weight of the processing aid.

The higher the molecular weight of the processing aid, the more pronounced the die swell effect is. Increasing the dosage of the flow modifier also results in an increased die swell.

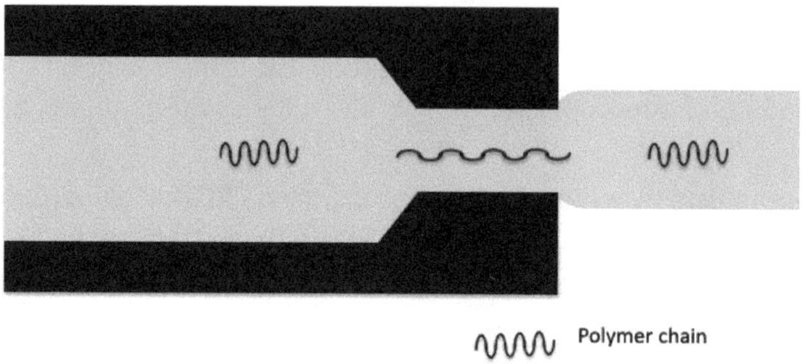

Figure 6.3 Formation of die swell

6.2.3.4 Process-specific Benefits

Further advantages of using processing aids in PVC processing become apparent within the context of individual processing methods, and are outlined below.

Extrusion

Processing aids:

- Increase ductility and toughness of the melt
- Shorten residence time in the barrel and promote faster conveying that may increase throughput
- Lower the thermal load on the compound and thus improve thermal stability of the extrudate
- Reduce scrap rates and the number of rejects
- Minimize plate-out on metal surfaces, cutting down cleaning-related downtime
- Stabilize melt flow and help suppress melt fracture
- Deliver a consistent appearance with smooth surfaces and high gloss on the finished profiles.

Foam extrusion

Processing aids:

- Prevent the collapse of the cellular material structure
- Achieve significant reductions in foam density due to the improved melt elasticity and melt strength.

Calendering/thermoforming

Processing aids:

- Reduce thickness variations due to even melt flow
- Increase production output due to the improved melt elasticity
- Result in uniform parts, and longer times between clean-outs due to better resistance to melt fracture.

Injection molding

The viscosity of a PVC compound is a key parameter for injection molding, since it affects how readily and how fast the molten PVC can fill the mold cavity. For parts with long flow paths, thin walls, or intricate geometries, compounds with very high flow (corresponding to low melt viscosity) are required, provided that the mechanical property targets of the application are still met.

Processing aids can also help to reduce jetting defects by increasing the elasticity and controlling swell of the polymer stream as it enters the cavity. Jetting describes a situation in which the incoming melt does not expand into a normal "fountain flow" at the gate, but instead shoots into the cavity as a narrow jet that solidifies irregularly, producing rough, worm-like surface streaks. By modifying the melt rheology and promoting more uniform spreading, processing aids lower the tendency for jetting. Figure 6.4 shows how jetting in a molded bar decreases with the use of high- and very high-molecular-weight processing aids.

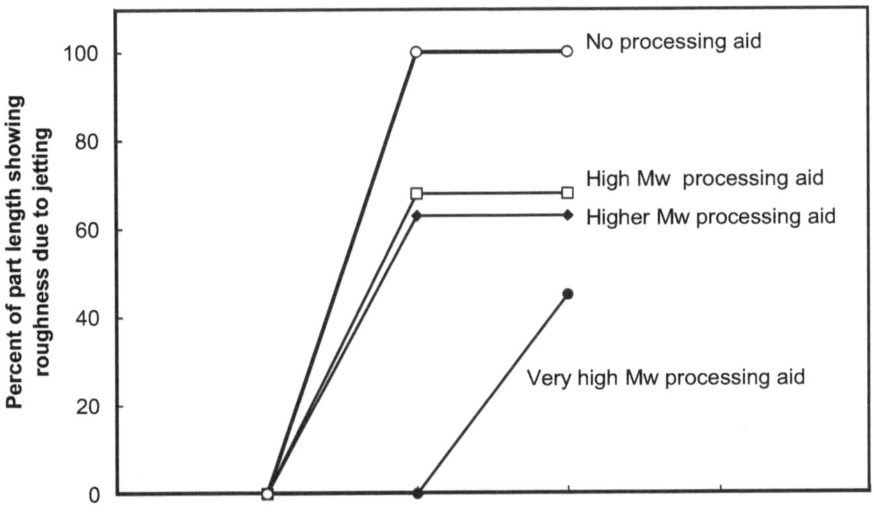

Figure 6.4 Effects of processing aids (1 phr) on the extent of jetting of a $K = 60$ PVC-based formulation[2)]

[2] The K value of a PVC resin is a measure of its average molecular weight, reflecting the polymer chain length, and influencing its processing and mechanical properties.

Fast gelation, low melt viscosity, and increased melt strength help to reduce "gate blush" around the injection point. Gate blush (also termed gate haze or gate frosting) describes surface defects that appear as pale rings, cloudy zones or flow lines in the gate region, when the incoming melt does not fill and cool uniformly. These imperfections are typically linked to very high shear stresses at an undersized gate and/or excessive injection speed, which cause strong chain orientation, local degradation, and ultimately melt fracture once the shear stress exceeds the melt strength. Under such conditions the polymer stream cannot form a stable, fountain-type flow pattern, and the resulting flow instability becomes visible as a whitish halo or rough zone around the gate.

The choice of processing aids can influence the final gloss and aesthetics of plastic articles. Processing aids enhance the gloss by improving surface structure and light reflectivity during processing.

Incorporation of processing aids into PVC compounds increases their melt viscosity, as illustrated in Figure 6.5 and expressed in mixing torque.

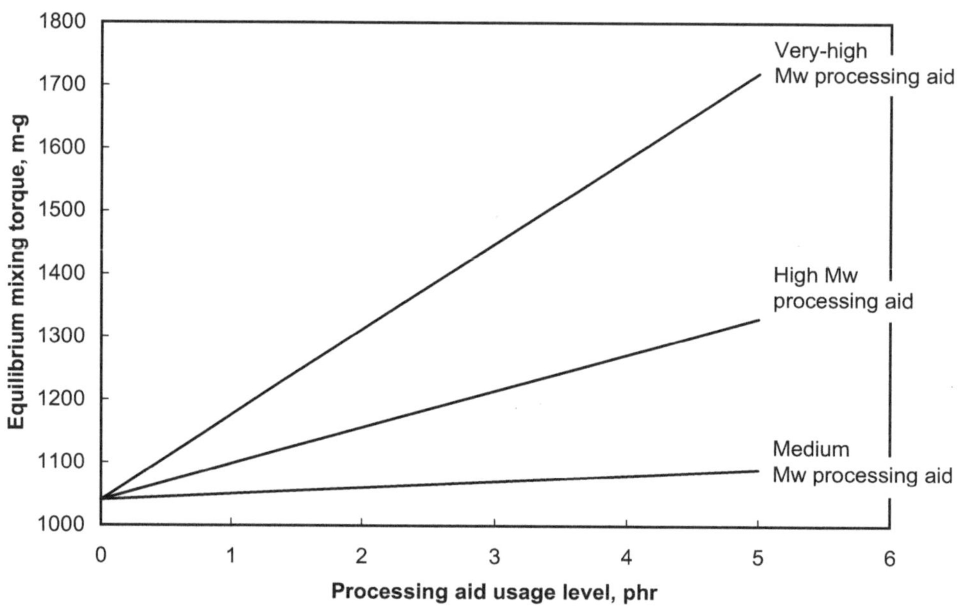

Figure 6.5 Effect of processing aids on the equilibrium mixing torque (PVC $K = 51$, Haake, 160 °C in a bowl)

The choice of molecular weight and dosage of a processing aid is critical for achieving the desired balance of melt properties. It is best evaluated by rheological (viscosity) measurements. In many cases, a relatively low addition level of a high-molecular-weight processing aid can provide a more favorable combination of elasticity, melt strength, and flow (for example, in reducing gate blush, jetting, and/or apparent viscosity) than a low- or medium-molecular-weight processing aid grade loaded at a higher dosage.

Blow-molding

Processing aids:

- Result in thinner and more uniform walls of articles (bottles) without tearing them due to higher melt elasticity
- Impart more effective mold release.

6.2.4 Processing Aids for Polyolefins

As in the case of PVC, processing aids in polyolefins are primarily employed to suppress melt fracture during extrusion. They are used at very low dosages, usually in the range 0.01–0.1% based on the polymer weight. Commercially available processing aids of this type can be used in LLDPE, LDPE, and HDPE, as well as in PP.

6.2.4.1 Mechanism of Action

Since these processing aids are largely immiscible with the base resin, they remain as a dispersed phase consisting of droplets with a typical size of about 1 µm. The droplets reach the die wall with the polymer flow. Over time, the die surface becomes fully covered with the processing aid, creating a lubricating layer along the die interior. Once this coating is formed, melt fracture becomes eliminated from the extrudate.

Figure 6.6 depicts a typical LLDPE response when 0.05% processing aid is added, in comparison with an unmodified control. It indicates that within roughly 45 min a critical coating develops on the die, and at that point melt fracture is essentially no longer an issue.

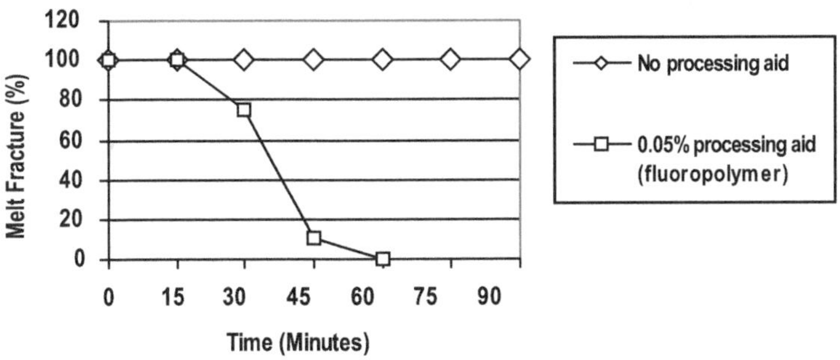

Figure 6.6 Effect of 0.05% processing aid on melt fracture elimination in LLDPE

As the die surface becomes fully covered with the processing aid, the back-pressure in the extruder gradually drops to a low, steady value, and an equilibrium is reached between the rate at which the processing aid is deposited on the metal and the rate at which it is removed. Operating conditions can shift both the deposition and removal rates and therefore change the processing aid dosage needed to reach this equilibrium, as well as the time required until the die is completely coated.

Several variables govern this deposition/removal balance:

- Processing aid concentration (higher levels generally increase the deposition rate and shorten the coating time)
- Dispersion of the processing aid in the melt
- Throughput, since higher output usually reduces the time needed to form a continuous coating
- A suitable match between the viscosity of the processing aid and that of the base resin.

Additionally, the chemical structure of the processing aid polymer, which includes the nature and number of its end groups, can influence how effectively it functions at the polymer–metal interface.

The optimum conditions are achieved at the highest shear rate possible that can be maintained without the melt fracture.

The shear rate also depends on temperature, and reducing the temperature may help to increase the shear stress to its optimum value. The optimal temperature depends on the target output, the equipment, and the polymer.

6.2.4.2 Main Types

Historically, low-molecular-weight fractions ("tails") that were present in polyolefin resins served as processing aids. When narrow-molecular-weight PP and PE became standard in the industry as a result of improved catalyst technologies, the processability of these polymers was adversely affected in certain applications. This led to the development of processing aids outlined below.

Fluoropolymer-based processing aids

Fluoropolymer-based processing aids (Figure 6.7) were added to PE for the first time in the early 1960s. This material is commonly described as a "fluoroelastomer", even though in its role as a processing aid, it is not crosslinked and therefore does not actually exhibit elastomeric behavior. This type of fluoropolymer has served as the main processing aid for polyethylene for more than 50 years, and is still widely used. It can be incorporated either as a single additive or as part of processing aid blends, in combination with other polymers and/or additives.

$$\begin{array}{l} -(CF_2\text{-}\underset{\displaystyle CF_3}{\underset{|}{CF}})_n\text{-}(CF_2\text{-}CH_2)_m\text{-} \end{array}$$

Figure 6.7
Structure of fluoropolymer-based processing aids

Silicone-based processing aids

Poly(dimethylsiloxane) or PDMS oil (Figure 6.8) and other silicon-containing polymers have been used as polymeric processing aids in polyolefins since the early 1970s. These additives, often referred to as "silicone processing aids" (SPAs), are incompatible with polyolefins such as PE and PP, and therefore tend to migrate or bloom to the surface. For optimal and sustained performance, they should be well-dispersed throughout the base polymer. In the past, this blooming behavior limited their use in applications where the finished parts needed to be printed, painted on, or used in contact with food. Newer SPA grades exhibit lower mobility within the polymer matrix, and therefore no longer impair printability or paint adhesion.

$$-(O\text{-}\overset{\displaystyle CH_3}{\overset{|}{\underset{\displaystyle CH_3}{\underset{|}{Si}}}}\text{-}O)_n\text{-}$$

Figure 6.8
Structure of silicone-based processing aids

6.2.4.3 Performance Attributes

The primary use of processing aids in polyolefins is in LLDPE blown film. Other typical applications for these additives include pipes and tubes, fibers, blow-molded parts, and profile extrusions produced from LDPE, HDPE, metallocene-catalyzed LLDPE, PP, thermoplastic elastomers, polystyrene, polyamides, acrylic polymers, PVC, and related materials. To obtain optimal dispersion of processing aids in polyolefins, the dispersed additive domains in the matrix should generally be kept below about 2 µm in diameter.

The main performance advantages are outlined below.

Melt fracture elimination

Historically, eliminating melt fracture has been the main motivation for using fluoropolymer-based processing aids. In the absence of such additives, linear polyethylene grades develop melt fractures on the surface when they exit the die above a certain critical shear stress. This defect appears as surface roughness, "sharkskin", and related imperfections on the extrudate. When processing aids are incorporated, the polymers can be extruded at higher shear rates without exhibiting melt fracture.

Reduction of plate-out in the die

During extrusion, deposits can build up on metal surfaces both at the die exit and along the internal die walls where the polymer flows. Processing aids help suppress this die build-up by forming a thin layer between the metal surface of the equipment and the polymer melt. As this coating develops, the processing aid dislodges and replaces degraded material adhering to the metal.

Once a fluoropolymer-rich layer has formed at the die exit, the emerging melt is less prone to swelling and attaching to the die surface. Within the die, the processing aid establishes and regulates a slip velocity at the polymer–metal interface, which reduces stagnant zones in the flow and lowers the risk of thermal degradation associated with excessively long residence times.

Improved production capacity

This improvement encompasses increased process output and reduction in equipment maintenance and clean-outs.

When using a processing aid, the resistance to the polymer melt from the extruder die is reduced. Consequently, both die pressure and the torque required to drive the extruder motor are reduced. These reductions indicate, and also contribute to, the enhanced productivity.

To assess improvements in productivity, comparisons were made between extrusion processes run with and without a processing aid under the same conditions of temperature and production rate. One measurable outcome of these comparisons was the decrease in energy consumption by the extruder motor (Figure 6.9).

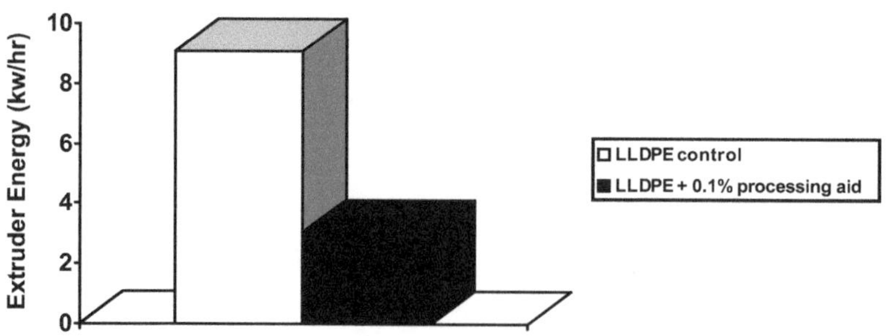

Figure 6.9 Reduction in extrusion energy after adding fluoropolymer processing aid to LLDPE

Figure 6.10 compares the extrusion throughput of LLDPE with and without a fluoropolymer processing aid under identical extrusion pressure and temperature conditions. When the processing aid was incorporated and the extrusion rate was increased accordingly, the process throughput rose by about 83% compared to the unmodified reference.

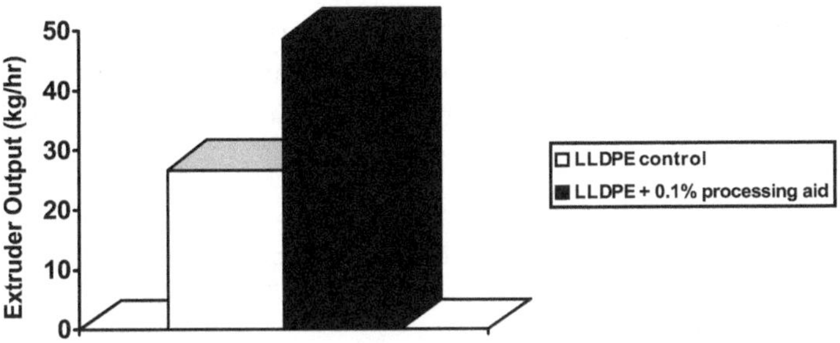

Figure 6.10 Output increase after adding fluoropolymer processing aid to LLDPE

Reduction in gel formation

Reduction in unmelted or poorly mixed gels is linked to avoiding break-up of the solid bed in the feed section of the extruder, since such interruptions can cause inadequate mixing of the still-unmelted particles further downstream. Fluoropolymer processing aids can help delay premature melting in the feed zone, allowing the material to move as a continuous, coherent solid bed. This type of gel formation is particularly associated with metallocene-catalyzed LLDPE.

6.2.4.4 Performance Effects

In pipe and tubing extrusion, processing aids are commonly used to reduce the apparent viscosity of the melt, eliminate die build-up, and achieve a smoother, glossier surface on the final article. These additives have proven effective in HDPE, PP, and LDPE for such purposes. Fluoropolymers and other processing aids also find use in extrusion and blow-molding of HDPE, where they help eliminate melt fracture, boost surface gloss and melt strength, and enable processing at lower temperatures. Silicone-based processing aids are used in injection molding to enhance mold-fill efficiency, lower the melt temperature, ease the mold release, and increase scratch and mar resistance of molded parts.

Additive interactions

Other additives may affect the performance of processing aids, by:

- Becoming involved in interactions on the surface (due to adsorption)
- Removing the processing aid from the die

- Reacting chemically with the processing aid
- Competing with the processing aid for metal surfaces.

Solid antiblocking agents and pigments that have an affinity for the processing aid can interfere with its action by adsorbing onto the equipment surfaces. When the solids have a high specific surface area or strong affinity with the processing aids, the effective concentration of the aid in the polymer matrix can decrease, diminishing its performance. In practice, such interactions often make it necessary either to increase the dosage of the processing aid to compensate for the loss in efficiency or switch to a different type of processing aid.

On the other hand, the effect of processing aid on the anti-blocking agent or pigment may rather be positive, since the processing aid may facilitate better dispersion of these additives in the polymer matrix.

Using additive concentrates that include processing aids can significantly reduce the effectiveness of those aids. This effect is observed with many inorganic additives, and it is particularly pronounced in the presence of silica or talc.

Hindered amine light stabilizers (HAS or HALS; see Chapter 3) are likewise known to interact with processing aids. It is still not fully understood whether this is driven mainly by an acid–base reaction, by competitive adsorption on the surfaces, or by a combination of the chemical and physical effects. Such interactions can be mitigated by operating the process at lower temperatures.

6.3 Impact Modifiers

6.3.1 Polymer Fracture and Functions of Impact Modifiers

In order to toughen a brittle polymer by adding an elastomeric phase, the rubbery modifier must interact well with the matrix, but not to the point that it dissolves in it like a plasticizer. In thermoplastics, two principal deformation mechanisms control how the matrix can be toughened: (a) crazing and (b) shear-yielding.

In practice, both mechanisms can act concurrently during a polymer fracture. When the matrix is crystalline, failure tends to proceed via crazing. In this case, finely dispersed, discrete rubber particles are needed, so that they can absorb impact energy and arrest the advance of craze-initiated cracks. Ductile amorphous polymers are deformed mainly by shear-yielding. For these systems, effective toughening requires that the yield stress is reduced before crack propagation can occur. This is typically achieved by uniformly dispersing a large number of rubbery particles, ranging in size from 0.1 to about 4 µm, throughout the polymer.

Crazing occurs when a polymer subjected to a sudden load forms tiny voids or microcracks aligned roughly perpendicular to the direction of the applied stress. The faces of the microcracks grow apart from each other, while remaining connected by numerous fine fibrils that bridge the gap. The resulting fibril-stabilized microcracks are referred to as crazes. The fibrils continue growing as long as sufficient concentration of molecular entanglements is maintained. Typical diameter of the fibrils is about 20 nm (Figure 6.11).

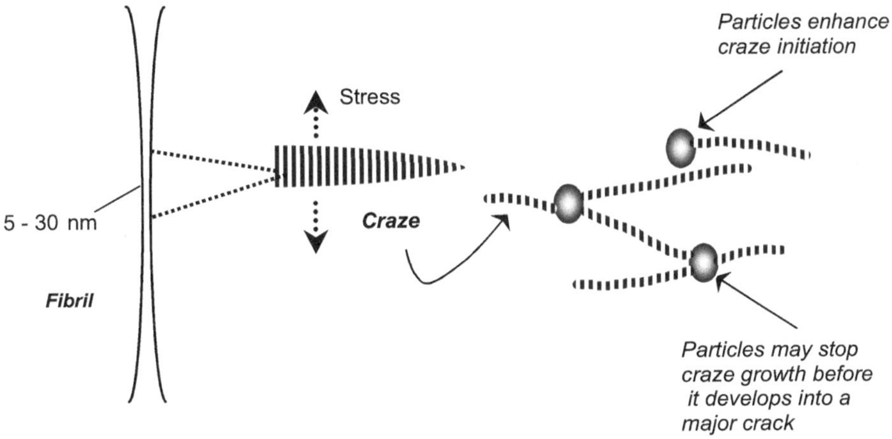

Figure 6.11 Schematic representation of crazing in a toughened polymer matrix

The initial formation of the craze faces and subsequent fibril drawing constitute energy absorption mechanisms that help the polymer to sustain the applied load and maintain craze enlargement under control.

However, if the load is large enough, the faces of a craze may continue developing up to the point where a crack forms, propagates through the material, and leads to rapid crack growth and brittle failure of the material. Several growing crazes may also link up and produce a fast-growing crack.

Shear-yielding occurs when the polymer responds to the impact load by deforming and flowing. This deformation process is oriented in a direction that is approximately 45° to that of the crack plane, absorbing large amounts of energy without creating microvoids or microcracks. As a result, the material volume changes very little, in contrast to crazing. Shear-yielding tends to occur in more ductile semicrystalline thermoplastics, such as polyamides, polyesters, PP, and PVC, whereas bisphenol A polycarbonate is a notable example of a glassy polymer.

Generally, the energy absorbed by plastics when failing in a shear-yielding mode is larger than that absorbed by crazing.

Therefore, control of crazes is required before they develop into cracks, and it is accomplished with the use of particles large enough to arrest craze growth.

Shear-yielding can be better controlled by incorporating small, rubbery modifier particles, typically well below 1 µm in diameter. The efficiency of such toughening agents depends on their particle size. For PVC, for instance, the optimum impact performance is often achieved with particles about 200 nm in size, where smaller particles (below roughly 200 nm) result in crazing being the dominant energy-absorbing mechanism. Larger particles result in a shift to shear-yielding as the mechanism for absorbing impact energy.

Impact resistance (or impact strength) describes the ability of a material to withstand a sudden, high-rate load or shock without cracking or breaking. It is commonly quantified as the energy required to fracture a specimen under standardized impact test conditions.

Impact modifiers are designed to enhance or improve the toughness, thermoplasticity, processability and use of thermoplastics. The toughness of inherently brittle polymeric materials can be effectively increased by lowering their modulus through the incorporation of elastomeric modifiers. These modifiers promote energy absorption throughout the material by altering the local stress distribution and triggering better-controlled deformation.The effectiveness of impact modification depends on several factors:

- Type of plastic
- Type and rate of impact stress/load
- Temperature and other process conditions
- Type and amount of the impact modifier.

The viscoelastic nature of polymers makes the rate of impact loading a key variable that differentiates them from metals. Polymers respond to low loading rates in a more ductile fashion, while high load rates result in more brittle behavior.

6.3.2 Classification and Architecture

The impact strength of rigid PVC can be increased by incorporation of an elastomeric phase into the PVC matrix. Impact modifiers can be classified into the following three types according to the architecture of the elastomers and their chemical composition. If brittle fracture is to be prevented, the energy must be transferred to the added elastomeric phase.

Type I: Core–shell copolymers with a rubber core and a polymer shell (Figure 6.12). Typical examples are methacrylate–butadiene–styrene (MBS), acrylic (also known as all-acrylic impact modifiers or AIMs), acrylonitrile–butadiene–styrene (ABS), and acrylic–styrene–acrylonitrile (ASA).

Modifiers with high rubber content impart greater toughness (or impact resistance) when added to brittle polymers, provided they are compatible with it. Hardness, tensile, and flexural strength decrease somewhat as impact strength increases.

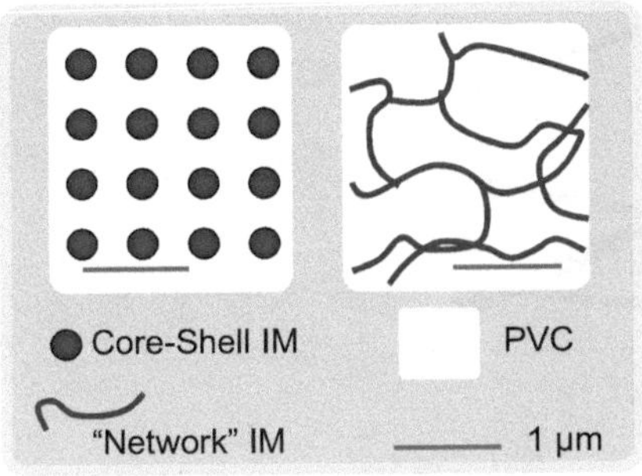

Figure 6.12
Impact modifiers with core–shell and network structures

Type II: Amorphous (co)polymers with a high degree of compatibility with the modifying polymers. These kind of impact modifiers are also called network polymers (NPs) (Figure 6.12). The main examples of NPs are chlorinated polyethylene (CPE), ethylene vinyl acetate (EVA), and acrylonitrile butadiene rubber (NBR). They are more shear-sensitive and effective within a narrower processing window compared to Type I. Type II impact modifiers encapsulate the primary particles during processing in such a way that they form a honeycomb-type network within the PVC matrix.

Type III: Inorganic modifiers of small particle size and large surfice. Examples are precipitated calcium carbonate (PCC; see also Chapter 12), certain synthetic alumo-silicates, spherical silicon dioxide, and fly ash.

6.3.2.1 Core-Shell Impact Modifiers

AIMs are produced by emulsion polymerization. Crosslinking agents are added to form the crosslinked rubbery phase of the core, which absorbs the impact and generally has a T_g of −45 to −60 °C. The grafted shell serves two purposes: it prevents the AIM particles from sticking to each other, and enables their dispersion and compatibility with polymers, such as PVC.

AIMs are suitable for outdoor, weatherable applications. With some exceptions, these impact modifiers are not suitable for applications where transparency is required.

In addition to increasing the impact resistance at optimum loadings, AIMs also reduce tensile strength and increase melt viscosity and cost of the compounds (Figure 6.13).

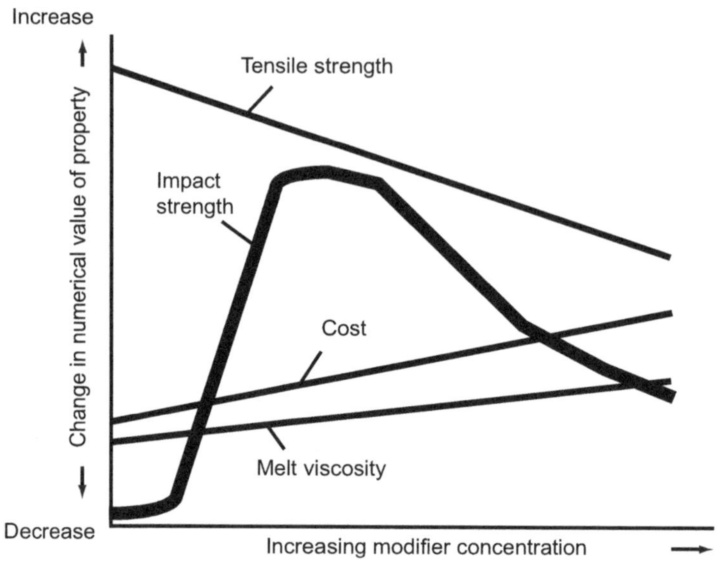

Figure 6.13 Effect of the AIM loading on cost–performance characteristics

MBS impact modifiers are produced by emulsion polymerization. It was established experimentally that the optimal shell thickness for these core–shell particles is typically in the ranges 4.2–9.8 nm, imparting high efficiency in terms of increased toughness. A significant benefit of MBS modifiers is the ability to adjust their refractive index during the production process to make them comparable to that of PVC, and even enabling certain grades to be employed in transparent PVC compounds without compromising clarity. The T_g of MBSs varies from −55 °C to −70 °C. This is notably lower than that of AIMs, making them particularly well-suited for PVC compounds designed for use at low temperatures. However, MBSs exhibit limitations for use in outdoor/weatherable applications. Commercially, MBSs are predominantly used to improve the toughness of PVC in clear packaging film and to modify performance of other polar thermoplastics.

ABS copolymer impact modifiers are similarly synthesized via emulsion polymerization. With a low T_g of about −70 °C, ABSs are effective for low-temperature uses. Transparent ABS grades are used for increasing the impact resistance of clear PVC bottles and films. Beyond imparting toughness to rigid PVC, ABSs improve processability and melt strength, facilitating effective thermoforming and increasing the chemical resistance needed for food packaging. In plasticized PVC, it boosts hot strength, embossing retention, and low-temperature flexibility of the compounds.

ASA, like ABS, is compatible with PVC, PC, SAN, and polyesters, imparting excellent heat and UV-light stability. In PVC, ASA is used in extruded and molded items where resistance to UV light and heat is essential. It is also used in flexible PVC films produced by extrusion and calendering..

6.3.2.2 Network-structured Impact Modifiers

Chlorinated polyethylene (CPE) is obtained by chlorinating high-density polyethylene (HDPE), a process that reduces the crystallinity of the original polymer. The resulting CPE typically contains about 25–50% chlorine, with around 36% generally regarded as an optimal level for impact modification. At higher chlorine contents, CPE becomes too compatible with PVC and behaves more like a strongly bound plasticizer, while at lower chlorine levels, both its compatibility with PVC and its impact efficiency decrease. The T_g of CPE is about -16 °C (compared to about -45 °C for AIMs). CPE forms a network-like structure within the PVC matrix. It is more easily deformable than PVC itself, which enables it to elongate substantially before fracture. It acts as an effective impact modifier due to its high flexibility at low temperatures, very low crystallinity, and effective impact energy dissipation required for protecting the rigid PVC phase. Both CPE and AIMs are suitable for weatherable including low-temperature outdoor applications.

The advantages of CPE-based impact modifiers are a combination of adequate impact resistant performance, good processability, and relatively low cost. However, compared to AIMs, CPEs are typically required to be used at higher loadings to achieve equivalent impact resistance; they also are sensitive to the presence of zinc-containing components (so heat stabilizers used in combination with CPE must have low zinc content).

Ethylene–vinyl acetate (EVA) copolymers are produced by high-pressure copolymerization of ethylene with vinyl acetate. They can be used in transparent PVC compounds and to a limited degree in weatherable formulations. For any given EVA grade, the specific weight ratio of the monomers determines the balance between T_g, compatibility with PVC, and crystallinity, and thus defines the compromise between toughness, clarity, and mechanical rigidity.

6.3.2.3 Inorganic Impact Modifiers

Since the early 1990s, ultrafine precipitated calcium carbonate (UFPCC) has been known to impart significant improvements in the impact strength of rigid PVC. The additive has been further optimized in regard to particle size, shape, and coating. It was found that only milled and surface-treated calcium carbonate improves impact strength up to a dosage of 10–20 phr (Figure 6.14). The increase in impact strength is a result of the formation of microcracks (crazing) in the immediate vicinity of the calcium carbonate particles.

The finer the particles of the impact modifiers, the more impact energy is absorbed before craze formation, reducing the likelihood of a macroscopic fracture.

In addition to increasing the impact strength, the calcium carbonate impact modifier grades added at 10–20 phr increase the modulus of elasticity and density of the PVC profile, but these effects come with potential for surface erosion of PVC articles under weathering conditions.

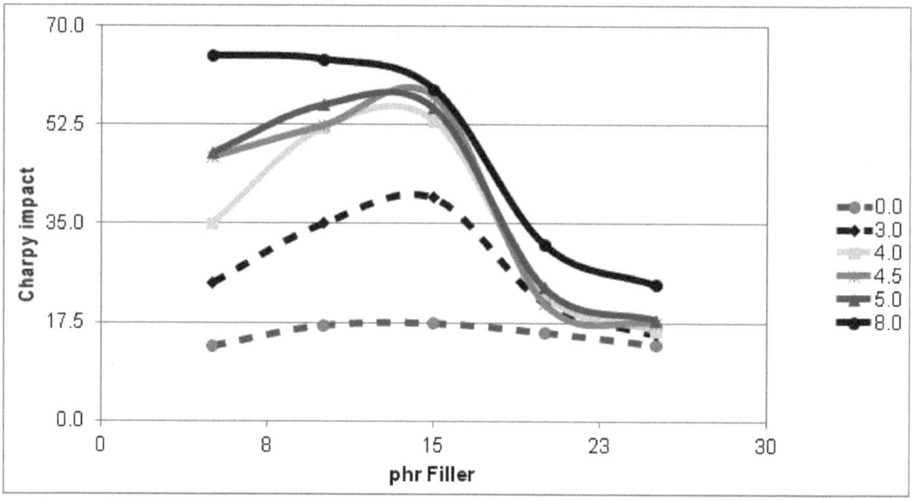

Figure 6.14 Notched Charpy impact strength as a function of the level of milled, surface-treated calcium carbonate filler; the dosages of the acrylic impact modifier were varied from 0 to 8 phr (key at right)[3)]

An amorphous, spherical silicon dioxide additive marketed under the trade name Sidistar™ has been shown to act as an inorganic impact modifier and processing aid in PVC. Its particles have a hollow, core–shell-like morphology with an inorganic outer phase and an empty middle. In rigid PVC, this additive can partially or completely replace conventional polymeric impact modifiers, delivering comparable impact strength while simultaneously increasing stiffness, improving melt flow, and raising softening temperatures. Achieving a very fine and uniform dispersion is crucial when using this additive. In practice, this is achieved by adding the silica early in the mixing sequence, before other components are introduced. Once well-dispersed, the additive not only contributes to impact modification itself but also helps to better disperse the other components. Typically, formulations that contain this type of impact modifier exhibit a broader and more forgiving processing window.

[3] M. Schiller, A. Sahin, B. Aldanmaz, and A. Elitez, *Rubbers Fibres Plastics International* (RFP), March 2017, p.188.

6.3.3 Performance Attributes

The main performance attributes of impact modifiers for applications in PVC are summarized in Table 6.2.

Table 6.2 Performance Attributes of Impact Modifiers for PVC

Impact modifier type	Typical use level (phr)	Performance attributes	Limitations
AIM	4–8	High impact efficiency High-temperature processing Weatherability Wide processing window Glossy surface	Transparency
MBS	6–10	High impact efficiency Crease-whitening resistance Transparency Improved low-temperature performance	Weatherability
ABS	2–6	High impact efficiency Crease-whitening resistance Low temperature impact Chemical resistance	Weatherability Transparency
ASA	4–8	Weatherability	
CPE	1–5	Moderate impact efficiency Weatherability Transparency Low cost Low temperature impact	Narrow processing window Sensitive to zinc
EVA	6–10	Transparency	Weatherability
UFPCC	5–10	Low impact efficiency	Surface erosion

6.3.4 Effect of Fusion and Process Temperature

The degree of fusion of PVC compounds strongly effects both the effectiveness of impact modification and its measurements. Therefore, care should be taken when comparing impact results for the specimens produced under identical or differing processing conditions. At processing temperatures above the optimum range, PVC can become over-fused, which promotes crack growth without sufficient energy-absorbing mechanisms, and can lead to melt fracture that is visible as rough surfaces on extruded articles. In a twin-screw extruded sheet, impact strength increased markedly as the melt temperature was increased to the 190–203 °C range, reaching a maximum of about 195 °C (Figure 6.15). This impact modifier-containing sheet was found to have a rough surface when extruded at temperatures above 205 °C.

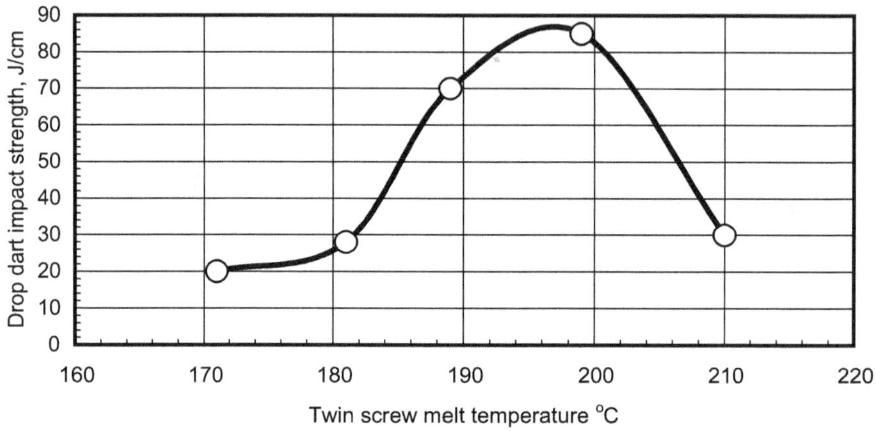

Figure 6.15 Impact strength of the extruded sheet over melt temperature (the compound also contained 1 phr calcium stearate and 1 phr paraffin wax)

6.3.5 Use of Impact Modifiers

6.3.5.1 Use in PVC

PVC, as a compounded homopolymer, possesses a certain degree of impact resistance. However, this inherent toughness is often inadequate for many applications, therefore additional impact modification is needed. Processes such as grafting or copolymerization can be used to lower its T_g and increase toughness, but these methods tend to reduce the material's heat deflection temperature. In most cases, rigid PVC compounds are made tougher by adding small amounts of impact modifiers, which increase its impact resistance without noticeably changing other essential performance characteristics.

In order to achieve a specific impact strength, the higher the molecular weight of PVC, the less impact modifier of any specific type is needed.

For a given application, formulators generally prefer the most efficient impact modifier, provided that other key properties remain unchanged. Using higher dosages of any additive can unintentionally alter certain characteristics of the compound. In rigid PVC, impact modifiers are commonly incorporated into window and interior profiles, sidings, foamed profiles, thin-wall small diameter pipes, electrical conduits, highly filled drain and sewer pipes, irrigation pipes, rigid sheets, and calendered films.

For rigid PVC used in transparent bottles, packaging films, and the outer layers of various cards, high optical clarity is essential. Under mechanical stress such as sharp folding or tight bending, certain clear PVC grades develop a white crease. These materials are described as exhibiting undesirable "crease-whitening". Transparent PVC compounds can be formulated either with modifiers that strongly suppress the crease-whitening effect or with types that are more prone to it. MBS and ABS impact modifiers are widely used in clear PVC, because they offer a good balance of toughness, transparency, and resistance to crease-whitening. In opaque rigid PVC, the appearance of creases is less critical. Therefore, at least from the standpoint of crease-whitening, essentially any established impact modifier family can be employed here.

In plasticized PVC, impact modifiers are selected for applications that must maintain flexibility and are tough at lower temperatures, especially when the plasticizer content is relatively low. Their effect is particularly beneficial in formulations containing polymeric plasticizers that are chosen for their permanence, low volatility, and reduced migration. Typical end uses that can take advantage of such modified plasticized PVC compounds include protective flooring, chair mats, automotive interior trim (such as instrument panel skins), durable book covers, and wear-resistant films.

6.3.5.2 Use in Non-PVC Polymers

In addition to PVC, impact modifiers are used in engineering polymers known for their mechanical strength and resistance to heat and solvents (Table 6.3). The melt-processing temperatures of these non-PVC polymers are generally higher than that of PVC.

Table 6.3 Impact Modifiers and Their Use Level (%) in Non-PVC Polymers[4]

Type	PBT	PET	PC	PB/PC	PA	Performance attributes
AIM	15–25	15	2–5	10	30	High-temperature processing, good aging
MBS	15–20	15	2–5	10	—	Improved heat stability, low temperature impact

With the exception of polyacetal (POM), most non-PVC polymers have processing temperatures of above 200 °C. For those polymers, the impact modifiers should be thermally stable up to 300 °C.

[4] J. Edenbaum, *Plastics Additives and Modifiers Handbook*, Van Nostrand Reinhold, 1992, p. 583.

Due to their relatively low cost compared to the non-PVC polymers, impact modifiers can also be used as extenders for reducing the overall formulation cost of other polymers, in addition to their primary function as toughening agents. In more recent developments, ethylene/propylene rubbers (EPM and EPDM) have become the main tools for enhancing the impact strength of polypropylene. Blends of PP with EPDM are widely employed in manufacturing automotive parts, such as bumpers, with typical modifier contents in the range of about 10–40%.

6.3.6 Impact and Toughness Testing

For impact modifiers, their defining characteristic is the improvement of impact resistance or strength. There are several methods used for testing impact strength. However, none of the methods described below correlates with any other. Furthermore, both notched tests like Charpy and Izod measure the impact strength after mechanical damage, while the falling weight test records the impact strength without causing any damage.

6.3.6.1 Charpy and Izod Impact Tests

For the Izod impact strength measurements, the testing equipment consists of a pendulum with a hammer that is released from a specific height, having to deliver specified impact energy. The test specimen with a single notch is clamped into a solid block, ready for "the hammer to fall". Once the hammer is released, it hits the sample on the side with the notch and breaks it (Figure 6.16). Based on the energy absorbed by the sample without the break, impact energy is calculated. A similar procedure is used for the Charpy impact test (Figure 6.16). In this case, the absorbed impact energy is correlated to the notch toughness.

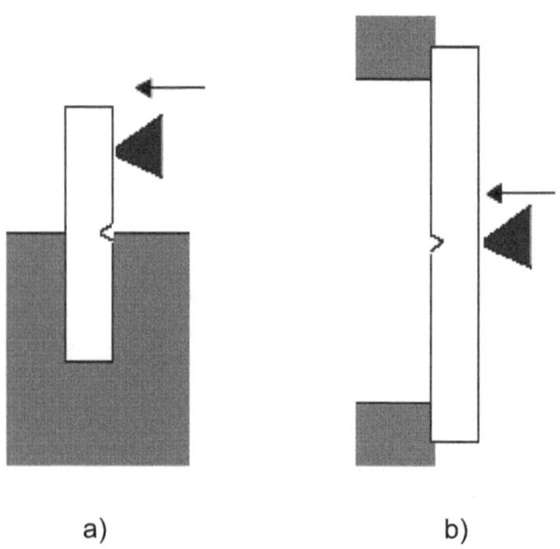

Figure 6.16 Specimen configuration in (a) Izod and (b) Charpy notched impact tests

6.3.6.2 Falling Weight Tests

For the free-falling dart impact test, the plastic specimen is placed on a base plate over an opening. An impactor sits on the top of the specimen and is in contact with the unsupported center of the specimen. A specified weight is raised inside a guide tube to a predetermined height, and then it is released/dropped onto the impactor, forcing the impactor nose through the test specimen. The drop height, drop weight, and the test result (pass/fail) are recorded.

A different principle is used in Gardner's free-falling dart impact test, which is a common method for evaluating the impact strength of plastic materials. The plastic specimen is placed on a base plate over an opening. An impactor sits on the top of the specimen and is in contact with the unsupported center of the specimen. A specific weight is raised inside a guide tube to a predetermined height and then released to drop onto the impactor, which forces the impactor nose through the test sample. The drop height, drop weight, and the test result (pass/fail) are recorded.

A variation of this method, useful for a rapid qualitative assessment of impact resistance, is to drop a weight from a defined height onto a (window) profile specimen at room temperature or at sub-zero temperatures, and note the result (pass/fail).

In Australia, PVC pipes are tested for their toughness in a similar manner according to AS-NZS 4765:2007 by using a torpedo-shaped drop weight of up to 20 kg, hitting the pipe from a height of 20 m. In this test, the 20 kg torpedo always penetrates the pipe. According to this standard, the pipes must fail in a totally ductile manner, which is not to crack or split around the impact hole.

6.3.6.3 Bottle-drop Impact Tests

Bottle-drop impact tests (ASTM D2463) are used to determine the maximum height from which a filled bottle can be dropped and remain intact. Both contact angles and impact plate angles can be varied to accommodate different bottle geometries.

6.4 Questions

Processing aids

1. What are the main types of processing aids used in PVC?
2. What monomers are commonly used for manufacturing processing aids for PVC?
3. How are the processing aids classified based on their molecular weight?
4. What is the mechanism or mode of action of processing aids used in PVC?

5. What is the typical composition of lubricating processing aids for PVC?
6. How do processing aids affect the fusion time of PVC compounds?
7. What are the main benefits of using processing aids in PVC extrusion?
8. What are the main benefits of using processing aids in PVC injection molding?
9. What is the mechanism of action of processing aids in polyolefins?
10. What are the main types of processing aids used in polyolefins?
11. What are the performance attributes of processing aids used in polyolefins?

Impact modifiers

12. What are the two main fracture modes in polymers?
13. What are the three commonly distinguished types of polymer modifier?
14. What are the main functions of impact modifiers in plastics?
15. What are the key performance attributes typically imparted by impact modifiers?
16. What types of impact modifiers are weatherable? What are their applications in PVC?
17. What types of impact modifiers are suitable for transparent applications?
18. What are the purposes of using impact modifiers in rigid PVC compounds?
19. What performance attributes are imparted by impact modifiers on plasticized PVC?
20. What is crease-whitening? How can it be reduced or prevented?
21. In what non-PVC polymers are impact modifiers used and why?
22. What are the standard test methods for measuring the impact resistance of polymers?

7 Lubricants and Lubricators

Content of this chapter was adapted from:

- E. Richter, "Lubricants" (Chapter 5), in: *Plastics Additives Handbook* (6th edition), ed. H. Zweifel, R. D. Maier, and M. Schiller, Hanser, 2009, pp. 539–580.
- T. Kromminga and G. Van Esche, "Anti-blocking Additives" (Chapter 7), in: *Plastics Additives Handbook* (6th edition), ed. H. Zweifel, R. D. Maier, and M. Schiller, Hanser, 2009, pp. 613–628.
- F. Wylin, "Slip Additives" (Chapter 8), in: *Plastics Additives Handbook* (6th edition), ed. H. Zweifel, R. D. Maier, and M. Schiller, Hanser, 2009, pp. 629–635.

7.1 Introduction

7.1.1 Lubricants for Plastics Processing

The processing of polymers occurs at temperatures above their crystalline melting points or, in the case of amorphous polymers, above their glass transition temperatures. The macromolecular structure of these materials results in high melt viscosities. Processing these polymer melts, e.g., conveying or shaping them, requires the application of pressure, and the resulting shear leads to dissipation of mechanical energy. At points of high flow resistance in particular, the heat generated can accelerate polymer degradation reactions or adversely affect the dispersion of multiphase systems.

The possibilities for optimizing the processing properties of plastics through appropriate monomer selection, molecular weight control, and macromolecular structure are limited. The primary considerations in solving problems involve not only the selection of the combination of machine and mold, but also lubricants that affect the rheology of the polymer melt in the desired way.

Besides their direct results on melt rheology, substances commonly used as lubricants have other beneficial effects on the properties of compounds and finished articles. For example, the lubricating action of these substances on machine parts and the mold results in finished articles with smooth, glossy surfaces, and the interfacial properties of certain products facilitate dispersion of fillers and pigments.

Most plastic lubricants are waxes; most of the non-waxy lubricants are fat or oil derivatives.

7.1.2 Lubricant Definitions

For polyolefins, the term lubricant means an additive that reduces the frictional resistance between the surfaces of finished parts at service temperatures.

Terms often used in connection with lubricants require clarification, with the commonly used terms internal lubricant and external lubricant being simplified descriptions of their different effects. Internal lubricants are incorporated into the polymer. In some instances, external lubricants can be applied directly onto the metal surfaces of the processing machine or the mold to prevent the polymer melt from sticking.

Polymeric processing aids for PVC are discussed in Chapter 6. The synonymous term "flow aids" indicates that these substances overlap with lubricants in their effects. Their extremely high molecular weight and different modes of action distinguish them from lubricants.

7.1.3 Commercial Viability of Lubricants in Plastics

Rigid PVC processing, for example, is impossible without lubricants. With other plastics, lubricants are often added only to certain grades, depending on the processing method or desired surface effects. Polymer blends (alloys), a rapidly growing segment of the plastics industry, as well as reinforced and filled thermoplastics, require relatively high lubricant concentrations.

The distribution of lubricant consumption among the most commercially viable plastics (Figure 7.1) shows the overriding demand by the PVC processing industry. The "miscellaneous" group, which includes engineering thermoplastics in particular, is also substantial. Lubricant consumption is expected to grow in the future, although at a slower rate than the overall plastics market.

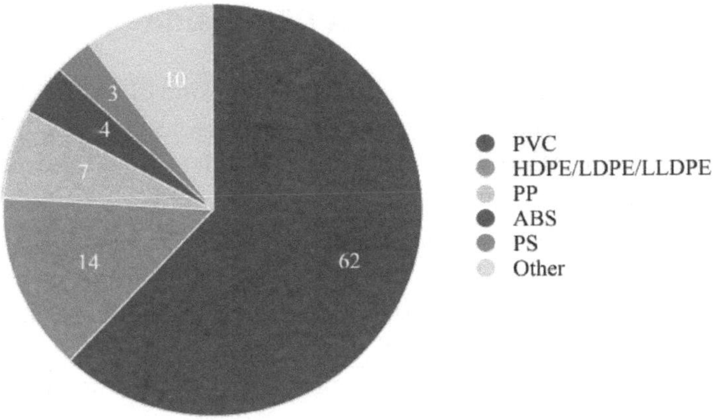

Figure 7.1 Worldwide lubricant consumption by type of plastics

7.2 Effects of Lubricants

7.2.1 Understanding the Lubricating Effect

Theories concerning how lubricants work have been primarily developed from studies on PVC. To summarize, easily soluble lubricants are known as internal, melt viscosity-reducing lubricants, while nearly insoluble ones are referred to as external lubricants. Paraffin waxes are consequently classified as external lubricants for PVC. In non-polar polymers, the situation is reversed. Hydrocarbon waxes, for example, are readily soluble in PE and function there as internal lubricants, while polar esters are incompatible with PE and hence, should be referred to as external lubricants.

> The higher the internal lubricating effect is, the higher the resulting maximum flow rate (assuming other processing parameters remain constant).

External lubricants provide a sliding film, a boundary layer, between the melt and the equipment wall. Even internal lubricants have a solubility limit beyond which its undissolved portions may act as an external lubricant. To achieve balanced processing properties, different lubricants are often combined. In discussing the nature of the lubricating effect, the term "compatibility" is preferred to "solubility".

7.2.2 Internal Lubricating Effects and Heat Dissipation

A broadly accepted definition of internal lubrication may be the sum of effects that change the rheological processes inside the polymer phase. Classification of lubricants according to the ratio of internal and external effects provides useful information for formulating PVC compounds.

The significant effects exerted by these lubricating additives on melt rheology, despite their low concentrations, can be theoretically explained as follows: lubricants do not coat the whole surface of the PVC macromolecules, but only much smaller parts of it. Typical viscosity-reducing lubricants for polar plastics are the fatty acid polyol partial esters. Similar effects are achieved in polyolefins by adding long-chain hydrocarbons, such as PE and PP waxes.

Energy dissipation can be defined as a conversion of mechanical energy into heat. High-speed processing equipment exerts high shear forces on polymer melts.

The higher the melt viscosity, the more mechanical energy is converted into heat (dissipated). This can lead to local overheating or undesirable "shear-burning" of the polymer.

Typical lubricants reducing heat dissipation are montan wax acid esters (Section 7.4.6) and metal soaps (Section 7.4.5). Their longer-chain aliphatic groups are believed to prevent solvation of the polymer but have a marked lubricating effect on internal interfaces. These lubricants, like the viscosity-reducing internal lubricants, also exhibit good compatibility with the polymer, and at the concentration they are normally used, little or no clouding in the polymer occurs.

7.2.3 External Lubricating Effects

External lubricating effects determine the behavior of hot plastic compounds at interfaces with other materials, such as processing equipment.

7.2.3.1 Release Effect

The extent to which plastic melts adhere to hot metal surfaces is of great importance in many processes. At low temperatures, adhesion of the melt to the screw and barrel of the processing equipment promotes plastification. Adhesion to the barrel wall is essential for the conveying action of single-screw extruders. Similarly, adhesion to the rolls is vital for calendering operations. Suitable lubricants can counter an excessive increase in adhesion with rising temperature.

Lubricants with a positive release effect can be chemically characterized by polar groups in the molecule. Hydrocarbon waxes, which are incompatible with polar plastics, may be classified as typical external lubricants that have only a moderate release effect. This underscores the need to differentiate lubricating effects more precisely than the general terms "internal" and "external" allow.

Reduction of the adhesive force between the polymer melts and solid surfaces is called the release effect. Certain lubricants can change the dynamic friction between the polymer melt and the walls of a processing machine.

Lubricants are also able to change the surface properties of finished articles. The smoothness and gloss of finished parts are determined by the processing behavior of the thermoplastic phase, and the lubricant type used in it. Sometimes, special effects are produced by the accumulation of lubricating additives on the surface.

7.2.3.2 Mold-release Effect

Mold-release effects are attributed to the demolding forces in the case of injection molding and similar processes that require ejecting a solidified part from the mold. The specific reduction of the demolding force is a high-priority target in many applications.

In processing reactive multicomponent polymer systems, it is still good practice to coat molds with release agents in the form of solutions, pastes, or emulsions. Among them, silicon-based chemicals are the most effective external anti-adhesives. However, residues potentially left on the surface of finished plastic articles may cause problems during further processing; for example, they may reduce the adhesion of paints or printing inks and impair bonding. For this reason, wax-based preparations are often preferred over silicon-based additives.

When injection molding technology became automated, there was a precondition: the demolding step had to be optimized by incorporating internal release agents so that the molded parts would uniformly eject in each cycle. Therefore, the forces required to eject the finished article needed to be as low as possible. Taking into account the cooling time thus saved, the resulting total cycle time was shortened.

Consequently, internal mold release agents must be very compatible with the polymer. It was also demonstrated that the demolding force reaches its maximum after several cycles. Therefore, the buildup of the release layer over time is critically important. Figure 7.2 illustrates the effects of different lubricants on the relative release force added to two polyesters (PET and PBT).

The number of additives that can meet these conflicting requirements is limited. Among them are fatty acids and wax acids, their soaps, amides and fully esterified polyol esters. Fatty acid amides, and especially bis(stearoyl) ethylenediamine (EBS or "amide wax"), have a tendency to exude during the storage of finished articles. For engineering thermoplastics, use of lubricants based on high-molecular-weight esters, such as pentaerythritol tetra-stearate, montan wax acid triglyceride, and other polyol esters is preferred. Hydrocarbon (paraffin) waxes that form lubricating films are no longer used as internal mold release agents; their metal-wetting effect is insufficient, and their residues may adversely affect the mold surface.

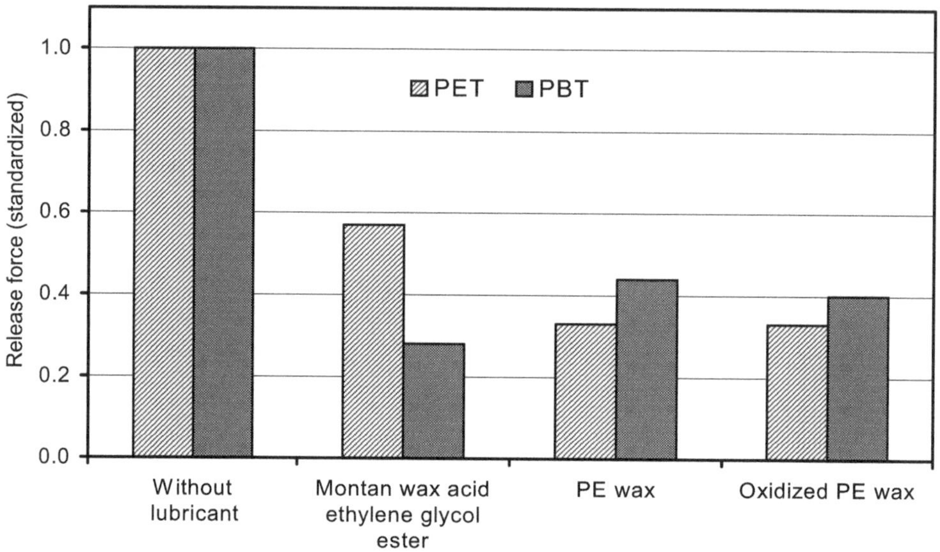

Figure 7.2 Release force as a function of different external lubricants added at 0.2% to two polyesters

7.2.4 Slip Effect

For certain applications, it is desirable for surfaces of finished parts to have good slip properties.[1] Polyolefin (and to a lesser extent PVC) films have a high coefficient of friction (COF), which leads to their tendency to adhere to themselves and to metal surfaces during processing. Polyolefin films with high COF can exhibit sticking when extruded without slip additives, resulting in wrinkled and unusable materials, and leading to lower production and a high scrap rate.

To be easily converted into bags or similar items, films must have a COF near 0.2.

---

[1] "Slip" refers to the ability of a surface to allow materials to move or slide across it without much resistance due to low friction.

Incorporation of slip additives prevents these problems. Slip additives modify the surface properties of polyolefin films and reduce the friction between layers of the film and between the film and other surfaces, with which they come in contact with. Lower friction also facilitates easier handling of the film in various packaging applications.

7.2.4.1 Slip Additive Action

As a requirement, slip additives must exhibit low compatibility with the polymer that facilitates the migration of the slip additive to the surface of the polymer film (see Figure 7.3).

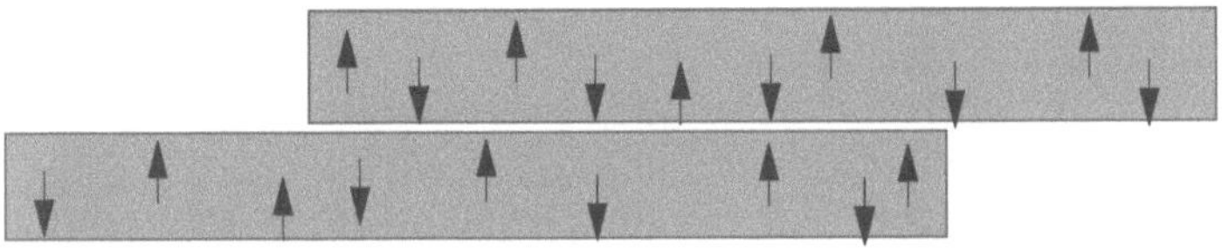

Figure 7.3 Migration of slip additives to the film surface

Fatty acid amide additives, such as erucamide and oleamide, function as slip agents in the production of polyolefin films. They migrate to the surface and form a very thin layer on the surface to reduce its friction shortly after exiting from the die. During processing, the fatty acid amides are solubilized in the amorphous melt, but as the polymer cools and begins to crystallize, the additives are squeezed out of the solidifying polymer matrix. Once on the surface, the slip additive forms a lubricating layer that effectively separates adjacent film layers. In addition to lowering surface friction, slip additives may also reduce surface resistivity and melt viscosity, and impart mold release and anti-block characteristics.

7.2.4.2 Slip Additives

The two most commonly used slip additives are erucamide and oleamide. Erucamide is derived from a mono-unsaturated C_{22} acid (erucic acid), while oleamide is the amide derived from a mono-unsaturated C_{18} acid (oleic acid). Oleamide is often referred to as the "fast-blooming" slip agent, because it migrates quicker to the film's surface than erucamide. However, although erucamide migrates slower than oleamide, after a certain period of time films treated with erucamide have lower COF than those treated with oleamide (Figure 7.4).

Due to its lower efficiency, stearamide, another slip additive, is used when the fatty acid amide also needs to provide an anti-block characteristic, and this is done primarily in combination with either oleamide or erucamide. However, most often, inorganic anti-block additives such as silica are used, with fatty acid amides being employed only in applications where the film's transparency is required.

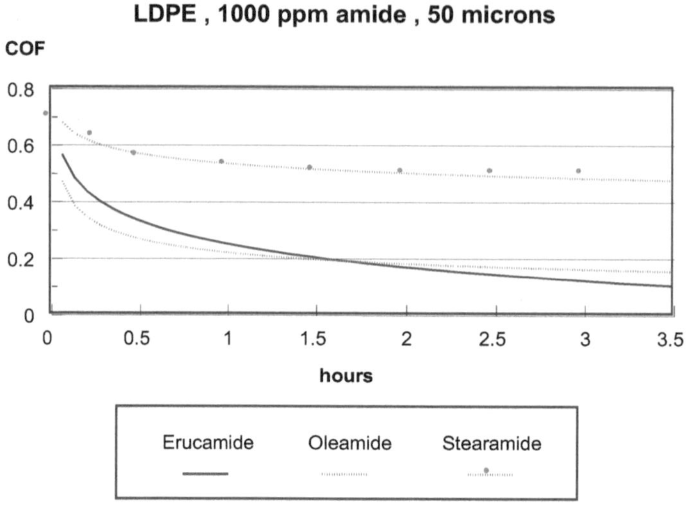

Figure 7.4 Reduction in COF as function of time resulting from the use of 1000 ppm amide slip agents in 50 μm LDPE

Erucamide is used most often in applications requiring high-temperature processing conditions, and is commonly incorporated in PP films. It has lower vapor pressure and therefore is less volatile than oleamide, which also means that it stays longer on the surface and does not vent off. Oleamide is often used for in-line bag converting operations, where low COF is required in a short period of time. It is commonly used in PE films. The structural formulae of erucamide and oleamide are shown in Figure 7.5 and Figure 7.6., and their main performance attributes are compared in Table 7.1.

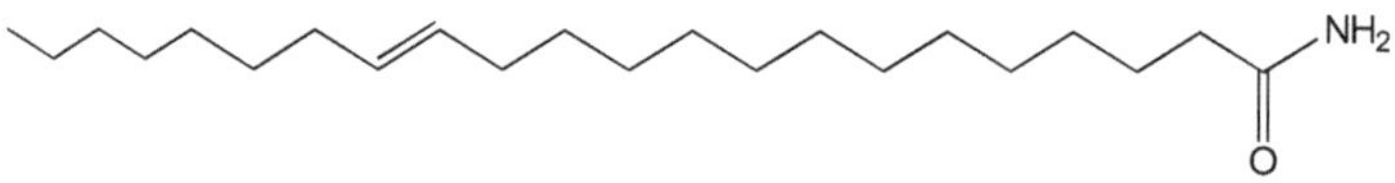

Figure 7.5 Chemical structure of erucamide

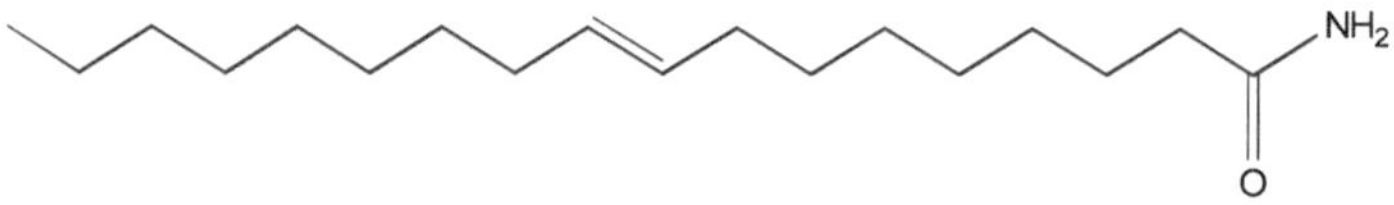

Figure 7.6 Chemical structure of oleamide

As well as the chemical composition of the slip additives, their concentration also affects the slip performance. The relationship between the COF imparted by the slip agents added at various concentrations is shown in Figure 7.7. Starting from a zero-concentration base, the COF drops sharply when the concentration of the slip additives increases. Further increases in slip additive concentrations result only in marginal decreases of COF.

Table 7.1 Comparison of Performance Attributes for Slip Additives

Oleamide	Erucamide
Faster migration	Slower migration
Lower cost-in use	Higher cost-in-use
More volatile	Less volatile
Animal-sourced	Vegetable-sourced
Higher final COF	Lower final COF

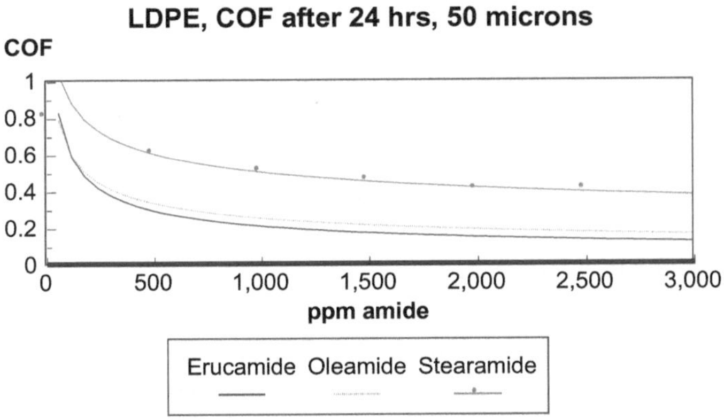

Figure 7.7 Effect of adding different concentrations of amide slip additives on COF of 50 µm LDPE after 24 h

7.2.4.3 Incorporation of Slip Additives in Polymers

For film producers, the use of erucamide or oleamide slip additive masterbatches is strongly recommended to achieve a uniform additive distribution at the lowest concentration levels. The masterbatches may also include an anti-block additive.

7.2.4.4 Effect of Other Additives

Slip additives, such as erucamide and oleamide, are in most cases used in concentrations between 500 and 1500 ppm. The exact test results depend on film thickness, the slip additive type used, and the presence of other additives. At typical concentrations, anti-block additives present in combination with slip agents may result in a slight further decrease of COF. However, high loadings of anti-block additives cause a very rough surface and will therefore increase COF. It is therefore recommended to always evaluate additive combinations based on end-use applications to determine their overall effect on COF.

Metal stearates may cause faster blooming of the fatty amides to the polymer film surface.

7.2.5 Anti-blocking Effect

The effect of anti-blocking agents is very similar to that of slip agents. These two groups of additives are often used in combination, and some special lubricants may also serve as anti-blocking agents.

PE or PP and other plastic films tend to stick together, often making it difficult to separate film layers. This is best illustrated by the common difficulties we experience when trying to open clear plastic bags used to package groceries. Problems also occur when plastic films are wound onto and off film rolls.

This adhesion between the film layers, usually called blocking, is an inherent property of some polymers such as LDPE, LLDPE, PP, PET, and PA. The most widely accepted explanation of this phenomenon is that low-molecular-weight fractions or waxy additives present in these polymers produce an adhesive layer at the film's surface. Therefore, a certain force is needed to separate the film layers from each other, and this is called the anti-blocking force. Anti-blocking additives added to the plastic film formulations minimize the adhesion and thus lower the blocking force between the film layers. When the right anti-blocking additive is used, film handling is significantly improved.

7.2.5.1 Mechanism of Anti-blocking Action

The anti-blocking force, defined as the force needed to separate two film layers, is a function of adhesion between the two plastic layers. Anti-blocking agents added to polymer film formulations create a micro-rough surface that reduces the adhesion between the film layers. As a result, the anti-blocking force is lowered. Anti-blocking additives simply act as "spacer-bars" between film layers. It is evident that two factors dominate the anti-blocking effect:

- The number of particles (concentration) of the anti-blocking agent on the film surface
- The size of the anti-blocking particles.

> The more anti-blocking particles that are distributed on the film, the more pronounced the imparted surface roughness.

> The coarser the anti-blocking particles are, the farther two film layers are kept apart from each other. Anti-blocking additives with small particle sizes are preferred for thin films (e.g., 20–30 µm), whereas coarse particles are used for thicker films (e.g., >30 µm).

Figure 7.8 shows that there is indeed a correlation between surface roughness and the anti-blocking performance. High surface roughness lowers the blocking force in the film and results in a pronounced anti-blocking effect.

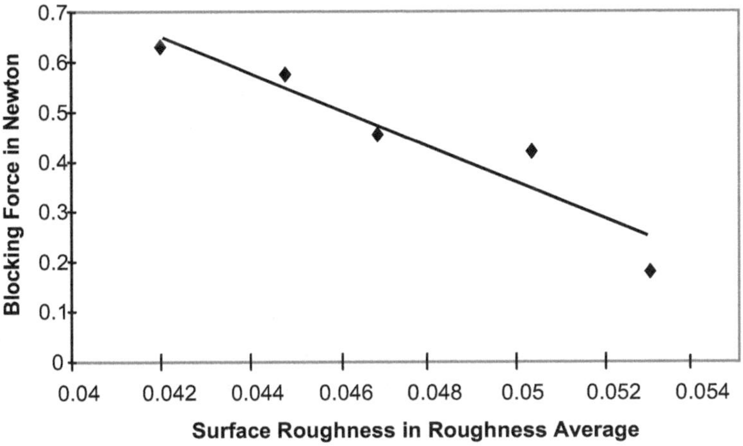

Figure 7.8 Blocking force versus surface roughness

7.2.5.2 Characteristics of Anti-blocking Additives

Several of the following anti-blocking additives are also used as functional fillers. For more detailed information, please see Chapter 12.

Synthetic silica gel

Synthetic silica gel (nominal chemical formula $SiO_2{\cdot}(H_2O)_x$) is a solid, amorphous form of hydrated silicon dioxide. The silica gel particles are spheroidal and 2–10 nm in diameter, resulting in a high surface area of approximately 300–1000 m^2/g (see Table 12.1 and Table 12.2).

The refractive index of silica gel is about 1.46, almost matching the refractive index of PP (ca. 1.49) and PE (ca. 1.50), enabling the films to maintain their high transparency and clarity.

Synthetic silica gel is 99.5% pure (by weight) silicon dioxide, based on the dried material. This high purity prevents quality deterioration when used in polymer films.

Natural silica

Natural silica mined as sedimentary rocks is mainly composed of the skeletons of single-celled diatoms, which are constituted from an amorphous form of SiO_2. The particles can be of various sizes, reaching up to 2 mm. Apart from water and organic substances, the most common other components of natural silica are quartz (crystalline SiO_2), calcium carbonate, and clay minerals, which need to be removed before use of the natural silica as a polymer additive. While drying and calcination can remove water and organic substances, the removal of quartz requires air classification. Natural silica's quartz content varies, depending on the deposit.

Talc

Compositionally, talc is a magnesium hydrosilicate that belongs to the group of phyllosilicates with a layered structure. Talc has the chemical formula $Mg_3(Si_2O_5)(OH)_2$. Its refractive index is about 1.57, which is close to the refractive indexes of PE and PP (see Table 12.1). As a natural product, talc contains certain levels of impurities that can be minimized by refining. Talc is used in many applications, including as an anti-blocking additive in plastics.

Zeolites

Zeolites are crystalline, hydrated alumosilicates of the general chemical formula $(M^{n+})_{y/n}[(SiO_2)_x(AlO_2)_y]\cdot(H_2O)_z$, where M is an alkali or alkaline earth metal. Because of their unique pore structure and excellent ion exchange capacity, zeolites are mainly used as detergent builders and adsorbents. As anti-blocking additives, zeolites are used to lesser extent compared to other anti-blocking agents.

Limestone

Limestone is a naturally occurring mineral that mainly consists of calcium carbonate ($CaCO_3$) and may also contain magnesium carbonate ($MgCO_3$). As a low-cost material, limestone is used as an anti-blocking aid primarily in film applications that do not require very high quality (see Table 12.1).

7.2.5.3 Applications for Anti-blocking Additives

Anti-blocking additives are used in a wide range of plastic films, e.g., PE, PP, flexible PVC, PET, ABS, and PA. Attributes critical for selecting anti-blocking aids are cost, efficiency, low abrasion and equipment wear, low health hazard, optical properties, and potential adverse interactions with other polymer additives. Table 7.2 shows the main uses of anti-blocking additives and their typical concentrations.

Table 7.2 Anti-blocking Aids and Their Typical Concentrations in Various Polymeric Films

Anti-blocking additive	Polymer films	Selected performance attributes	Typical concentrations
Synthetic silica	PP, LDPE, LLDPE, PET	High efficiency, low abrasion, higher price; used where high transparency is required	1000–2000 ppm
Natural silica	LDPE	Moderate efficiency, high abrasion and scratch marks, health risks; used where transparency is not critical	2500–4000 ppm
Minerals (clay, talc, chalk	LDPE, LLDPE	Low health risks, low efficiency	3000–6000 ppm

Anti-blocking additive	Polymer films	Selected performance attributes	Typical concentrations
Zeolites	LLDPE, PVC, PET	Low cost; may result in scratch marks	2000–4000 ppm
Organics (waxes and fatty acid amides)	PP	Low efficiency, slip effect; used where a slip/anti-block balance, as well as high transparency, are required	2000–4000 ppm

7.2.5.4 Incorporation of Anti-blocking Additives

For anti-blocking additives to be effective in polymer films, they are added at 0.1–0.5% and should be well-dispersed. Therefore agglomerates of the fine additive powders must be broken down to their primary particles, and these should be are evenly distributed within the polymer. Good dispersion of anti-blocking additives can be achieved by using twin-screw extruders, for example.

The additives can be incorporated into the polymer streams as powders or concentrates in polymer carrier resins, called masterbatches. These masterbatches can contain the anti-blocking additives, such as natural silica, talc, or zeolites, at up to 50 wt%. Porous materials, such as synthetic silica, have a more pronounced effect on the melt flow index and therefore on the output rates, and can be present in masterbatches at a maximum of about 20 wt%.

The manufacture of PET occurs by reacting ethylene glycol with dimethyl terephthalate or terephthalic acid to produce PET (plus methanol or water from the condensation reaction). The method of producing film-grade pellets containing anti-blocking agents is completely different from that used for polyolefins. In PET production, silica is initially added to the ethylene glycol as a slurry. The silica is then often ball-milled to the required particle size. The slurry is then passed to a filter to remove the oversized particles and from there into the reactor containing terephthalate or terephthalic acid for conversion to PET pellets. Concentrations of silica in these final pellets are in the range 0.1–0.3 wt%.

Plastic films are extensively used for packaging food and pharmaceuticals. These applications require that not only the polymer but also all the additives in the polymer be authorized for use in these applications. Most grades of anti-blocking additives, such as silica, talc, zeolites, and limestone, are suitable for use in food contact applications. However, it has been proven that crystalline silica, containing quartz, may cause silicosis, depending on exposure. Crystalline silica has been classified as carcinogenic to humans when inhaled.

7.2.5.5 Technological Trends

Since anti-blocking additives may adversely affect the optical properties of plastic films, there are constant efforts to minimize that impact. There is a trend to offer anti-blocking additives in narrower particle size distributions that also help to maintain film's optical properties.

Much developmental work has been focused more on improving their physical characteristics for safer handling, improved flow properties, more accurate dosing, and lower cost. This has specifically resulted in low-dust grades. An attractive option is to combine anti-blocking aid powders with other, mainly organic additives to form additive combinations. Multi-component additive packages are already commercially available in powder and pelletized forms. They contain only the active ingredients (no carrier is necessary), and exhibit the same performance properties as the individual components. Combination products can also lower the cost of dosing and mixing equipment.

7.2.6 Dispersion Effect

Metal soaps, fatty acid esters, and montan wax acid esters are also added as dispersion agents to pigment concentrates for polar plastics, but amide wax is preferred for styrenic polymers. The concentrations of these substances are usually lower than in color concentrates for non-polar polyolefins. The lubricating properties of the additives used in these concentrates normally improve processing significantly.

Because of this better dispersion of the filler, lubricants used in reinforced thermoplastics help create smoother finished article surfaces.

7.3 Testing Lubricants

Lubricant testing can be divided into two major areas:

- Physical and chemical measurements and specifications to characterize the lubricants
- Lubricant testing in formulations under specified processing conditions to evaluate their performance.

7.3.1 Two-roll Mill Tests

The heated two-roll mill is the most common testing equipment for determination of the release effects of lubricants in PVC and other polymer melts that tend to stick. This is a particularly suitable preliminary test for calendaring processing. The laboratory two-roll mill is also very useful for the preparation of PVC samples when rheological, mechanical, color, and transparency testing is needed.

To determine the release effect, the powdery compound is fed into the rotating rolls, adjusted to a narrow slit. The sheeted, rolled-out compound is cut by hand or automatically, then turned and put onto the rollers again. This procedure is repeated until the release is assessed.

7.3.2 Torque Rheometer

The torque rheometer (also called a laboratory kneader, plastograph, or plasticorder) is a versatile testing instrument widely used in the polymer and plastics industry to characterize and optimize the flow and processing behavior of thermoplastics, elastomers, and compounds under realistically modeled processing conditions. The device essentially consists of a heated mixing chamber equipped with rotating kneader blades. As the material is processed, both the torque (rotational force) and temperature are continuously recorded, providing a direct insight into melting, plastification, and material stability during mixing.

A key application of the torque rheometer is the evaluation of lubricants that are added to polymers to improve their processing and flow properties. In a typical test, a defined amount of lubricants is mixed with the polymer, and the compound is processed in the torque rheometer. The maximum torque and the overall "torque curve" are recorded and analyzed during the process.

An effective lubricant (or a lubricant system) is *critical for reducing friction and enhancing the performance of polymer-processing machinery. It* usually decreases and accelerates plastification by lowering frictional heat build-up and improving melt flow, and the polymer melt becomes easier to process, the mechanical load on the mixing blades is reduced, and both shear heating and energy input are minimized. Differences in lubricant performance are directly visible in the shape and height of the torque curve, and the instrument can also detect undesirable effects such as material degradation. A typical plastogram is shown in Figure 7.9.

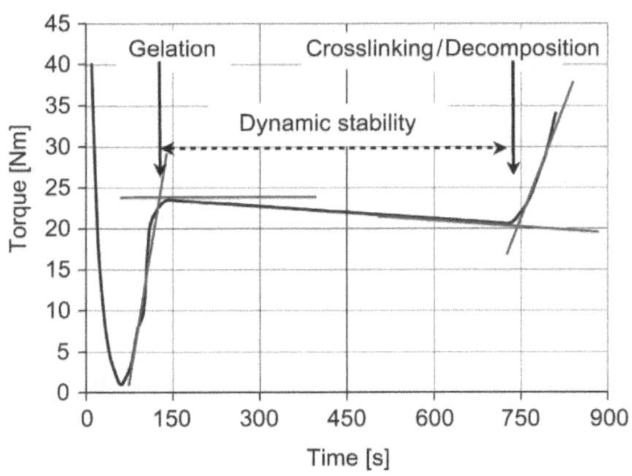

Figure 7.9
Example plastogram showing the changes in torque during a dynamic heat stability test in a laboratory kneader

7.3.3 Measuring Gloss

The gloss, color, and smoothness of finished part surfaces are substantially affected by lubricants. Gloss and color are measured by reflectometers.

7.3.4 Testing for Slip

A method commonly used for determining the slip performance of a plastic film is ASTM D1894 (Static and kinetic coefficients of friction (COF) of plastic film and sheeting).

7.3.5 Anti-blocking Test Methods

During film processing, winding, or storage, films stick together or "block". Because these conditions are neither controllable nor reproducible, films are de-blocked before conducting measurements. Once films have been de-blocked, they only are blocked again under the set artificial conditions, such as exposure to heat and pressure. Despite the fact that no standard method exists for the pretreatment of films prior to the blocking measurements, there are international standards for measuring the blocking force itself. They are described in ASTM D1893 and ASTM D3354.

7.4 Lubricant Groups and Their Characteristics

7.4.1 Lubricant Structure–Effect Relationships

Several experience-based points are summarized in the following rules:

Aliphatic chains of 12 carbon atoms or more have an effective lubricating action in polymers.

With increasing chain length, their compatibility decreases in polar plastics and increases with non-polar ones.

Compatibility of lubricants with polar polymers is increased by using additives with polar functional groups.

Polar groups on lubricant molecules delivered to the polymer surfaces result in significant effects even at very low concentrations.

Carboxylic acids and some of their derivatives wet metal surfaces, and are therefore effective release agents.

Amide waxes often impart marked slip properties on surfaces of finished articles.

Lubricant groups that are commercially viable in the plastics industry are discussed below, in order of decreasing polarity.

7.4.2 Fatty Alcohols and Their Dicarboxylic Acid Esters

Alcohols with chain lengths from C_{12} to C_{22} are effective internal lubricants for polar polymers, such as PVC, but have the disadvantage of high volatility. By esterification with dicarboxylic acids, highly polar lubricants with excellent compatibility are obtained.

7.4.3 Fatty Acids, Alcohols, Esters, and Complex Esters

The carboxylic acids mostly used in PVC applications are saturated or unsaturated C_{16}–C_{18} fatty acids such as stearic acid, oleic acid, and 12-hydroxystearic acid. Commercial stearic acid contains a mixture of C_{14}–C_{20} fatty acids, and is suitable for direct use in PVC compounds with limited compatibility.

Renewably sourced esters are derived from animal fats or vegetable oils. These natural triglycerides have poor hydrolytic and oxidative stability, leading to rancidity and hydrolysis. Therefore, synthetic esters are manufactured to fit the specific PVC processing requirements.

Stearyl stearate shows a very balanced internal/external behavior, and is used in combination with distearyl phthalate. Glycerol esters used in PVC compounds are glycerol mono-, di- or tristearate. The OH group in glycerol monostearate (GMS) contributes to its polarity, and GMS is an effective internal lubricant. Its derivative, the glycerol monoester of 12-hydroxystearic acid, provides higher clarity and stronger stabilizer coordination. Glycerol mono-oleate (GMO) is used in clear semi-rigid PVC packaging cling film that must have low fogging. It is also used as a lubricant in plasticized PVC compounds.

Other esters used in PVC include pentaerythritol stearates of variable composition, as well as complex esters prepared from polyols and mixtures of carboxylic acids, such as pentaerythritol adipate–stearate and pentaerythritol adipate–oleate. The polymeric character and multiple ester groups impart a very high viscosity index, low tack, and exceptional boundary lubrication. Complex esters that contain unsaturation contribute to higher clarity than the saturated esters, indicating increased compatibility for lubricants that contain double bonds.

7.4.4 Fatty Acid Amides

Primary amides used as PVC lubricants are stearamide, oleamide, and erucamide. Ethylene bistearamide, $(CH_2NHC(O)C_{17}H_{35})_2$ or EBS, is a diamide and called "amide wax". In amide waxes, amide groups impart polar characteristics, and long-chain fatty acids (like saturated stearic and mono-unsaturated erucic or oleic acids) give non-polar molar segments. EBS is the most viable amide lubricant that is used in PVC compounds and other plastics. Primary amides adversely affect PVC heat stability. The most common applications for EBS are in rigid siding, profiles, and molded fittings. A combination of EBS with internal ester lubricants is used in PVC films, where clarity is required. While in service, it serves as a slip agent without forming oily or waxy surfaces.

7.4.5 Metal Soaps

Metal soaps of calcium, barium, magnesium, and zinc provide lubrication as well as heat stability during PVC processing (see also Section 4.3 and Table 4.1). Water-soluble soaps of lithium, sodium, potassium, etc. are not used as lubricants. The lubricating effect of a metal soap depends upon the chain length of the fatty acids – for example, lubricity increases from laurates (C_{12}) to stearates (C_{18}).

7.4.6 Montan Waxes

Montan waxes, sometimes called mineral waxes, are derivatives of lignite wax, which is a fossilized vegetable wax. This wax is hard, having a melting range of 95–100 °C. Montan wax is a mixture of chemical compounds that can be divided into three groups: waxes, resins, and asphaltic substances. It is used in rigid as well as plasticized PVC. Its composition is non-glyceride long-chain (C_{24}–C_{30}) carboxylic acid esters (62–68 wt%), free long-chain organic acids (22–26 wt%), long-chain alcohols, ketones, and hydrocarbons (7–15 wt%), and resins. Over the years, montan acid, its esters, and its soaps have found wide use in PVC processing.

7.4.7 Polar PE Waxes

Polar, oxygen-containing groups are introduced into hydrocarbons, such as PE waxes and to some extent paraffin waxes, by non-selective air-oxidation. The degree of oxidation is expressed in the "acid value" (AV) based on the content of carboxylic acids formed. Because ketone, alcohol, and ester groups are also formed along with the acid groups, the increase in polarity is much greater than might be expected based on the acid value.

Polar waxes with acid values of 15–25 mg KOH/g are most common, but there are special grades with acid values of up to 70 mg KOH/g. These oxidates exhibit a strong release effect and provide reasonable transparency in PVC.

Polar polyolefin waxes are preferred for extrusion processes, particularly in blow-molding and in manufacturing tin-stabilized PVC pipes, profiles, and sheets. Plate-out on on the screw, inside the barrel, in the adapter, and in the tooling of extrusion lines is observed from time to time.

7.4.8 Non-polar PE Waxes and Paraffin Waxes

The molecular weight of non-polar PE waxes ranges from about 1000 to 10,000 Da. Similarly to all other hydrocarbons, non-oxidized polyolefin waxes are highly incompatible with polar plastics and very compatible with non-polar plastics, such as polyolefins. They are used especially as friction-reducing lubricants in rigid PVC to adjust plastification time and to produce smooth surfaces. Large quantities of polyolefin waxes are used as dispersion-promoting carrier materials in pigment masterbatches for thermoplastics processing.

Petroleum waxes are low-cost external lubricants for rigid and plasticized PVC, and are especially suitable for improving the surface smoothness of extrudates. Their friction-reducing effect increases with molecular weight. This effect is particularly pronounced with the synthetic paraffin waxes produced by the Fischer–Tropsch process ("FT waxes").

7.5 Practical Use of Lubricants

A large number of lubricants are used to cover the wide range of manufactured plastics, but an overview of the main choices is summarized below.

7.5.1 Lubricants for PVC

PVC, especially the rigid polymer, can rarely be processed without lubricants. PVC applications are the largest consumers of these additives, accounting for about 80% of overall demand. While rigid PVC normally requires about 1% lubricant (with some special cases needing up to 4%), in plasticized PVC the dosage of lubricants is 0.5% or less.

Unlike other plastics, PVC does not melt completely during processing, but instead undergoes fusion at a primary particle level, with a fusion level of 65–70% imparting optimum engineering properties to the final product. Lubricants play a major role in optimizing the extent of fusion during PVC processing, and (depending on the additive type) can influence the following aspects of the product and process:

- Fusion
- Viscosity, flow, melt pressure, melt temperature, melt elasticity, and its consistency, as well as the extrusion torque
- Frictional heat
- Mobility of stabilizers and in turn effectiveness of stabilization
- Die swell and variation in product dimensions
- Surface finish, gloss, haze, and printability
- Mechanical and thermal properties of the finished articles.

A single lubricant cannot fulfill all the requirements related to processing, quality, and service life. Dozens of natural and synthetic lubricants belonging to different chemical groups, and having different molecular weights, chain lengths, polarities, and physical properties are available on the market, and processors globally are using combinations of them, according to their experience, to achieve the desired quality in the final products. Therefore, the selection of lubricant combinations to attain specific processing parameters and article quality is quite complex.

Any lubricant for PVC must be at least slightly soluble in the polymer. If the solubility is zero, the lubricant will bloom to the surface ("plate-out"). If the solubility is too high, the lubricant will act as a plasticizer and adversely affect the article's flexibility, hardness, and T_g, among other parameters.

7.5.1.1 Classification of PVC Lubricants

Different and sometimes contradictory theories have been developed over time regarding the classification of lubricants for PVC and their mechanism of lubrication.

External and internal lubricants

The simplest way to classify lubricants is on the basis of their compatibility with PVC, defining two functional lubricant types: internal and external.

An *internal lubricant* is readily miscible with PVC, reducing its viscosity and the internal friction between the molecules during processing, facilitating movement of PVC chains to undergo fusion and reducing shear-burning and back-pressure. Internal lubricants are either liquid or (in most cases) low-melting polar solids, having relatively low molecular weight. Hence, they are added at the beginning of the hot mixing cycle. Depending on the polar groups, their dosage, and melting point, a certain reduction in T_g and the Vicat softening temperature (VST) of the final article can be expected.

> The lower the melting point of a lubricant is, the more substantial is the decrease in VST.

> Lubricants with a melting point >100 °C have a very limited or no effect on VST.

An *external lubricant* forms a film on the surface of PVC particles, reducing friction between PVC and the metal surface of the production equipment, and between PVC particles and $CaCO_3$ particles during processing. They do not reduce the T_g of PVC compounds.

> Double bonds, epoxy groups, hydroxyl groups, ester groups, and ketone groups contribute predominantly to internal lubricating properties. Amides, amines, fatty acids, ethers, and metallic ions contribute predominantly to external lubricating properties of PVC.

The influence of internal and external lubricants on the speed of melt front is illustrated in Figure 7.10.

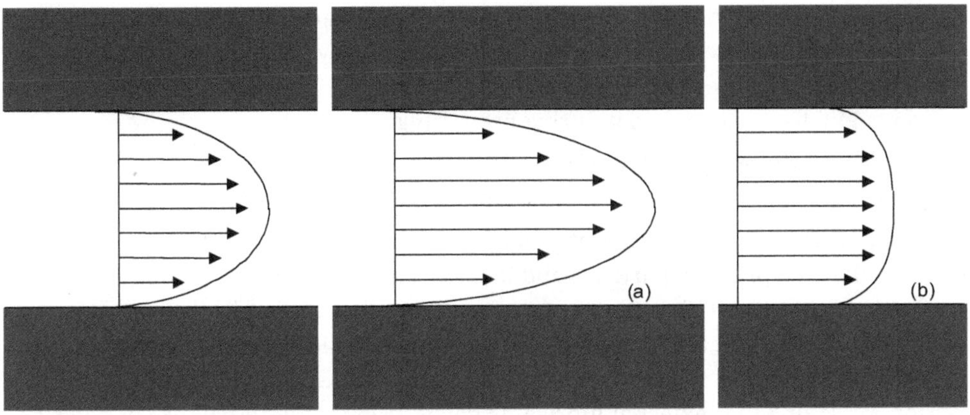

Figure 7.10 Influence of (a) internal and (b) external lubricants on the speed of a melt front

By definition, non-polar lubricants control the degree of particulate flow, whereas polar lubricants control the melt flow.

Fisch and Bacaloglu[2)] proposed that during blending, lubricant molecules are attached to PVC particle surfaces and distributed between primary particle surfaces, interstitial spaces, and a separate phase. They concluded that no single material functions entirely as an internal or external lubricant, but usually exhibits a property combination of the two, with one being predominant.

The traditional classification considers that the lubricant systems used in PVC processing do not undergo chemical reactions and remain in the compound like a lubricated ball-bearing. However, this classification fails to explain the behavior of certain lubricants, like calcium stearate, that react chemically during processing.

Modern classification of PVC lubricants

Currently, PVC lubricants are classified into three types based on their polarity, rather than their internal or external functionality:

a) **Non-polar lubricants** (instead of external lubricants): Examples include paraffin oil, paraffin waxes, Fischer–Tropsch waxes, and polyethylene waxes. They have no affinity with either PVC or metal surfaces of the manufacturing equipment.

b) **Polar lubricants** (instead of internal lubricants): Examples include stearyl alcohol, glycerol monostearate, stearyl stearate, butyl stearate, distearyl phthalate, pentaerythritol tetra-stearate and montan ester wax. They are polar, usually low-molecular-

[2] M. H. Fisch and R. Bacaloglu, Study of additive compatibility with poly(vinyl chloride), *Journal of Vinyl and Additive Technology*, 1998, 4: 4–11.

weight liquids or low-melting solids, capable of diffusion within PVC grains. Due to their moderate polarity, they have affinity with PVC and partial compatibility with it.

c) **Film-formers** (a new category): These are reactive polar lubricants, and examples include stearic acid, calcium, barium, neutral lead and zinc stearates, high-molecular-weight oxidized PE waxes, montan waxes. They have affinity towards very polar metal surfaces and form a monolayer film on the metal surface, which prevents PVC from sticking to it during processing.

In practice, all three types of lubricants are necessary for PVC processing. As a result, they are more commonly used in a combination deployed as a "lubricant system".

During processing, polar heads of polar (internal) lubricants orient themselves towards polar PVC primary particles and lubricate the released PVC chains, facilitating its faster movement. This is the reason for saying that internal lubricants promote fusion. On the other hand, non-polar lubricants stay outside the primary particles and coat the emerging PVC chains, causing them to slip while retarding fusion. This is the reason for saying that external lubricants delay fusion. Thus, a balance of polar and non-polar lubricants is required.

Film-forming lubricants, such as calcium stearate and oxidized PE, have polar heads and non-polar tails. The very polar heads of the molecules of film-formers are attracted to polar metal surfaces as well as moderately polar PVC, while the tails are attracted to themselves, forming a reduced-friction surface to ensure sliding of the moderately polar polymer melt. Thus, PVC is displaced from the metal surface, which ensures that it does not stick.

Film-formers are further classified into two categories: one that promotes fusion and another that does not (with an example being stearic acid).

7.5.1.2 Effect of Overlubrication and Underlubrication

For optimum output, non-polar lubricants must not be added at concentrations that are so high that there is slippage on the screw, preventing the screw from pushing the molten polymer mass forward. This "surging" also takes place when a non-polar lubricant is used at too high a level.

Overlubrication can result in low production output, poor impact strength and poor surface quality, and burning (scorching), which is a consequence of material slippage meaning that it remains in the equipment for an overextended residence time.

Underlubrication may result in frictional heat and in thermal degradation of PVC, poor fusion, and loss of mechanical properties.

7.5.2 Lubricants for Styrenic Polymers

Styrenic polymers are supplied almost exclusively as ready-to-process granules containing all the required additives. If they are to be prepared by compounders (for example, for coloration or special modification with a flame retardant or an antistatic agent), it would be possible to optimize processing behavior by adding a lubricant at this stage, for which ABS is a good option.

Processes involving ABS use large quantities of fatty amide wax and metallic soaps, mainly zinc stearate. The effects of the lubricants are similar to those in PVC: glycerol mono-stearate and the amide wax act as typical internal lubricants. Stearic acid and its metal soaps occupy an intermediate position, while oxidized PE waxes, pentaerythritol fatty acid esters, and montan waxes function as release-effective lubricants, causing only a slight deterioration in heat deflection temperature.

7.5.3 Lubricants for Polyolefins

Generally, polyolefins do not require a lubricant to be processed. However, calcium stearate or zinc stearate used as acid acceptors (see Section 4.7) in HDPE and PP, as well as free stearic acid formed as a result of the neutralization reaction with chloride ions present in Ziegler catalysts, affect flow and the demolding force.

Low-molecular-weight PE waxes or PP waxes (with a molecular weight of about 2000 Da) are sometimes used as internal lubricants in PE or PP to raise their melt index. However, PE waxes are not fully compatible with PP and can therefore act as strong external lubricants, causing overlubrication. To avoid these issues, PP waxes have to be used instead in PP. Amide waxes are used as an external lubricant in HDPE tapes, normally at concentrations of 0.3–0.5%.

In PP, external lubricants are sometimes used as demolding aids, and erucamide is very suitable for this purpose. Slip agents (oleamide and erucamide) are added mainly to film grades at a concentration of 0.1%. The additive is either incorporated by the polymer manufacturer in a pelletizing extruder or by the processor via a masterbatch.

7.5.4 Lubricants for Engineering Thermoplastics

Engineering thermoplastics such as polyamides, polycarbonate, polyalkylene terephthalates, polyoxymethylene, polyphenyleneoxide, and thermoplastic polyurethane are processed mainly by injection molding. Therefore, mold release effects are of prime consideration. Additionally, improving flow is highly important. Lubricant requirements in engineering thermoplastics may differ considerably depending on the lubricant properties needed in each polymer.

The chemical reactivity of these engineering thermoplastics, and particularly their susceptibility to hydrolysis, make them sensitive to the presence of acidic and/or alkaline components in the corresponding formulations. This restricts the use of metal soaps and carboxylic acids in some cases. Due to the high processing and service temperatures involved, the only additives suitable for engineering thermoplastics are those with good thermal stability and low volatility at high temperatures. Low migration is another important requirement for lubricants in this case, taking into consideration the higher service temperatures and the intended long-term use of the finished plastic parts. Table 7.3 provides examples of lubricants that are suitable for use in engineering thermoplastics.

Table 7.3 Lubricants Used in Selected Engineering Thermoplastics to Achieve the Target Performance

Thermoplastics	Flow improvers	Release agents	Dispersants
Styrenics	Montan wax Oxidized PE wax Wax esters Polyol wax	Amide wax Montan wax PE wax Polyol esters Wax esters	Amide wax Montan wax Polyol esters
PP (reinforced)	Montan wax Copolymer wax Polyol esters	Montan wax Amide wax Wax esters Metal soaps	Montan wax Polyol esters
PET/PBT	Montan wax Polyol esters Wax esters	Montan wax PE wax Wax esters	Montan wax Polyol esters
POM	Montan wax Amide wax Complex esters Polyol esters Wax esters	Montan wax Amide wax Oxidized PE-wax Polyol esters Fatty acids	Amide wax Polyol esters
PC	Montan wax Wax esters	Montan wax PE wax Oxidized PE wax Polyol esters Wax esters	Montan wax Polyol esters

Thermoplastics	Flow improvers	Release agents	Dispersants
PA	Montan wax Amide wax Copolymer wax Wax esters Polyol esters	Montan wax Copolymer wax Metal soaps Polyol esters Wax esters	Montan wax Polyol esters Wax esters
TPU	Montan wax Complex esters Polyol esters	Montan wax Amide wax Complex esters Polyol esters	

The preferred lubricants for certain engineering thermoplastics are montan wax acid derivatives. These provide good release and flow properties, adequate compatibility, very high thermal stability, and superior migration resistance. However, their use is limited by their intrinsic color, which is not pure white. Pentaerythritol esters are also preferred in engineering thermoplastics because of their thermal stability.

In polyamides, metal stearates and amide wax dominate as the major lubricant additives. In their glass-fiber reinforced grades, montan wax acid diol ester works well to enhance the mechanical properties of these composites.

In reinforced PC, a similar effect is achieved with stearyl stearate. Pentaerythritol stearate is generally used in transparent PC. Oxidized PE waxes also impart good release characteristics in PC. The partial esters of sorbitol are credited with a color-improving effect.

In PET and PBT, as well as fatty acid esters of trimethylol propane or pentaerythritol and montan wax acid esters, metal soap-containing montan wax acid esters are also used particularly in glass-fiber reinforced grades. Sodium montanate displays slip and anti-blocking effects in PET films in addition to its nucleating effect. In polyacetals (POM), besides amide waxes, metallic soaps are mainly used as co-stabilizers due to their acid scavenging effect. For the highly transparent PMMA, stearic acid polyol esters are preferred over montan waxes, which impart an opaque appearance. In thermoplastic polyurethanes, amide wax is used for its efficient release effect.

7.6 Outlook for Lubricants

As discussed earlier, some stabilization problems are solved with blends or combinations of lubricants and other additives. So, in many cases, formulations are improved and optimized by using conventional additives with the aim of attaining lower-cost additive combinations, thus avoiding the time-consuming and expensive development of entirely new chemicals. Improving the thermal stability of lubricants and other additives would enable processing of plastics at higher temperatures for increased output.

7.7 Questions

1. What is the mechanism of action of internal lubricants in plastics?
2. What is the mechanism of action of external lubricants in plastics?
3. What are the functional differences between processing aids and lubricants?
4. What commercial polymers consume most lubricants?
5. What is the correlation between the internal lubrication and maximum flow rate?
6. What is the effect of lubricants on heat dissipation?
7. What is the mold release effect?
8. What is the slip effect? What is the mechanism of slip action?
9. What are the commonly used slip additives?
10. What is the anti-blocking effect? What is the mechanism of anti-blocking?
11. What anti-blocking additives are commonly used in plastics?
12. What are the main applications of anti-blocking additives in plastics?
13. What are the main lubricant groups and their characteristics?
14. What is the ideal solubility of lubricants in polymers?
15. What are the effects of lubricants on the processing of PVC materials?
16. What is the modern classification of PVC lubricants?
17. What are the adverse effects of overlubrication and underlubrication in polymer processing?
18. What is plate-out?
19. What lubricants are used in polyolefins? What lubricants are used in engineering thermoplastics?

8 Antimicrobials

Content of this chapter was adapted from:

- D. Ochs, "Antimicrobials" (Chapter 11), in: *Plastics Additives Handbook* (6th edition), H. Zweifel, R. D. Maier, and M. Schiller, Hanser, 2009, pp. 675–700.

8.1 Terms and Definitions

Antimicrobial agents	Chemicals that inhibit or prevent microbial growth on material surfaces.
Inorganic chemicals	Compounds not primarily based on carbon, and often taking the form of minerals or salts.
Organic chemicals	Carbon-based compounds containing hydrogen, oxygen, nitrogen, or other elements.
Natural chemicals	Substances derived from plants, animals, or minerals without significant modification.
Esters	Organic compounds formed from acids and alcohols, often used as plasticizers for plastics.
Composites	Materials composed of two or more distinctly different material components to achieve enhanced properties.

8.2 Functions of Antimicrobials

Antimicrobials are usually low-molecular-weight substances that kill or suppress the growth of viruses, bacteria, algae, and/or fungi (molds) that feed on plastics and their additives, often leading to a potential loss of product integrity. This is particularly true for outdoor applications, where environmental effects dramatically reduce the service life of untreated plastic articles. Antimicrobials are incorporated into plastics during the manufacturing process to create articles that are able to:

- Improve robustness towards weather variations.
- Resist microbial growth throughout their lifespan
- Help reduce the risk of cross-contamination in healthcare, food processing, and public spaces
- Maintain their aesthetic appeal
- Prevent growth of odor-causing bacteria and fungi, as well as staining by mold and mildew.

There are two different types of antimicrobial activity:

- **Microbiostatic activity** involves only the inhibition of *reproduction* of the microorganisms, i.e., the cells are not killed but only suppressed in growth. Depending on the target organisms, this mode of activity is described as bacteriostatic or fungistatic.
- **Microbicidal activity**, in contrast, involves the death of bacteria or fungi, i.e., it significantly reduces the number of microorganisms within a short period of time. Depending on the target microorganisms, this mode of activity is described as bactericidal or fungicidal.

Having made this distinction, the terms microbiostatic and microbicidal describe two extremes in a spectrum of antimicrobial activities. The same antimicrobial substance can act as both a microbicidal or a microbiostatic agent depending on the conditions and the target microorganisms. The activity of an antimicrobial agent depends on many parameters, e.g., their concentration, pH, temperature, the polymer matrix into which it was incorporated, the mode of incorporation of the antimicrobial, and the sensitivity of the target microorganism, including the amount of time over which the microorganisms were exposed to the antimicrobial.

Antimicrobials are chosen depending on their intended function in plastic applications. In many cases, a mixture of antimicrobials is recommended to protect against a broader target range of microorganisms.

8.3 Factors Affecting Performance of Antimicrobials

Most microorganisms need some basic conditions for their survival and growth. Water is one of the most critical environmental factors, but at the same time, is often a limiting parameter for microbial growth. The so-called "water activity", a_w (i.e., the bioavailable water) in or on a given polymer matrix is defined by the following:

$$a_W = \frac{\text{Vapor pressure of the matrix}(p)}{\text{Vapor pressure of water}(p_0)}$$

As a common rule, bacteria need more water, usually a_w = 0.99–0.93, than yeasts (a_w = 0.91–0.88) or molds (a_w = 0.83–0.73).

Most bacteria have an optimal temperature range of 30–37 °C; yeasts and molds usually prefer temperatures between 25–30 °C. Microorganisms whose growth is limited to typical environmental temperatures (20–40 °C) are called mesophilic microorganisms.[1] Although temperatures outside this range reduce their growth significantly or eliminate growth completely, extremophilic microorganisms can survive and even grow at temperatures far removed from this moderate range. With regard to acidity, pH values below 3 and above 9 make most bacterial growth impossible.

Oxygen is a key element for all aerobic microorganisms. Most bacteria and yeasts have the potential to cause degradation of plastics, and all molds need oxygen for growth. In contrast, anaerobic microorganisms grow only in the absence of oxygen, and oxygen can even be toxic to them.

Sunlight is usually not essential for microbial growth, with the exception of phototrophic microorganisms. Nevertheless, sunlight is necessary for the growth of algae on plastic surfaces. It leads to increased growth of molds and bacteria on the plastic articles, because the algae provide conditions that are necessary for microbial growth in terms of humidity and organic matter.

1 "Mesophilic" refers to organisms, especially microorganisms, that reproduce or grow best at moderate, non-extreme temperatures – usually 20–45 °C. In microbiology, for example, the term is commonly used for mesophilic bacteria that require these optimal temperature conditions, in contrast to thermophilic (heat-loving) or psychrophilic (cold-loving) organisms.

8.4 Types of Antimicrobials

Antimicrobial additives can be organic, inorganic, or natural.[2],[3] Organic antimicrobials include phenolic compounds, isothiazolinone, thiabendazole, and zinc pyrithione. Inorganic antimicrobials are silver, copper, and zinc compounds, which generate metal ions. Natural antimicrobials include plant extracts, essential oils, and enzymes.

Organic antimicrobials used in a wide range of plastic articles have limited effectiveness, since they are relatively small molecules that are incompatible with the polymer matrix. This results in their leaching to the surface, where they prevent reproduction of microorganisms. The leaching phenomenon limits the useful life of the additive, especially on regularly cleaned surfaces. This leaching can also cause discoloration of the material and an unpleasant taste in applications associated with food and/or water contact. Therefore, organic antimicrobial agents fit better for use in disposable items and applications rather than durable goods. Organic antimicrobials are also sensitive to temperature, and may decompose at typical polymer processing temperatures, leading to high losses of antimicrobial agents from the plastic during article production.

Inorganic antimicrobials are well-suited for incorporation into molded plastics. Many of these additives contain metal ions that are non-reactive until released in contact with moisture or other reagents. These antimicrobials are released gradually to the surface of polymers and provide continual, long-lasting protection. These additives offer both biocidal and biostatic effects. Inorganic antimicrobials do not migrate much, since they are immobilized in the polymer matrix. The antimicrobial effect of silver ions has long been recognized, dating back to ancient times.[4]

To be incorporated into the polymer matrix, highly mobile metal ions are bonded to the delivery systems, allowing small quantities of ions to be released through a process of ion exchange at the plastic's surface (Figure 8.1). The main advantage of such a mechanism is that the ions are made available over the lifetime of plastic articles. Moreover, inorganic systems are also characterized by high thermal stability, minimizing any losses during polymer processing and making them suitable for use in a

[2] M. L.Kurtz, "Antimicrobial additives" [White paper], The Madison Group, *https://madisongroup.com/antimicrobial-additives/* [accessed on 30 June 2025].

[3] V. Marturano, P. Cerruti, and V. Ambrogi, "Polymer additives" (Chapter 5), in: *Polymer Engineering*, ed. B. Tylkowski, K. Wieszczycka, and R. Jastrzab, De Gruyter, 2017, pp. 162–166, *https://doi.org/10.1515/9783110469745-005*.

[4] Silver-containing substances, frequently used in polymers as antimicrobial additives, may result in discoloration when exposed to light, which is a result of the silver ions (Ag^+) being reduced to elemental silver (Ag^0). This photochemical process is similar to what is known from classic silver halide photography. In polymers, the resulting metallic silver appears as a fine, gray-to-black deposit, leading to unwanted graying – an effect that is especially pronounced under intense or prolonged light exposure.

wide range of durable applications of polymer articles containing them. Silver-based compounds dominate the inorganic antimicrobial markets, especially in food packaging and healthcare.

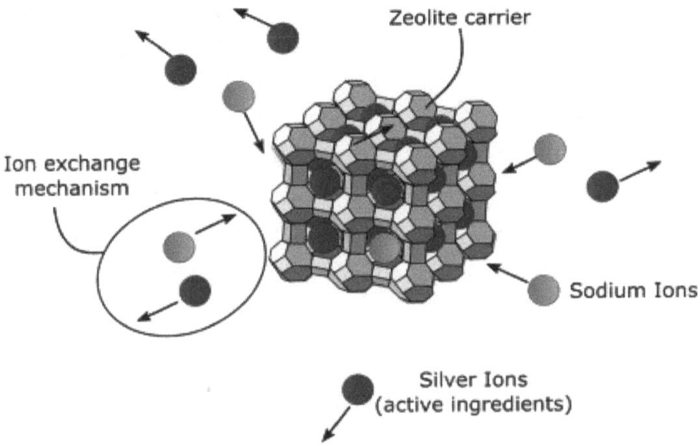

Figure 8.1 Example of an ion exchange mechanism operating in an antimicrobial zeolite containing silver ions. Reproduced with permission from *Polymer Engineering*[3)]

The desire to reduce risks to health and the environment related to the use of synthetic antimicrobial additives in plastics, especially in food-packaging applications, has led to the rediscovery of plant extracts as natural antimicrobial agents. These include basil, a popular culinary herb used extensively in food flavoring. However, the current market share of natural antimicrobials is small.

8.5 Plastics and Incorporation of Antimicrobials

Typical plastic materials used for incorporating antimicrobials include ABS, PP, PS, HDPE, LDPE, PVC, PC, PA, polyesters and elastomeric polymers and rubbers such as TPEs and TPVs.

Antimicrobial additives intended for use in plastics can be added directly into the polymer formulations (dry blends), or as an antimicrobial masterbatch, during the manufacturing process. Antimicrobial additives are typically incorporated into the plastic matrix during the manufacturing process either during polymer processing or compounding. Loading levels of antimicrobials vary depending on the additive type and the polymer, but typically are in the range 0.5–5%.[5)]

5 “Antimicrobial additives”, PRIMEX, *https://www.primexplastics.com/antimicrobials/* [accessed on 30 June 2025].

8.6 Biodeterioration of Plastics

The undesired effects of microorganisms on plastic materials are often called "biodeterioration". Particularly in wet or humid environments and in outdoor applications, the risk of microbial growth and subsequent damage to plastic articles increases significantly. Examples of plastic applications with enhanced risks for biodeterioration include:

- Building and construction (inflatable buildings, barrier films, roofing membranes, electrical cables, vinyl hoses, awnings)
- Household and recreation products (floor coverings, wall coverings, shower curtains, bathmats, shoes, pool and pond liners, waterproof clothing, greenhouses)
- Transportation (automobile tops, truck covers, marine upholstery and foam packing/cushioning).

Biodeterioration of plastics by microorganisms can occur by three modes of action:

- Direct degradation using plastics components as nutrients
- Indirect degradation or change in appearance by microbial metabolites
- Microorganism accumulation on the plastic surface without degrading it chemically.

Initially, plastics are often attacked by fungi, which may cause major degradation of the polymers and/or the additives. Once biodegradation starts by microorganisms capable of using these components as nutrient sources and converting them into smaller organic molecules, the deposition of other microorganisms is facilitated, resulting in increased material biodeterioration.

The main effects caused by microorganisms are:

- Staining – pink, white, orange, green, red, blue, and black discolorations
- Deterioration of electrical insulation properties
- Deterioration of mechanical properties (brittleness, shrinkage, surface roughness, loss of tensile strength, increased permeability of solvents and gases)
- Enhanced dirt uptake
- Disagreeable odor – unpleasant odors can be generated by microbial metabolites, such as amines, ammonia, or hydrogen sulfide.

The main factors involved in microbial adherence to polymers are surface tension, surface texture, and charge on the polymer surface.[6)]

---

[6] The charge of a polymer can be quantified by the zeta potential, which determines the behavior of bacteria, yeasts, or other particles on the polymer surface.

The higher the surface tension of the plastic, the lower the rate of bacterial deposition.

Surface irregularities and imperfections promote microbial adhesion.

Since most bacteria carry a negative surface charge, polymers with an anionic surface charge show reduced bacterial adherence.

8.7 Use of Antimicrobials

The main markets for antimicrobials in plastics include healthcare, packaging, foods & beverages, construction, and automotive. In the polymer industry, antimicrobials are used for two major purposes:

- As a biostabilizer (i.e., preservative)
- As an antimicrobial active ingredient.

8.7.1 General Requirements

General requirements for antimicrobials used as either biostabilizers or active ingredients are:

- Low toxicity to humans, animals and the environment
- Ease of use/applying
- Compatibility with other additives
- No adverse effect on the performance, processability. and appearance of plastic articles
- Storage stability and long-lasting efficacy.

Depending on the intended effect, the active ingredients should have an activity spectrum against the special group of undesirable microorganisms, often originating from humans, e.g., malodor-generating bacteria and fungi for malodor-prevention, or activity against pathogenic microorganisms for sanitizing or hygiene efficacy. In some cases, antimicrobials exhibit an activity profile that makes them suitable for use as both biostabilizers and active ingredients.

8.7.2 Antimicrobials as Biostabilizers

Biostabilizers in plastic articles protect them against deterioration due to organic matter build-up on their surface by microorganisms during use, especially outdoors.

In the recent past, the biostabilizer market for plastics was dominated by arsenic-containing antimicrobials, particularly 10,10′-oxybisphenoxarsine (OBPA). About 70% of the demand for biostabilizers in plastics was met by this type of additive. However, due to toxicological and ecological concerns as well as new regulations adopted in Europe, arsenic-free products have gained a significant market share. The most common additives used now to manufacture antimicrobial-containing plastics include various isothiazolinones, zinc pyrithione, and silver ion-containing antimicrobials.[7]

Plasticized PVC, polyurethanes, LDPE and polyesters demonstrate particular susceptibility to microbial attacks. Flexible PVC is by far the main plastic in which biostabilizers are incorporated, followed by polyurethane foams and other resins. While PVC itself is resistant to microbial attacks, the plasticizers, fillers, pigments, lubricants, and other additives present in soft PVC compounds can be used by microorganisms as nutrients. Of the additives in PVC, the plasticizers can comprise up to 50% of the total compound formulation.

The plasticizers (mostly ester types) can be grouped in terms of their susceptibility to microbial attack as follows:

- High susceptibility: Sebacates, epoxidised oils, polyesters, and glycolates
- Moderate susceptibility: Adipates, azelates, and pentaerythritol esters
- Low susceptibility: Phthalates, phosphates, and chlorinated hydrocarbons.

In most cases, biostabilizers are formulated with a liquid carrier, usually a plasticizer, with concentrations of active ingredient in the range 2–10%.

Polyurethanes, especially their blown foams, are also susceptible to microbial attacks because of their open cell structure, which allows air to flow through it, resulting in the deposition of soil, dust, and fungal and bacterial spores. Combinations of soil and oxygen advance microbial growth on the surface of these materials.

In general, materials based on esters are very susceptible to microbial deterioration compared to others. That is the reason as to why polyester polyurethanes are considerably more easily degraded by microorganisms than polyether polyurethanes. The susceptibility of the polyurethanes is increased further by the presence of aromatic diisocyanate groups.

PE is generally less susceptible to microbial deterioration than plasticized PVC and polyurethanes.

---

[7] "Antimicrobial plastic & polymers", Ultra-Fresh, *https://www.ultra-fresh.com/antimicrobial-applications/antimicrobial-plastic* [accessed on 30 June 2025].

8.7.3 Antimicrobials as Active Ingredients

Antimicrobial additives used as active ingredients in plastics provide protection against pathogenic microorganisms capable of resulting in health risk in hospital and other primarily indoor settings.

Over the last 30 years, applications of antimicrobials as active ingredients have significantly increased. Starting with medical applications, such as gloves, catheters, wound bandages, and bedding, the trend has also moved into homes and offices, resulting in antimicrobial-treated garbage bins, pails, kitchen utensils, bathtubs, sinks, tubes, air filters, containers, and mattresses, among others. This trend is driven by changes in consumer demands for more leisure time and comfort.

The following aspects of using antimicrobials as active ingredients, as well as the associated growing awareness about health, hygiene, and cleanliness, should be kept in mind:

- As adhesion to surfaces by microorganisms is the first step in the contamination of those surfaces, preventing adhesion is highly effective to prevent the undesirable effects on plastics. The anti-infective properties of these antimicrobials do not decrease over time, because their anti-adhesive activity remains consistent over time and does not leach out of the plastics.
- As well as the infection risk in hospitals, there has been an increasing awareness of allergic reactions to mold spores in recent years. These reactions are substantially reduced by polyurethane foams used in cushions and mattresses.
- Improved hygiene outside hospital settings. Institutional and industrial (I&I) applications are traditional markets for antimicrobials due to the reduced risk of transferring infectious diseases and avoiding economic damage to the producers in the presence of these additives.
- Antimicrobial-containing plastics can further improve hygiene in I&I areas. Examples of such anti-microbially treated articles include conveyor belts, flooring, surface coatings, packaging materials, tubing, and cleaning sponges and brushes, among others.
- Odor control is usually associated with textile protection. Also, plastic articles exposed to food or that come in direct contact with human skin (gloves, bathmats, mattresses, apparel, etc.) provide ideal habitats for microbial growth. Therefore, contaminations by skin scales, hairs, and especially sweat and skin oils contribute to the environment conducive to microbial growth.

8.8 Modes of Action

Antimicrobial additives inhibit or prevent growth of microbes by interfering with their cellular processes, damaging their structure, and continually disrupting their growth and reproduction on a treated plastic surface.

It is necessary to point out that not all antimicrobial additives are biocides; many of them are simply biostatic agents that instead hinder microorganism reproduction.

According to Vigo's model for textile applications, the efficacy of most antimicrobial active ingredients in plastics can be described as a *controlled release mechanism*. The antimicrobial active ingredients are slowly released from the polymer matrix, and migrate to the surface. Under ideal conditions, the polymer matrix can serve as a reservoir of the antimicrobial over a long period of time.

Another mode of action called the *regeneration method* has lately gained importance for synthetic textile fibers. According to this method, the active ingredient can be regenerated by surface treatment with bleaching agents during laundering or by UV light during drying.

8.9 Test Methods

For a general assessment of an antimicrobial substance, the minimum inhibitory concentration (MIC) of the substance against microorganisms is determined. The MIC is defined as the lowest concentration of an antimicrobial (in parts per million within the resin) that results in full growth inhibition of a test organism. While the MIC gives only information about the microbiostatic activity of a substance, other tests evaluate the microbicidal activity of the antimicrobials.

Knowledge of the antimicrobial MIC and bactericidal activity data is valuable in assessing the activity spectrum and antimicrobial potential of a substance and to compare it to other antimicrobials. To evaluate the efficacy of antimicrobial-containing plastics, *in situ* microbiological tests are necessary.

A variety of test methods are used to determine the susceptibility of a plastic material to microbial degradation, as well as its antimicrobial resistance. Standardized test methods like ASTM G21, AATCC Method 30, ISO 22196, and JIS Z 2801 are used to evaluate the performance of plastic materials containing antimicrobials.

8.9.1 Agar Plate Tests

The test specimen is placed on (or immersed in) a medium containing test microorganisms. After solidification of the agar and incubation of the agar-containing plates until the microbial growth is visible, the growth of the test organisms is assessed on and around the test specimen. This test, and ones similar to it, depend on limited diffusion of the antimicrobial into the water-based agar medium. Since the agar medium contains high concentrations of water, water-soluble substances show superior diffusion compared to lipid-soluble agents, and therefore result in better efficacy assessment results.

Agar plate tests are suitable only for investigation of microbiostatic activity. It is not possible to differentiate between microbiostatic and microbicidal activity in this test. However, despite these limitations, these tests provide reasonable results, they are fast and easy to perform, and enable screening of antimicrobial-containing plastics against a range of test microorganisms.

Figure 8.2 shows the antibacterial activity of an HDPE sample in an agar plate test with the test organism. Even the sample with a low concentration of antimicrobial (B) showed strong antibacterial activity, resulting in a large growth inhibition zone around the plastic disc. The plate with the sample without any antimicrobial (A) was totally overgrown by the bacteria.

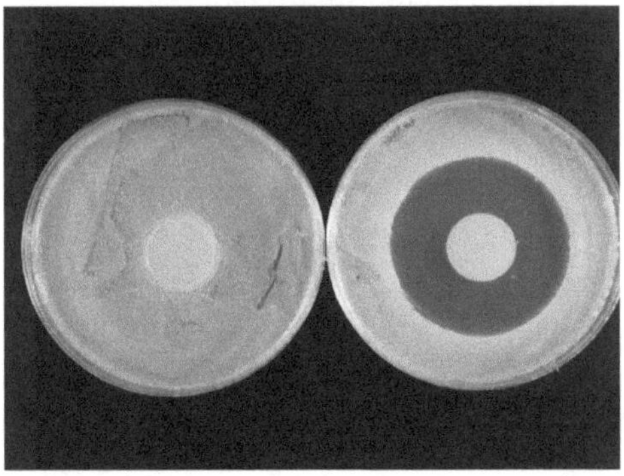

Figure 8.2
Agar plate test against *Staphylococcus aureus* ATCC 9144 with HDPE samples: (left) containing no antimicrobial; (right) with 0.25% of an antimicrobial

In a modified agar diffusion test method, the microbiostatic efficacy of a plastic article is investigated by applying an agar disc containing microorganisms to the surface of the test material. As shown in Figure 8.3, the antimicrobial-containing material showed full growth inhibition, while the antimicrobial-free material has no bacteriostatic activity. To increase visibility of the bacterial growth, the agar contained 0.01% potassium tellurite, resulting in black-stained microorganism colonies.

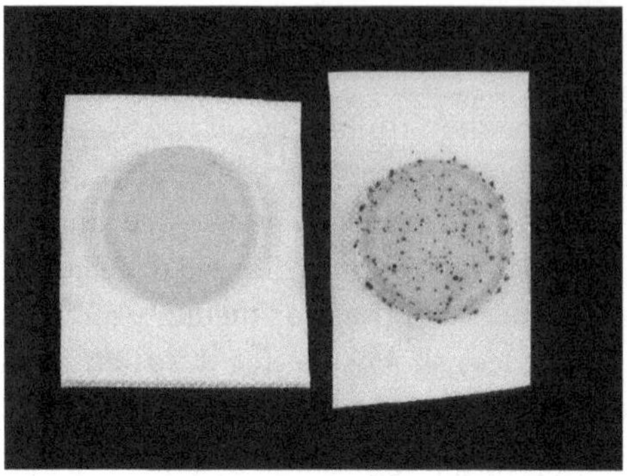

Figure 8.3
Antibacterial activity on an antimicrobial-containing TPU conveyor belt. Discs of agar medium containing *Staphylococcus aureus* were laid on the surface of the conveyor belt containing: (left) no antimicrobial agent; (right) 0.25% of a bacteriostatic antimicrobial. Subsequently, the plates were incubated for 24 h at 37 °C

8.9.2 Direct Contamination of the Test Specimen

In this test, the specimen is directly contaminated with test organisms in the absence or presence of additional contamination (e.g., proteins, lipids, etc.) to simulate antimicrobial activity under soiled conditions. The survival rate of the microorganisms on the test specimens is investigated after the set time intervals by rinsing off the test organisms from the plastic, diluting the microorganism suspension, and subsequently plating out aliquots on agar plates for determination of the colony-forming units.

In contrast to the agar plate tests, tests involving direct contamination of the test specimen are to a large extent independent of migration (diffusion) of the antimicrobial agent. Therefore, these methods are also suitable for testing samples containing water-insoluble antimicrobials or substances fixed in the plastic matrix (e.g., inorganic antimicrobials based on silver or zinc ions that have been incorporated into the plastic during processing).

Moreover, these kinds of test methods are suitable for the assessment of both microbiostatic and microbicidal activity of the antimicrobials, giving them an advantage over the simple agar plate tests.

8.9.3 In-use Tests

In these tests, an antimicrobial-treated material is subjected to realistic conditions. According to ASTM D3083, the plastic material is buried for 90 days to investigate its resistance to rotting.

In the modified test, physico-chemical parameters, such as tensile strength and integrity, are investigated as well as microbial contamination. To increase visibility of the microbial colonies, dyes, such as Methylene Blue, can be used to stain the living cells. Field tests lasting for one year or more are usually performed as outdoor-panel exposure studies for a better assessment of how well the plastic withstands outdoor uses.

8.9.4 Other Test Methods

ASTM G21-90 is entitled "Standard practice for determining resistance of synthetic polymeric materials to fungi." In this test, the material is placed on nutrient-free, sterile agar (mineral salt agar without a carbon source), which is subsequently inoculated on its surface (including the surface of the test specimen) with a fungal spore suspension. The plates are incubated for at least 21 days at not less than 85% relative humidity and 28–30 °C.

Provided that the surface of the plastic specimen is free of organic contamination, fungal growth on the test material is only possible if the fungi metabolize in the plastic specimen, since the nutrient agar does not contain any carbon source. After the 21-day incubation and microbiological assessment of the plastic samples, the test specimen can be used to evaluate the effects of the fungi on the mechanical, optical, and electrical properties of the plastic material.

A comprehensive test method for the assessment of the resistance of plastic materials to fungal and bacterial attacks has been recently approved by the European Committee for Standardization (CEN). This method, EN ISO 846, is a combination of (a) the agar plate tests for investigation of fungi and bacteria growth on plastic materials in nutrient-free and nutrient-containing agar media, and (b) soil burial tests for the determination of plastic's resistance to attack by soil microorganisms, particularly aimed at plastic materials intended for outdoor uses.

Another in-use method for evaluating effect of microbes on plastic materials under realistic field conditions is the IEC 68-2-10 method. In this test, specimens are inoculated with mold spores, followed by a period of incubation under conditions that promote spore germination and growth of molds (28–84 days at 97% humidity). As in EN ISO 846, microbial growth is tested in the presence and absence of nutrients to determine whether the additives in the plastic serve as nutrients for microorganisms.

8.10 Regulations Related to Antimicrobials

Regulations vary by country and require antimicrobials to be tested and registered. The manufacturer that uses antimicrobial agents must maintain up-to-date knowledge of the relevant regulations pertaining to the industry/region.

8.10.1 USA

All biocides must be registered by the US Environmental Protection Agency (EPA) for intended use in plastics. Under the Federal Insecticide, Fungicide, and Rodenticide Act (FIFRA), companies cannot make public health-related pesticidal claims for any plastic product distributed or sold unless the plastic product has been approved and registered (licensed) by the EPA or is covered by an exemption from registration. EPA created this limited exemption to allow pesticide-treated articles to be sold without an EPA registration so long as product claims are solely limited to protection of the treated product itself (i.e., when the biocide acts as preservative/biostabilizer).

A prerequisite for any public health claim on plastic products, such as "protects against microorganisms", "controls fungus", or "kills germs", is registration of the product at EPA. To obtain FIFRA registration, EPA requires specific efficacy data to support the product's public health labeling claims for the patterns of use and other data in support of product safety. Antimicrobials used in plastics intended to come in contact with food have to be registered as indirect food additives by the US Food and Drug Administration (FDA).

8.10.2 Europe

Antimicrobial agents in non-food contact applications are not regulated in the European Community by any specific directive, although some countries have national regulations for antimicrobials in plastics (e.g., the Netherlands, Belgium, and UK).

The Biocidal Products Directive (BPD) regulates the use of antimicrobials in plastics. The directive aims to introduce a harmonized European system for biocidal products, replacing various national regulations, and to provide greater protection for both the human population and the environment. The directive covers 23 product types and fills in the gaps between the existing European legislation on products, such as pesticides, medicines (both human and veterinary), and cosmetics. The directive covers both biocidal substances and active ingredients.

For antimicrobials used in plastics intended to come in contact with food, the indirect food contact additive registration is required.

8.11 Future Trends

There is certainly a trend toward more environmentally friendly antimicrobials of low toxicity. Arsenic- and heavy-metal-containing antimicrobials are being replaced by isothiazolones or silver- or zinc-based substances, among others.

Increased demand for plastic products with antimicrobial efficacy beyond preservation is expected. Antimicrobials with a favorable toxicity profile and high efficiency when used on the surface of plastics, such as triclosan, which is already approved for skin care applications and food contact in the EU, are likely to increase their market share over the agents that cannot make similar claims. For food contact-suitable plastics and applications, research on natural antimicrobial substances is ongoing.

Two main alternatives for obtaining 3D-printed, antimicrobially active articles are: (a) use of natural polymers that already exhibit intrinsic antibacterial performance, and (b) use of synthetic polymers produced with monomers that contain antimicrobial functional groups. These 3D-printing methodologies enable manufacturing and fabrication of materials with molecular architectures similar to biological tissues, such as bones, cartilages, or heart valves.

Use of naturally occurring antimicrobial polymers and their derivatives has been increasing in recent years. In this context, cellulose, cellulose acetate, and nanocellulose are the most extensively employed natural polymers due to their intrinsic antimicrobial properties. It makes these polymers attractive either for use as antimicrobial polymer additives or for 3D-printing of biomedical polymer articles.

Composites have also been explored for the preparation of antimicrobial 3D-printed parts. These composites have been fabricated using graphene oxide, metal ions of zinc, copper, or silver, and TiO_2 in combination with a thermoplastic polyurethane (TPU) and poly(lactic acid) (PLA).[8)]

8.12 Commercially Available Antimicrobials

Selected commercially available antimicrobials used for protecting plastics are listed in Table 8.1.

Table 8.1 Selected Types of Antimicrobial Agents Used in Polymers

Abbreviation	Chemical composition/name	CAS no.
Inorganic		
Ag (Ag/Zn, Ag/Cu)	Silver ion on a carrier (zeolite, glass)	—
Zn	Zinc ion on a carrier	—
Cu	Copper ion on a carrier	—
Organic		
OBPA	10,10′-Oxybisphenoxarsine	58-36-6
IPBC	3-Iodo-2-propynyl butylcarbamate	55406-53-6
ZNP	Zinc pyrithione	13463-41-7
TSC (triclosan)	2,4,4′-Trichloro-2′-hydroxydiphenyl ether	3380-34-5
OIT	2-n-Octyl-4-isothiazolin-3-one (in DOP, DINP, ESBO)	26530-20-1

[8] C. M.González-Henríquez, M. A.Sarabia-Vallejos, and J. R.Hernandez, "Antimicrobial polymers for additive manufacturing", *International Journal of Molecular Sciences*, 2019, 20: 1210.

Table 8.1 Selected Types of Antimicrobial Agents Used in Polymers (*continued*)

Abbreviation	Chemical composition/name	CAS no.
DCOIT	4,5-Dichloro-2-octyl-4-isothiazolin-3-one	64359-81-5
TCPP	Tris(chloropropyl) phosphate	13674-84-5
TCMS	2,3,5,6-Tetrachloro-4-(methyl sulfonyl)pyridine	13108-52-6
Chlorothalonil	Tetrachloroisophthalonitrile	1897-45-6
BCM	Benzimidazole carbamic acid methyl ester [carbendazim]	10605-21-7
BIT	1,2-Benzisothiazolin-3-one [benzisothiazolinone]	2634-33-5

8.13 Questions

1. What is called microbiostatic activity?
2. What is called microbiocidal activity?
3. What outdoor conditions facilitate microbial growth?
4. What are the main three types of antimicrobials?
5. What is the mechanism of action of organic antimicrobials?
6. What is the mechanism of action of inorganic antimicrobials?
7. What are the disadvantages of organic antimicrobials?
8. What are the advantages of inorganic antimicrobials?
9. What are the modes of biodeterioration of plastics? Give examples.
10. What plastics require the incorporation of antimicrobials?
11. What microorganisms and their functions are targeted by biostabilizers?
12. What microorganisms and their functions are targeted by antimicrobial active ingredients?
13. What components of flexible PVC compounds are most susceptible to microbial attacks?
14. What are the modes of action of biocides?
15. What is the controlled release mechanism? How is it applicable to antimicrobials?
16. Give examples of antimicrobial polymers and their uses.
17. Give examples of antimicrobial composites and their uses.

9 Colorants and Fluorescent Agents

Content of this chapter was adapted from:

- S. Scherrer and R. L. Sykes, "Colorants" (Chapter 15), in: *Plastics Additives Handbook* (6th edition), ed. H. Zweifel, R. D. Maier, and M. Schiller, Hanser, 2009, pp. 831–899.
- A. G. Oertli, "Optische Aufheller" (Chapter 8), in: *Handbuch der Kunststoff-Additive* (4th edition), ed. R. D. Maier and M. Schiller, Hanser, 2016, pp. 813–832.

9.1 Key Terms and Definitions

Colorants	Substances like dyes or pigments added to plastics to impart or modify color, affecting appearance.
Dyes	Substances that dissolve in plastics, giving transparent materials.
Pigments	Substances that remain as solid particles, providing opacity.
Organic pigments - properties	Substances that have chemical structures based primarily on carbon; they are bright, pure but less stable.
Inorganic pigments	Substances that have chemical structures that do not contain carbon; they are duller but highly durable, heat- and light-stable.
Pigment dispersion	The process of evenly distributing pigment particles in plastic to achieve a uniform color and to avoid streaks or clumps.

Processing challenges in plastics	Difficulties like poor dispersion, thermal degradation, or uneven coloration during plastic manufacturing.
Fluorescent dyes	Additives that absorb UV light and emit visible light, producing intense, glowing colors.
Optical brighteners	Chemicals that absorb UV light and re-emit blue light, making plastics appear whiter and brighter.
Color strength	The intensity of a colorant's effect; higher strength means less additive is needed to achieve the same shade.
Metamerism	The phenomenon where two samples match under one light source but differ under another.
Migration testing	The process of assessing if colorants or additives move from plastics into other substances (e.g., food, skin).
Environmental impact	The potential effects of colorants and additives on ecosystems, recyclability, and sustainability.
Heavy-metal-free colorants	Colorants formulated without toxic metals like lead, cadmium, or chromium in order to meet safety standards.

9.2 Introduction

Colorants change the appearance of plastics to the human eye, enabling them to appear yellow, green, blue, orange, black, or another color when viewed in the light.

The term "colorant" generally encompasses pigments and dyes. In this chapter, the definition will be expanded slightly to include fluorescent dyes and optical brighteners, because, as their names suggest, they also affect the perceived color. Colorants influence the properties of a formulation and have their own characteristics, such as weather resistance, lightfastness, transparency, and, in some cases, approval for food contact packaging.

Color is an individual perception triggered by the light that reaches our eyes. Physically speaking, light consists of electromagnetic waves perceived by the human eye in the wavelength range 400–700 nm. The color stimulus triggered by an object is a dual function of illumination and observation. First, the distribution of luminous intensity across the wavelength spectrum determines the illumination. Observation then encompasses the influence of the object on the proportion of light it reflects from the light source to the observer's eye.

It is important to note that black and white are not, strictly speaking, "colors". Black is the absence of light, and an object appears black when no light is reflected or emitted.

Therefore, black is considered a "non-color" or a "hue without color". White, on the other hand, is the sum of all visible light colors, and occurs when all wavelengths of visible light are reflected equally. White is therefore not a single hue, but the result of all colors combined.

Colorants and additives absorb energy depending on their function, and energy absorption is substance-specific. A particular colorant absorbs visible light, and the remaining light we see is perceived as the color of the object. The corresponding terms are reflection, transmission, and absorption (see also Section 3.2). There is a special feature called metamerism, which occurs when two colors look the same under a certain light source but appear different under a different light source. This happens because the colors have different compositions, but our eye perceives them as identical under certain lighting conditions. As soon as the light changes (e.g., from daylight to artificial light), the differences become visible. This phenomenon is particularly important when color-matching materials, for example, in the textile or plastics industries.

9.3 Colors, Classifications, and Measurements

In the past, there have been various approaches to defining colors, three of which are discussed below.

9.3.1 Color Index

The Color Index (C.I.) is published by the British Society of Dyers and Colorists and the American Association of Textile Chemists and Colorists, and is a reference work for all common dyes and dye-based chemicals that has existed since 1925. It is considered the standard work in the field of pigment and dye chemistry, and contains information on the chemical structure and common names of dyes, as well as on their synthesis, reactivity, solubility, and patent literature. Manufacturers voluntarily include information about dyes in the Color Index, but this does not generally extend to information about the chemical structures, which are not publicly available for many dyes.

The C.I. incorporates a system for classifying the very large number of known pigments and dyes, with each product offered on the market being classified by a "generic name" that describes the product. The corresponding generic names for colorants used in plastics are called "pigment" and "solvent". The generic name consists of the colorant type, the hue, and a special number. For example, C.I. Pigment Red 254 (commonly abbreviated as P.R. 254) is a red diketopyrrolopyrrole pigment (DPP pigment), while an example of a polymer-soluble dye is C.I. Solvent Yellow 21 (abbreviated S.Y. 21), a chromium complex of an azo dye.

9.3.2 RAL Colors

RAL Colors refer to standardized colors used by RAL gGmbH (a subsidiary of the RAL German Institute for Quality Assurance and Labeling). This means it is a series of world-proven color systems and color catalogs, each with a palette of standardized colors (digital and owned). Each color is associated with a unique number, which can be exchanged between a customer and supplier, avoiding the need to create a color sample on a defined material.

The best known RAL color collection is "RAL Classic", which was first created in 1927 by the previous organization, the Reichs-Committee for Delivery Conditions (RAL). The color collection contains 213 shades, originally for water-soluble paints, and is used by many manufacturers for products and brands in order to facilitate color communication and to ensure a long-lasting color impression. Each color in the color catalog is assigned a four-digit color number. This standardization goes back to a table of 40 colors that was drawn up in 1927.

Various criteria must be met for another color to be included in the catalog. On the one hand, the color must be timeless and subject to an overriding public interest. Furthermore, the color must be environmentally friendly, weather-resistant, high-opacity, and with a visible difference to existing colors.

The "RAL Design" color system was introduced in 1993. It comprises 1625 standardized colors, which are described with coordinates in the CIELAB color space (see below). In contrast to the "RAL Classic" color collection, the colors here are systematically classified according to hue (H), brightness or lightness (L) and chroma (C).

In October 2010, "RAL Plastics", a separate color collection for plastics, was introduced. It comprises the 100 most frequently requested colors from the "RAL Classic" color collection, and is to be gradually expanded. The color samples consist of DIN A6-sized polypropylene panels, which have three surface structures and three levels of panel thickness.

A software platform ("RAL Digital") makes the color catalogs available digitally, and RAL gGmbH publishes the color trend book "RAL Color Feeling" at regular intervals, along with "The Colour Dictionary", "Colours of Health & Care" and "Colours for Hotels".

In the color table according to RAL Classic, the arrangement is made according to the color groups, as marked by the leading digit (1000 = yellow, 2000 = orange, 3000 = red, 4000 = violet, 5000 = blue, 6000 = green, 7000 = grey, 8000 = brown, 9000 = white/black).

9.3.3 CIELAB Color Space

Colors can also be measured. Color spaces like CIELAB (also called CIE Lab) are mathematical models used to describe colors precisely and independently of devices. CIELAB was developed in 1976 by the Commission International d'Eclairage (CIE) to represent colors as perceived by the human eye. It is three-dimensional:

- L^* represents lightness (from 0 = black to 100 = white).
- a^* represents the color axis from green (negative values) to red (positive values).
- b^* represents the color axis from blue (negative values) to yellow (positive values).

Compared to other color spaces like RGB or CMYK, CIELAB has the advantage of being device-independent – meaning that colors can be communicated and compared accurately across different devices (e.g., monitors, printers). It also covers the entire range of human color perception and allows precise calculation of color differences. In practice, CIELAB is widely used for quality control, color measurement, and color matching between different media.

9.4 Transparency, Opacity, and Dichroism (Optical Anisotropy)

Pigments are relatively inert particles that are physically enclosed in the polymer matrix. Particles smaller than approximately 0.2 µm scatter light very little. Therefore, objects containing these particles are transparent. Pigment particles with a size of approximately 0.2 µm and larger, on the other hand, scatter light, and the objects colored with them are opaque or semi-opaque. When an object is colored with a dye, it appears transparent because the dye dissolves molecularly in the substrate. It does not form particles that cause light scattering.

Pigment particles can orient themselves in a preferred direction during processing, and the color then depends on the direction of observation. The cause of this phenomenon is the fact that certain pigments form crystal systems with low symmetry, and such crystals have direction-dependent properties. Typical examples are the needle-shaped iron oxide yellow or lead chromate pigments. Regarding their coloristic properties, the absorption and refractive indices differ between pigments with different crystallographic structures.

9.5 Definitions of Some Colorants

Colorants are divided into two classes based on their physical behavior: pigments and dyes. The key application properties are color strength, hue, transparency, heat resistance, lightfastness, weather resistance, migration behavior, dispersibility, and ease of incorporation.

Pigments are organic or inorganic, colored, white, or black particles that are practically insoluble in the medium into which they are introduced. These particles absorb and scatter light in the medium, with the smallest unit being called a primary particle. Primary particles with regular shape, such as needles or plates, can form aggregates, and due to strong cohesive forces, aggregates are difficult to break up. Primary particles have a crystalline character with a predetermined surface area. When they aggregate together, the inner surface is inaccessible. Therefore, the surface area of an aggregate is smaller than the total surface area of the individual particles that make up the aggregate. When irregularly shaped primary particles and/or aggregates bond loosely together, they form agglomerates, which can be broken up by applying high energy during the dispersion process.

Dyes are soluble in the application medium and their chemical structure is very similar to that of pigments. However, since dye materials, unlike pigments, dissolve in the medium, there are no visible particles, and the transparency of the medium remains unchanged. Both pigments and dyes cover a wide range of properties (Figure 9.1).

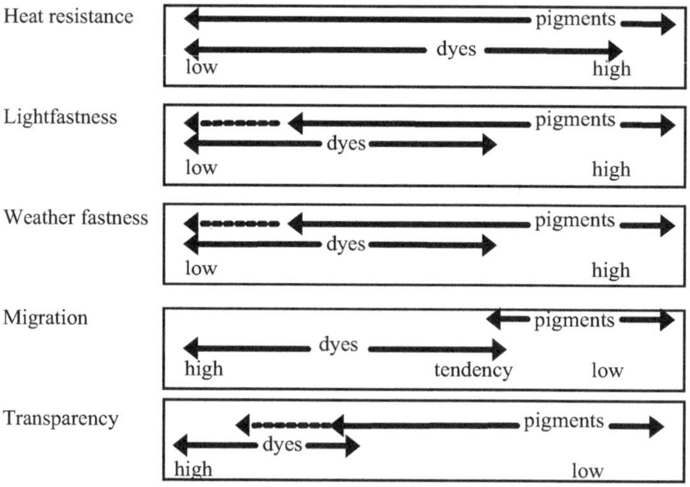

Figure 9.1 Overview of the properties of dyes and pigments

While regular dyes appear colored because they absorb certain wavelengths of visible light and reflect or transmit others, fluorescent dyes absorb higher-energy light (such as ultraviolet light) and immediately re-emit part of this energy as visible light – they actively "glow" under appropriate illumination. Fluorescent dyes are a special type of dye because they not only result in color by absorbing and reflecting light, but they also have the unique ability to absorb light of a certain (usually shorter) wavelength and then emit light of a different (usually longer) wavelength. This process is called fluorescence (see also Section 3.2).

In addition to the CAS registration number, there are many ways of referring to colorants. These include:

- C.I. Generic Name and C.I. Constitution Number
- Chemical structure or chemical class (e.g., azo pigment, phthalocyanine green, diketopyrrolopyrrole red)
- Trade names.

9.6 Properties, Test Methods, and Assessment of Performance

Colorant manufacturers test and publish the properties of their products to enable users to more quickly find the desired colorant in their product range. Due to the complexity of plastic formulations and the specific application conditions, the appropriate selection must be based on specific, individual formulations, taking into account the final processing conditions.

The assessment of a colorant's properties is performed colorimetrically using a spectrophotometer, or visually. The human eye and experience are still good means of making comparisons.

Increasingly high processing speeds, the use of thinner films and thin-walled injection-molded components, and rising expectations regarding the aesthetic appearance of plastic products are leading to higher demands on dispersion quality. Insufficient dispersion manifests itself in various ways:

- Fluctuations in color intensity and deviations in shade
- Streaking
- Appearance of visible pigment agglomerates and specks
- Blockage of melt filters in the extruder
- Tearing of the film during blown film extrusion due to unbroken pigment agglomerates
- Breakage of fibers, monofilaments, and ribbons

- Printing problems due to surface inhomogeneities
- Reduced mechanical strength (impact resistance, tensile strength, and elongation behavior)
- Severe abrasion on machine components due to inadequate coating of the pigment particles by the polymer
- Viscosity problems such as sedimentation and separation or premature thickening of pigment pastes.

Solving such problems requires a systematic approach to the entire dispersion process. This is often not possible for the processor, so they must rely on the pigment supplier to deliver the colorant in a form that prevents such difficulties from occurring. Colorants may change color when exposed to high temperatures during processing (see also Table 1.4). Heat resistance, thermal stability, or temperature resistance is always specified by two parameters: the temperature and the period of time for which the colorant is exposed to that temperature.

In many plastic applications, it is necessary that the original color does not change after exposure to daylight and artificial light. UV light is the most harmful component of the daylight radiation spectrum. ISO 105 B02 is the standard test method for lightfastness using xenon lamps, and the results are typically evaluated with the blue scale; this scale is also used in Atlas Xenotest® weathering test equipment (see also Section 3.5.2). Table 9.1 shows the lightfastness ratings and their meaning; note that these ratings are not absolute but indicate lightfastness under specific test conditions.

Table 9.1 Lightfastness Scale

Rating	Description
8	Outstanding
7	Excellent
6	Very good
5	Good
4	Moderate
3	Fair
2	Poor
1	Very poor

The best way to test the durability of a dyed material is to expose it to outdoor weather conditions. In addition to visible light and UV and IR radiation, rain, humidity, and temperature changes play an important role in this test. One of the most well-known locations for outdoor weathering tests is in Florida, USA (see Table 3.2).

Testing instruments used for artificial weathering include the Atlas Weather-Ometer®/Xenotest®, which determines weather resistance by applying light from a xenon arc lamp according to test method ASTM G155 or ISO 4892-2. After exposure, the sample is compared to the original, and fading is assessed using a gray scale. Other changes, such as a change in color, loss of gloss, darkening, and chalking, are also assessed, and the test methods are usually published along with the results.

Chalking is a phenomenon of weathering during which the surface of a plastic product turns white independently of the previous color. This occurs because, under long-term UV radiation during outdoor exposure, the outer polymer layer can degrade slightly, increasing the number of microparticles of pigments (such as titanium dioxide) and/or releasing inorganic functional fillers (such as calcium carbonate) from the surface, which produces a white, powdery deposit on the polymer. The higher the content of calcium carbonate in a PVC product, the earlier that chalking appears and the more visible it is.

When the filler content is increased, the resulting materials exhibit a distinct lightening in color.

Both outdoor weathering and artificial weathering were found to significantly affect product performance, and this influence was particularly pronounced in the presence of chalk loadings. The impact of solar irradiation was especially pronounced during outdoor weathering, with samples undergoing faster and more pronounced chalking when exposed to high solar irradiation compared with periods of lower solar exposure. Other factors, such as temperature, temperature variations, precipitation amounts, and the type of precipitation, also played a role. For instance, snow cover provided temporary relief from light stress during the period covered.

If the colorant is only partially dissolved in the plastic, it can migrate to the surface or into a system in contact with the plastic. Migration is also referred to as "bleeding" or "blooming". In pigmented plastics whose glass transition temperature is above the storage temperature, pigment migration generally does not occur.

Four types of migration can be distinguished:

- **Solvent bleeding** occurs when the colored plastic releases colorant into the solvent. For testing, the colored plastic is stored in the solvent for a defined period of time. If the solvent does not extract any colorant during the test, the colorant's resistance to the solvent is rated as 5. Any value below 5 means that the colorant is not resistant to the corresponding solvent.
- **Contact bleeding** occurs when the colorant migrates from the colored object to another object. Typical test conditions include, for example, migration into p-PVC. According to DIN 53775 Part 3, the colored PVC is in contact with the corresponding uncolored PVC for 24 h at 80 °C under a pressure of 0.1 MPa. The uncolored PVC is evaluated using the ISO 105 A03 gray scale (1 to 5), with the range between

4 and 5 divided into 0.2 increments. A rating of 5 represents no migration and 1 represents very strong migration.

- **Blooming** is the phenomenon of pigments migrating to the surface of the plastic. This is a very slow process. The result is a hazy surface that may appear bronze-like. The assessment is carried out in the same way as for contact bleeding (gray scale).
- **Plate-out** is defined as uncontrollable occurrence of often quite disruptive deposits or coatings, which form after a longer production period on the hot or cold metal surfaces of an extrusion line or on the calender rolls.

However, these incompatibilities could in principle be estimated by calculating the Hildebrand solubility parameters of the organic dyes from their molecular structure and density using the group contribution method (see Table 1.3) and then comparing the resulting values with the polymer solubility parameters listed in Table 1.1.

Colorants can change the flow properties of plastics. Soluble dyes generally have no significant effect when added in small concentrations, but pigments can change the viscosity in the same way as fillers. The particle size and distribution of the pigments play an important role. However, pigment concentrations used are usually low, and therefore their influence is relatively small, except in highly pigmented thin films and thin-walled parts.

Colorants can affect the final shape of extruded or injection-molded parts, especially in semi-crystalline plastics. When such deformations occur after processing, the phenomenon is called warpage, which can be tested on a flat plate. When warpage occurs, the part bends after the injection-molding process.

9.7 Colorant Classes

9.7.1 White Colorants

C.I. Pigment White 6, titanium dioxide (TiO_2) is the most widely used white pigment in the plastics industry. Its main characteristics are brilliance, weather resistance, opacity, and chemical resistance (see also Section 3.5.1.2). Two types of titanium dioxide are available on the market, anatase and rutile, which differ in their crystal form. Anatase is relatively soft, a photocatalyst that accelerates the decomposition of the polymer in sunlight and is used in fibers. Rutile is the most widely used white pigment in the plastics industry. The rutile form of titanium dioxide has very good opacity, excellent lightfastness, and excellent weather resistance. It is often used in combination with organic pigments to control opacity and create pastel shades. Other white pigments, such as zinc oxide and zinc sulfide, are occasionally used for plastics.

9.7.2 Black Colorants

C.I. Pigment Black 7, carbon black, is simple carbon. It is by far the most widely used black colorant for plastics applications (see also Section 3.5.1.2). Carbon black is available in many different particle sizes that determine its color strength and hue (brownish to bluish gray tones). Carbon black possesses outstanding chemical resistance, lightfastness, and weathering resistance.

C.I. Pigment Black 11, iron oxide (Fe_3O_4), possesses outstanding dispersibility and migration resistance. However, it is limited in terms of thermal stability and color strength. Under certain conditions, it accelerates the thermal decomposition of PVC.

There are also black pigments in the category of “complex inorganic colored pigments”, such as C.I. Pigment Black 26 (Mn/Fe oxide), C.I. Pigment Black 27 (cobalt/chromium/iron/manganese oxide), or C.I. Pigment Black 29 (Fe/Cr oxide). They are characterized by outstanding heat resistance, weather resistance, and chemical resistance, and high NIR reflectivity.

Specialty products such as the IR-transparent black pigments Lumogen® Black K 0087 and K 0088 are suitable for applications requiring high transmittance in the infrared range. Polymer-soluble dyes such as Oracet Black 007 are also available on the market, which are primarily used for PS, PMMA, and u-PVC.

9.7.3 Inorganic Pigments

Metal oxides, hydroxides, and sulfides are, in many cases, extremely stable colorants, and have been used for many years. Some of them are viewed critically from an environmental perspective, particularly those based on lead, mercury, hexavalent chromium, and cadmium. Recent market developments offer new possibilities for coloring plastics with inorganic pigments. These include, for example, the bright yellow bismuth vanadate and orange-to-red cerium-based sulfides.

Yellow pigments

Chrome yellow, nickel titanium yellow, chromium titanium yellow, iron oxides, and bismuth vanadates are the most important inorganic yellow pigments. C.I. Pigment Yellow 42 (iron oxide), can be used for rubber, and has a heat resistance typically around 160–180 °C. C.I. Pigment Yellow 34 (lead chromate) has been used in many plastics applications in the past, and has a heat resistance in the range 140–300 °C. Products with higher resistance to heat and light are coated to varying degrees with silicon dioxide. However, this product group is being replaced in many applications because it contains heavy metals. C.I. Pigment Yellow 34 also darkens when exposed to light.

C.I. Pigment Yellow 53 and C.I. Pigment Brown 24 contain oxides based on nickel/antimony/titanium (greenish) and chromium/antimony/titanium (reddish). They exhibit exceptionally good heat resistance up to 1000 °C. These pigments have relatively low color strength.

C.I. Pigment Yellow 184 is based on bismuth vanadate. This material replaces chromium- and cadmium-containing pigments. It exhibits excellent lightfastness and weather resistance, and its heat resistance is in the range 240–300 °C. Because it contains no heavy metals but has good properties, this pigment type is gaining increasing importance in the market.

Orange pigments

The only important inorganic orange pigments are molybdate orange, cerium sulfide, and orange pigments from the series of complex inorganic pigments.

C.I. Pigment Red 104 (lead chromate molybdate sulfate) exhibits high color strength, but is known to darken under the influence of light. Because it is inexpensive and readily dispersible, this pigment type was often used in plastic applications in the past, despite its lead content, but this product group is being replaced in many applications because it contains heavy metals.

C.I. Pigment Orange 75 and 78 (cerium sulfide), have high hiding power as well as outstanding heat and weather resistance. In combination with titanium dioxide, they can produce brilliant shades of salmon red and pink.

C.I. Pigment Orange 82 and C.I. Pigment Yellow 216 (zinc/tin/titanium oxides) are from the series of complex inorganic pigments, and are new to the Color Index. In addition to the inherent properties of this pigment class, they are characterized by brilliant shades and high color strength. They are particularly suitable for demanding plastics such as polyamide.

Brown pigments

C.I. Pigment Brown 29, C.I. Pigment Brown 33, C.I. Pigment Brown 35, and C.I. Pigment Brown 35 (all based on chromium or iron oxide) offer outstanding lightfastness and heat resistance (300 °C) combined with high opacity.

Red pigments

C.I. Pigment Red 101 (iron oxide), is used in a wide variety of products in opaque to transparent variants. This pigment does not produce a very pure color but offers outstanding heat resistance and lightfastness. Iron oxide serves as a versatile tinting pigment in many formulations.

C.I. Pigment Red 104 (lead chromate molybdate sulfate), is also known as a mixed-phase pigment. Coating (stabilization) with sodium silicate during production increases the pigment's lightfastness and weather resistance to a level sufficient for many applications. Heat resistance can reach 300 °C for highly stabilized grades. However, this product group is being replaced in many applications because it contains heavy metals.

C.I. Pigment Red 265 and C.I. Pigment Red 275 (cerium sulfide), offer high hiding power as well as outstanding heat and weather resistance. They are suitable for demanding polymers such as engineering plastics.

Violet pigments

C.I. Pigment Violet 15 (aluminium sodium thiosilicate), Ultramarine Violet, offers heat resistance up to 250 °C and outstanding lightfastness.

However, ultramarine pigments (including Ultramarine Violet and Ultramarine Blue) react with acids, causing the pigment to lose its color and release hydrogen sulfide; this takes place even with weak or diluted acids. This is why more acid-stable, coated grades have been developed for special applications. Surface-treated grades of this pigment with improved acid resistance are also available.

C.I. Pigment Violet 16, Manganese Violet, offers heat resistance up to 300 °C and good lightfastness.

Blue pigments

C.I. Pigment Blue 29 (aluminium sodium thiosilicate), Ultramarine Blue, has heat resistance in the range 300–400 °C and good lightfastness. However, it reacts with acids (see Ultramarine Violet). Therefore, special grades are available with surface treatments to increase acid resistance.

C.I. Pigment Blue 28 and C.I. Pigment Blue 36 are mixed metal oxides based primarily on cobalt aluminate. They are heat-resistant up to 1000 °C and offer excellent dispersibility.

Green pigments

C.I. Pigment Green 17, Chromium Oxide Green, offers outstanding lightfastness and heat resistance up to 1000 °C, but the pigment is highly abrasive.

C.I. Pigment Green 26 and C.I. Pigment Green 50 are cobalt-based mixed metal oxides. They exhibit outstanding heat resistance in the range 300–1200 °C as well as outstanding lightfastness.

9.7.4 Organic Colorants

Due to increasing environmental concerns about heavy metals, some important groups of inorganic pigments are being replaced by organic colorants. Inorganic pigments are particularly useful when high opacity and high processing temperatures are required. Organic colorants, on the other hand, are primarily used when high color strength and brilliant shades are needed. Organic colorants are often used in combination with inorganic pigments. This section describes organic colorants according to their chemical class, as shown in Table 9.2. This table will not be discussed further in the text. This list is intended to serve as a guideline for beginners, and no claim to the completeness of this information is made; product designations are registered trademarks.

9.7.5 Fluorescent Dyes

Fluorescent dyes are specialized organic colorants that emit intense, vivid light when excited by UV or visible light. In the plastics industry, these dyes are used to achieve brilliant, luminous colors and to incorporate functional features such as security markers or product tracers. Unlike pigments, fluorescent dyes dissolve completely in the polymer matrix, resulting in transparent and highly colored plastics.

Several classes of fluorescent dyes are commonly used in plastics, each listed in the Color Index (C.I.):

- Rhodamine dyes (e.g., C.I. Solvent Red 49, C.I. 45170) produce strong red-to-pink fluorescence and are widely used in PS, PC, and PMMA.
- Fluorescein dyes (e.g., C.I. Solvent Yellow 98, C.I. 56220) provide yellow–green fluorescence and are suitable for transparent plastics such as PS, PMMA, and PVC.
- Coumarin dyes (e.g., C.I. Solvent Blue 35, C.I. 61554) offer blue-to-green fluorescence and are also used in PS, PMMA, and PC.
- Benzoxazole-based dyes (e.g., C.I. Fluorescent Brightener 184, C.I. 367) act as optical brighteners and are applied in PE, PP, PVC, and polyesters.

Fluorescent dyes are especially effective in transparent and amorphous thermoplastics, where the dye molecules can dissolve well and the fluorescence is most visible. The most suitable plastics include PS, PMMA, PC, PVC (especially when unplasticized), polyesters such as PET, PE, and PP, with the latter two mainly for optical brightening. For technical applications, such as security or recycling, fluorescent dyes are also used as markers to uniquely identify or sort plastic materials. Fluorescent dyes are ideal for masterbatch production and for coloring transparent plastics via melt processing, and for demanding applications requiring high light or heat stability, specially stabilized dye types are available. It should be noted that complete dissolution of the dye is essential for uniform fluorescence and transparency.

9.7.6 Powdered Metal Pigments

Metal-effect pigments consist of flakes or lamellae of aluminium or other metals (copper or zinc). The aluminium-based pigments, called aluminium flakes or aluminium bronze, are the most widely used metallic pigments; their silver-metallic shade can be varied by adding transparent colorants. Copper/zinc bronze pigments are used for gold and copper shades.

9.7.7 Pearlescent Pigments

Pearlescent pigments based on metal oxides with mica are the major effect pigments on the market. Similar to the effect seen with pearls, light is reflected from different parallel layers. Because of the different refractive indices of the titanium dioxide layer compared to the mica, the angle of reflection changes and the typical pearl effect is observed. These pigments have excellent heat resistance and can be used in practically all plastic applications.

9.8 Coloring Technology

Colorants for a specific polymer and end-use are selected according to the following criteria:

- The desired color and effects – taking into account transparency/opacity and any requirements related to selective absorption/transmission of certain wavelengths of light (for example, for camouflage or infrared control)
- The requirements of the polymer and processing technology – taking into account the risk of interaction with other additives included in the formulation
- The performance requirements of the final product during use – taking into account the need to comply with legal regulations regarding the coloring of consumer goods, food packaging, and toys.

In recent years, two developments have led to higher demands being placed on colorants in plastics. First, the development of numerous polymer blends, combined with ever-increasing processing speeds for almost all polymers, has led to colorants having to meet increasingly stringent requirements regarding color tone and heat resistance during processing. Second, the use of pigments based on certain heavy metals, especially cadmium, lead, mercury, and hexavalent chromium, is increasingly subject to health, safety, and environmental restrictions. These pigments are continually being replaced by alternative products.

Table 9.2 lists the most important inorganic and organic color pigments and dyes used in plastics coloration, based on information available from manufacturers. Principal areas of suitability and use are indicated, together with examples of commercial products and leading suppliers. This list is intended to serve as a guideline for beginners, and no claim to the completeness of this information is made; product designations are registered trademarks.

Table 9.2 Inorganic and Organic Colorants

C.I. Designation	Chemical description	Light-fastness[a]	Heat-resistance (°C)[b]	Migration[c]	Principal polymers in which used[d][e]
Yellow inorganic pigments					
Pigment Yellow 53 (Nickel Titanium Yellow)	$(Ti,Ni,Sb)O_2$	8 (pure); 8 (1:4)	300	5	Generally applicable
Pigment Brown 24 (Chrome Titanium Yellow)	$(Ti,Cr,Sb)O_2$	8 (pure); 8 (1:4)	300	5	Generally applicable
Pigment Yellow 42	Iron oxide hydrate (FeO(OH))	8 (pure); 8 (1:4)	220 (PVC)	5	p-PVC
Pigment Yellow 119	Zinc/iron	8 (pure); 8 (1:4)	300	5	Generally applicable
Pigment Yellow 184	Stabilized bismuth vanadate	7–8 (pure); 7–8 (1:2)	280 (PO)	5	LDPE, HDPE
Pigment Yellow 184	Highly stabilized bismuth vanadate	8 (pure); 8 (1:2)	300 (PO)	5	HDPE, ABS, PA
Yellow organic pigments					
Pigment Yellow 81	Diazo diarylide	7 (pure); 6 (1:10)	230 (PO)	4–5	PVC, LDPE
Pigment Yellow 93	Diazo condensation compound	8 (pure); 7 (1:10)	280 (PO)	5	PVC, PS, LDPE, HDPE, PP
Pigment Yellow 95	Diazo condensation compound	6–7 (pure); 6–7 (1:10)	280 (PO)	5	PVC, PS, LDPE, HDPE, PP
Pigment Yellow 128	Diazo condensation compound	8 (pure); 7 (1:10)	240 (PO)	5	PVC, LDPE

Pigment Yellow 62	Monoazo Ca salt	7 (pure); 7 (1:10)	240 (PO)	5	PVC, LDPE, HDPE, PP, PS
Pigment Yellow 168	Monoazo Ca salt	7 (pure); 7 (1:10)	260 (PO); 240 (PO)	5	PVC, LDPE, HDPE, PP, PS
Pigment Yellow 183	Monoazo Ca salt	8 (pure); 5 (1:10)	300	4–5	PVC, PO, PS, SAN, PMMA, ABS, PPE
Pigment Yellow 191	Monoazo Ca salt	7 (pure); 6–7 (1:5)	300 (PO)	5	PVC, LDPE, HDPE, PS, PP
Pigment Yellow 191:1	Monoazo NH_4 salt	7–8 (pure); 6–7 (1:10)	280 (PO)	5	PVC, LDPE, HDPE, PP, PS
Pigment Yellow 180	Diazo benzimidazolone	7 (pure); 6 (1:10)	290 (PO)	5	PVC, LDPE, HDPE
Pigment Yellow 181	Monoazo benzimidazolone	8 (pure); 7–8 (1:10)	300 (PO)	5	PVC, LDPE, HDPE
Pigment Yellow 109	Tetrachloro-isoindolinone	8 (pure); 7–8 (1:10)	280 (PO)	5	PVC, PS, LDPE, HDPE, PP
Pigment Yellow 110	Tetrachloro-isoindolinone	8 (pure); 7–8 (1:10)	280 (PO)	5	PVC, PS, LDPE, HDPE, PP
Pigment Yellow 138	Quinophthalone	8 (pure); 8 (1:10)	280 (PS); 260 (PO)	4–5	PVC, PS, LDPE, HDPE
Pigment Yellow 139	Isoindoline	8 (pure); 7–8 (1:10)	220 (PVC); 250 (PO)	5	PVC, LDPE, PP
Pigment Yellow 147	Anthraquinone	8 (pure); 6–7 (1:10)	300 (PET); 300 (PC)	4–5	PMMA, PS, PC, PBT, PET

Table 9.2 Inorganic and Organic Colorants *(continued)*

C.I. Designation	Chemical description	Light-fastness[a]	Heat-resistance (°C)[b]	Migration[c]	Principal polymers in which used[d][e]
Yellow dyes					
Solvent Yellow 93	Pyrazolone	8 (0.05%, PS); 8 (1/3, SD-PS)	300 (0.05%, PS)	5	PMMA, U-PVC, PS
Solvent Yellow 68	Fluorescent polycyclic	7 (pure); 6 (1:5)	300 (PS)	4–5	PS, PMMA, U-PVC
Solvent Yellow 145	Fluorescent polycyclic	7 (0.1%, PS); 5 (1:20, PS)	300 (PS); 340 (PC)	4	PS, SAN, PC, ABS, PET
Solvent Yellow 163	Anthraquinone	8 (0.05%, PC); 6–7 (1:20)	360 (PC)	4	PMMA, PS, PC, PBT, PET, ABS, PPE
Disperse Yellow 201	Methine	8 (0.05%, PS); 7–8 (1/3, SD-PS)	300 (0.05%, PS)	—	PS, SAN, ABS, PMMA, PC, U-PVC, PET, PPE
Disperse Yellow 54	Quinophthalone	8 (0.05% PS); 8 (1/3 SD-PS)	300 (0.05%, PS)	—	PS, SAN, ABS, PMMA, U-PVC, PET, PPE
Solvent Yellow 130	Naphthazine	8 (0.05% PS); 8 (1/3 SD-PS)	300 (0.05%, PS)	—	PS, SAN, ABS, PMMA, U-PVC, PET, PPE, PA
Red inorganic pigments					
Pigment Red 101	Iron oxide (Fe_2O_3)	8 (pure); 8 (1:4)	300	5	Generally applicable, except U-PVC
Pigment Red 101	Semitransparent iron oxide (Fe_2O_3)	8 (pure); 8 (1:4)	300	5	Generally applicable, except U-PVC

Red organic pigments					
Pigment Red 208	Monoazo benzimidazolone	7 (pure); 6–7 (1:5)	220 (PVC)	4–5	PVC
Pigment Red 247	Ca salt of a monoazo acid	6–7 (pure); 6 (1:10)	300 (PO); 280 (PS)	4–5	LDPE, HDPE, PS
Pigment Red 144	Diazo condensation product	7–8 (pure); 7–8 (1:10)	300 (PO)	5	PVC, PS, LDPE, HDPE, PP
Pigment Red 166	Diazo condensation product	7–8 (pure); 7–8 (1:10)	300 (PO); 280 (PS)	5	PVC, PS, LDPE, HDPE, PP
Pigment Red 122	Dimethyl quinacridone	7–8 (pure); 7 (1:10)	240 (PS); 280 (PO)	5	PVC, PS, LDPE, HDPE, PP,
Pigment Red 202	Dichloro quinacridone	7–8 (pure); 8 (1:10)	300 (PO)	5	HDPE, PP, PA, ABS
Pigment Red 178	Perylene	8 (pure); 7 (1:10)	300 (PS); 300 (PO)	4–5	PVC, PS, HDPE, PP, ABS, PA
Pigment Red 149	Perylene	8 (pure); 7 (1:10)	280 (PS); 300 (PO)	—	PVC, PS, LDPE, HDPE, PP, PA
Pigment Red 177	Anthraquinone	8 (pure); 7 (1:10)	280 (PO)	5	PVC, PS, LDPE, HDPE, PP
Pigment Red 176	Monoazo benzimidazolone	7 (pure); 5–6 (1:10)	240 (PO)	5	PVC, PS, LDPE, HDPE
Pigment Red 185	Monoazo benzimidazolone	7–8 (pure); 7 (1:10)	250 (PO); 240 (PS)	5	PVC, PS, LDPE, HDPE
Pigment Red 187	Monoazo naphthol AS	7 (pure); 6–7 (1:10)	250 (PO)	5	PVC, LDPE, HDPE
Pigment Red 254	Diketopyrrolopyrrole	8 (pure); 8 (1:10)	300 (PO)	5	PVC, PS, LDPE, HDPE, PP
Pigment Red 264	Diketopyrrolopyrrole	7–8 (pure); 7 (1:10)	300 (PO)	5	PVC, PS, LDPE, HDPE, PP
Pigment Red 272	Diketopyrrolopyrrole	8 (pure); 7–8 (1:10)	300 (PO)	5	PVC, LDPE, HDPE, PP, PS

Table 9.2 Inorganic and Organic Colorants

C.I. Designation	Chemical description	Light-fastness[a]	Heat-resistance (°C)[b]	Migration[c]	Principal polymers in which used[d][e]
Red dyes					
Vat Red 41	Fluorescent	5–6 (pure); 3 (1:5)	280 (PS)	—	PS, PMMA, U-PVC
Solvent Red 111	Anthraquinone	7 (0.1%, PC)	340 (0.1%, PC)	4–5	PMMA, PS, PC
Solvent Red 135	Perinone	6–7 (1:10, PC)	340 (PC)	4–5	PMMA, PS, PC, PBT, PET, ABS
Solvent Red 179	Perinone	8 (0.05%, PS); 6 (1/3, SD-PS)	300 (0.05%, PS)	—	PS, SAN, ABS, PMMA, PC, PET, PA
Solvent Red 195	Azo	8 (0.05%, PS); 7 (1/3, SD-PS)	300 (0.05%, PS)	—	PMMA, PS, U-PVC
Orange organic pigments					
Pigment Orange 43	Perinone	7–8 (pure); 6 (1:10)	300 (PO); 240 (PS)	4–5	PVC, PS, LDPE, HDPE
Pigment Orange 64	Azo	7–8 (pure); 7–8 (1:10)	300 (PO)	5	PVC, PS, LDPE, HDPE, PP, ABS
Pigment Orange 72	Azo	8 (pure); 7–8 (1:50)	290 (PO)	5	PVC, PS, LDPE, HDPE, PP, ABS

Orange dyes					
Solvent Yellow 14	Azo	5–6 (0.05%, PS); 2–3 (1/3, SD-PS)	270 (PS)	5	PMMA, PS, U-PVC
Solvent Orange 63	Fluorescent polycyclic	7 (pure); 5 (1:5)	300 (PS)	5	PS, PMMA, U-PVC
Solvent Orange 60	Perinone	7–8 (pure)	360 (PC)	4–5	PMMA, PS, PC, PBT
Solvent Orange 114	Quinophthalone (quinoline indandione) type	8 (pure PS); 6–7 (1:20, PS)	300 (PS); 360 (PC)	4	PS, PMMA, PC, PET
Disperse Orange 47	Methine	7–8 (0.05%, PS); 6 (1/3, SD-PS)	300 (PS)	—	PS, SAN, ABS, PMMA, PC, PET, PPE
Violet inorganic pigments					
Pigment Violet 15 (Ultramarine Violet)	Aluminium sodium thiosilicate	8 (pure)	280–300	5	Used in PVC, PS, PP, PO. Not suitable for acidic media due to color loss
Pigment Violet 16	Manganese ammonium pyrophosphate	8 (pure)	~300	5	Widely used in PVC, PP, PS, ABS. Not suitable for alkaline media
Violet organic pigments					
Pigment Violet 23	Dioxazine	6 (pure); 5 (1:10)	250 (PS); 260 PO	3–4	PVC, LDPE
Pigment Violet 19	Quinacridone (β-modification)	7–8 (pure); 7–8 (1:10)	280 (PO); 260 (PS)	5	PVC, PS, LDPE, HDPE, PP

Table 9.2 Inorganic and Organic Colorants *(continued)*

C.I. Designation	Chemical description	Light-fastness[a]	Heat-resistance (°C)[b]	Migration[c]	Principal polymers in which used[d][e]
Violet dyes					
Solvent Violet 36	Anthraquinone	7 (0.05%, PS); 6–7 (1/3, SD-PS)	300 (PS)	—	PS, SAN, ABS, PMMA, PC, PET, PPE, PA
Solvent Violet 13	Anthraquinone	7 (0.05%, PS); 6–7 (1/3, SD-PS)	300 (PS)	—	PS, SAN, ABS, PMMA, PC, U-PVC, PET, PPE
Blue inorganic pigments					
Pigment Blue 29 (Ultramarine Blue)	Aluminium sodium thiosilicate	8 (pure); 8 (1:4)	300 (PS); 300 (PO)	5	PVC, PS, LDPE, HDPE
Pigment Blue 28	Spinel ($CoAl_2O_4$)	8 (pure); 8 (1:4)	300 (PS)	5	PVC, PS, LDPE, HDPE, ABS, PA
Pigment Blue 36	Spinel ($Co(Al,Cr)_2O_4$)	8 (pure); 8 (1:4)	300 (PS); 300 (PO)	5	PVC, PS, PO, ABS, PA
Blue organic pigments					
Pigment Blue 15:1	Cu-phthalocyanine (stabilized α-form)	8 (pure); 8 (1:10)	280 (PS); 240–280 (PO)	5	PVC, PS, LDPE, HDPE, PP, ABS
Pigment Blue 15:3	Cu-phthalocyanine (β-form)	8 (pure); 8 (1:10)	280 (PS); 280 (PO)	5	PVC, PS, LDPE, HDPE, PP, PET, PA
Pigment Blue 60	Anthraquinone (indanthrone)	8 (pure); 8 (1:10)	250 (PS); 270 (PO)	5	PVC, PS, LDPE, HDPE, PP, PET, PA
Blue dyes					
Solvent Blue 67	Phthalocyanine	7 (0.02%, PET); 4–5 (1:20, PS)	280 (PET); 220 (PS)	5; 4–5	PET, PS
Solvent Blue 97	Anthraquinone	7 (0.05%, PS); 6 (1/3, SD-PS)	300 (0.05%, PS)	—	PS, SAN, ABS, PMMA, PC, U-PVC, PET, PPE, PA

Green inorganic pigments					
Pigment Green 17	Chromium oxide (Cr_2O_3)	8 (pure); 8 (1:4)	300 (PS); 300 (PO)	5	Generally applicable
Pigment Green 26	Spinel (Co,Cr,Ti,Zn oxide)	8 (pure); 8 (1:4)	300 (PS); 300 (PO)	5	Generally applicable
Pigment Green 50	Spinel (Co,Cr,Ti,Zn oxide)	8 (pure); 8 (1:4)	300 (PS); 300 (PO)	5	Generally applicable
Green organic pigments					
Pigment Green 36	Halogenated phthalocyanine	8 (pure); 8 (1:4)	300 (PS); 300 (PO)	5	PS, PVC, LDPE, HDPE, PP
Green dyes					
Solvent Green 3	Anthraquinone	7–8 (0.05%, PS); 7 (1/3, SD-PS)	300 (0.05%, PS)	—	PS, SAN, ABS, PMMA, PC, PET, PPE, PA
Solvent Green 28	Anthraquinone	8 (0.05%, PS); 7–8 (1/3, SD-PS)	300 (0.05%, PS)	—	PS, SAN, ABS, PMMA, PC, U-PVC, PET, PPE
Brown pigments					
Pigment Brown 6	Iron oxide (Fe_3O_4)	8 (pure); 8 (1:4)	240 (PE)	5	PVC, PS, LDPE
Pigment Brown 29	Chrome iron ($(Fe,Cr)_2O_3$)	8 (pure); 8 (1:4)	300	5	Generally applicable
Pigment Brown 31	Zinc iron	8 (pure); 8 (1:4)	300	5	Generally applicable
Pigment Yellow 164	Manganese titanium	8 (pure); 8 (1:4)	300	5	Generally applicable
Pigment Brown 23	Diazo condensation product	8 (pure); 8 (1:10)	260 (PS)	5	PVC
Pigment Brown 25	Azo	8 (pure); 8 (1:10)	290 (PE)	5	PVC

[a] Generally carried out using xenon light according to DIN 53387/ISO 105 B02, and expressed using the "blue scale" rating, from 1 (very low) to 8 (very high).
[b] In most cases tested along the lines of EN 12877 part 2.
[c] Usually tested in contact with plasticized PVC film according to DIN 53775 part 3.
[d] Polymer types are abbreviated according to ISO 1043-1, 11987.
[e] Manufacturers should be contacted for information on shrinkage behavior of organic pigments in HDPE injection molding.

It should be noted that products with the same C.I. Generic Name may differ considerably in their properties, particularly regarding color strength and dispersibility. Leading colorant manufacturers often supply pigments controlled and adapted specifically for use in plastics, and it is usually advantageous to use such products instead of non-specific generic materials.

Lightfastness data is given for full-shade applications and in some cases, in combination with TiO_2. Testing is generally carried out using xenon light according to DIN 53387/ISO 105 B02. Heat resistance data is in most cases generated from tests carried out using an injection molding machine along the lines of EN 12877 part 2, in which the colored polymer is exposed to increasing test temperatures for a residence time of 5 min. Particularly with organic pigments, the test concentration and polymer type can significantly affect the results obtained. Furthermore, different suppliers use different definitions of the extent of acceptable color change. Because of these variations, figures given in these tables must be treated with caution. Migration resistance is usually tested in contact with plasticized PVC film according to DIN 53775 part 3.

9.9 Fluorescent Whitening Agents (Optical Brighteners)

The use of fluorescent whitening agents (FWAs), also known as optical brighteners (OBs), in plastics has been extensively described in the literature. This section contains technical data for various plastics.

FWAs have long been used to improve the appearance of end products in the paper, soap, detergent, and textile industries, and FWAs have now become widespread in thermoplastics. The reasons for this are the same as for paper and textiles:

- Improving or masking the initial color of plastics, which is often slightly yellowish
- Producing brilliant white end-products, such as fashion items made of artificial leather, packaging materials, etc.
- Increasing the brilliance of colored and black-pigmented articles.

In principle, FWAs can be used in all plastics. However, not all FWAs are suitable for all plastics due to limitations in certain applications. Therefore, the type and concentration of a FWA must be carefully selected. A technically suitable FWA exhibits high brightening power at low concentrations. In plastic applications, concentrations of 50–500 ppm are typically used, but in some specialty applications or when using recycled materials, concentrations of 1000 ppm and more are also possible.

The mechanism of action of FWAs is described in Section 3.2 and Figure 3.4.

9.9.1 Selection of Fluorescent Whitening Agents

The determining factors for selecting the most suitable FWA for common plastics are:

- Achievable whitening effect
- Compatibility
- Light stability.

The selection of FWAs is influenced to some extent by their color. In principle, neutral and neutral–blue colors are preferred over blue–green colors. However, reddish shades are sometimes preferred, especially in Asia. The achievable whiteness of an optical brightener depends on the carrier material (Figure 9.2). In addition, the whiteness can be influenced by both process conditions and possible reactions with other components of the formulation, such as pigments and UV absorbers.

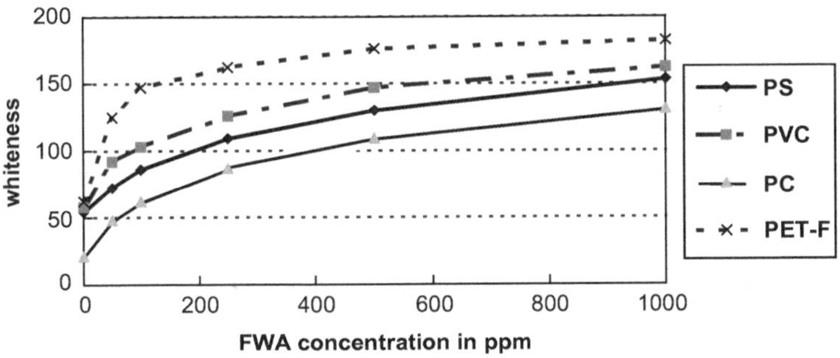

Figure 9.2 Whitening effectiveness (Ganz formula) of a bis-benzoxazole FWA in various substrates (with rutile-type TiO_2) as a function of FWA concentration

Titanium dioxide pigments absorb in the long-wave UV range, which is crucial for excitation of the FWA, and so they result in lower whiteness levels. Anatase pigments absorb approximately 40% of the incident radiation at 380 nm, while rutile pigments absorb approximately 90% (Figure 9.3).

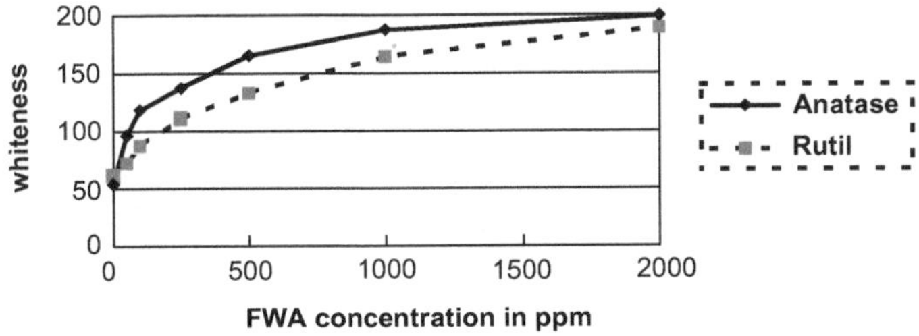

Figure 9.3 Whitening effectiveness of a bis-styryl biphenyl FWA in plasticized PVC as a function of whitener concentration in the presence of 5% anatase-type and rutile-type titanium dioxide

UV absorbers (Section 3.5.1) also absorb radiation in the range in which FWAs absorb, and can therefore lead to reduced whiteness. However, this loss of whiteness is minimal because the brightener can absorb sufficient energy at the polymer surface to be excited.

A key requirement for the technical suitability of FWAs is their light stability in the carrier material. FWAs are usually more-or-less light-sensitive in the wavelength range required for excitation. Therefore, the light stability of the FWA is limited and significantly lower than the light stability of pigments and the light stability of most plastics. Colored structures, which could lead to additional yellowing, should therefore be avoided at the end of the photochemical degradation of the optical brighteners. The light stability of FWAs is significantly influenced by the carrier material (Figure 9.4).

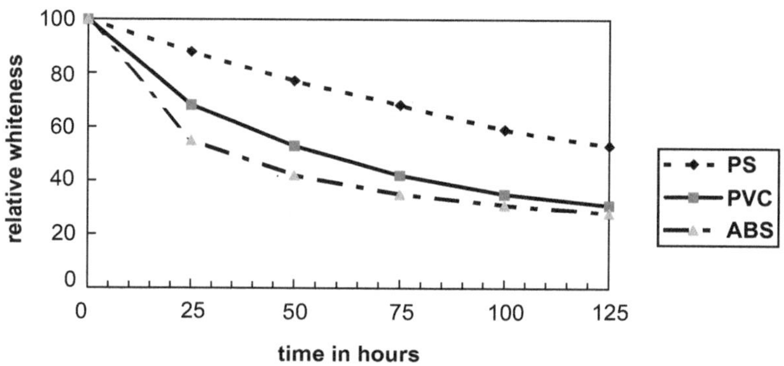

Figure 9.4 Lightfastness of a bis-benzoxazole-type FWA in various substrates, showing elative whiteness (Ganz formula) as a function of weathering time in a Xenotest® 150 with filtered xenon radiation corresponding to daylight behind window glass

To achieve maximum effectiveness of an FWA, it must be completely dissolved and homogeneously distributed in the final product, i.e., the distribution must be monomolecular. Otherwise, the finished product may exhibit bright spots or streaks. Adequate compatibility also prevents exudation. The requirements regarding thermal stability and low volatility of FWAs are higher for thermoplastics than for textiles.

Migration stability is particularly important for optical brighteners used in p-PVC and polyurethane coatings. An incompatible FWA can be transported to the surface, e.g., through plasticizer migration, which results in a partial loss of whiteness. As a result, the surface may become completely or partially discolored.

9.9.2 Structure of Fluorescent Whitening Agents

Only a few of the known chemical classes of FWAs (Figure 9.5) possess the properties required for mass whitening of plastics and fibers; Table 9.3 lists some key information, and various considerations are discussed in Section 9.9.3.

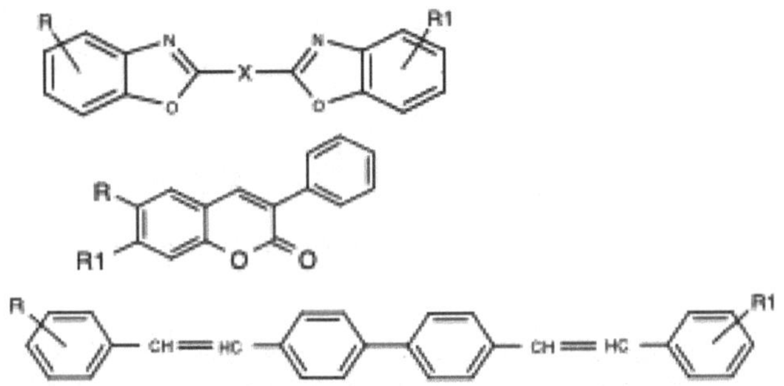

(1)

(2)

(3)

Figure 9.5 Chemical classes for the mass whitening of plastics and fibers: (1) bis-benzoxazoles; (2) phenylcoumarins; (3) bis-(styryl)biphenyls

Table 9.3 Fluorescent Whitening Agents

Abbreviation	CAS no.	Chemical name	Application
FWA-1	91-44-1	4-Methyl-7-diethylaminocoumarin	PVC, PS, PC, PU, polyolefins
FWA-2	53850-91-2	3-Phenyl-7-(4-methyl-6-butyloxybenzoxazole)coumarin	PVC, PS, PC, PU
FWA-3	1533-45-5	4,4′-Bis(benzoxazol-2-yl)stilbene	PVC, PS, PC, PMMA, PET, PA
FWA-4	91-44-1	4-Methyl-7-diethylaminocoumarin	PU (foamed), PMMA, unsaturated polyesters
FWA-5	3271-22-5	2,4-Dimethoxy-6-(1′-pyrenyl)-1,3,5-triazine	PVC, PS, PC, PU
FWA-6	5089-22-5	1,4-Bis(benzoxazol-2-yl)naphthalene	PVC, PS, ABS, PC, PU, PMMA, unsaturated polyesters, PA
FWA-7	5242-49-9	Mixture of 4,4′-bis(benzoxazol-2-yl) stilbene and 4-(benzoxazol-2-yl)-4′-(5-methylbenzoxazol-2-yl)stilbene and 4,4′-bis(5-methylbenzoxazol-2-yl)stilbene	PVC, PS, ABS, PET
FWA-11	3333-62-8	3-Phenyl-7-(2H-naphtho-[[1,2-d]]triazol-2-yl)coumarin	u-PVC, PS, ABS, PC, PU, PET

Table 9.3 Fluorescent Whitening Agents *(continued)*

Abbreviation	CAS no.	Chemical name	Application
FWA-12	40470-68-6	4,4′-Bis(2-methoxystyryl)biphenyl	PVC, PS, ABS, PC, PU, PMMA, unsaturated polyesters, PA
FWA-13	27344-41-8	4,4′-Bis(2-sulfostyryl)biphenyl disodium salt	PU (foamed)
FWA-14	7128-64-5	2,5-Bis(5-*tert*-butylbenzoxazol-2-yl)thiophene	PVC, PS, ABS, PC, PU, PMMA, unsaturated polyesters, PET, PA, polyolefins

9.9.3 Incorporation and Performance of Fluorescent Whitening Agents into Polymers

In **PVC**, bisbenzoxazoles or phenylcoumarins are typically used, both in p-PVC and u-PVC, and the addition of FWAs does not pose any particular difficulties. Sufficiently homogeneous distribution is ensured by hot-mixing FWA and PVC powder together with conventional thermal stabilizers. In p-PVC, the fluorescent brightener is often pre-dissolved or dispersed in the plasticizer. Nevertheless, the preferred method is to add FWA as a concentrate (masterbatch) or mixture. For transparent formulations, concentrates with plasticizer or PVC are used. For pigmented formulations, premixes with chalk are used.

Even very small amounts of FWA are sufficient to increase the basic whiteness of PVC. A concentration of 50–100 ppm – or less – is usually sufficient to compensate for the yellow tint. In transparent formulations, a water-clear appearance is achieved. Ultra-white PVC products require a concentration of up to 500 ppm of fluorescent whitening agent, and for applications with rutile-type whitening agents, up to 1000 ppm may be necessary. However, overdosage should be avoided, because some FWAs tend to have a green tint at higher concentrations, which can lead to a less pleasing white tone and a reduction in whiteness.

The same bis-benzoxazole and phenycoumarin types can be used in both PVC and **styrene-based plastics**. The light stability in crystal-clear PS is somewhat better than in PVC, but it is somewhat lower in other styrene polymers. Bis-styryl-biphenyls also achieve excellent whitening effects. However, their light stability is somewhat lower than expected for their chemical class.

Crystal-clear PS exhibits a slight inherent fluorescence, which can be quenched by traces of yellow decomposition products formed by thermal stress during processing. However, as with PVC, this yellowing can be compensated for with low concentrations (less than 50 ppm) of an FWA. In white-pigmented PS, which primarily uses rutile titanium dioxide, “ultra-white” products are obtained by adding approximately 500 ppm of an optical brightener.

Styrene copolymers such as HIPS, SAN, ASA, ABS, and SBS occasionally exhibit a distinct yellow tint, which can be compensated for by adding an FWA. FWAs are typically incorporated into the styrene polymer by dry-blending (mixing) the brightener with the polymer granules prior to processing.

Polycarbonates are processed at temperatures of 300 °C and higher, and yellowing occasionally occurs. However, these high processing temperatures place stringent demands on the thermal stability and volatility of FWAs. FWAs of the same chemical classes used in PVC and styrene polymers can compensate for the yellowing in unpigmented polycarbonates at concentrations of around 50 ppm; ultra-white pigmented products require concentrations of up to 500 ppm. The addition of FWAs to polycarbonate is handled in the same way as for PS, by "dry-blending" with the dried granules before processing. To improve the adhesion of the FWAs to the polycarbonate granules, e.g., butyl stearate is added in amounts of 50–100 ppm.

Typically, only one-component **polyurethane** systems are brightened, or "whitened", with FWA, and these products are used in films and coatings. The brightener is added directly to the polyurethane solution or, as in the case of PS or PC, mixed with the granules. Particular attention is paid to solubility and migration stability in cast films, as the FWA can reach the film surface during drying due to solvent evaporation. In this case, exudation occurs, and the FWA appears in the form of yellow specks on the surface.

FWAs are also used in polyurethane-based adhesives, and liquid brighteners or dispersions of solid brighteners are used for such applications. FWAs are also used in two-component polyurethane foams in athletic shoes, for which the FWA must be liquid or soluble in the polyol, and must not lead to any side-reactions with the isocyanates. Larger and more rigid FWA molecules are fundamentally incompatible and lead to a very pronounced green discoloration of the foam. The light stability of FWAs in polyurethanes is generally significantly worse than in other plastics. However, this is also due to the rapid discoloration of the pure polyurethane, which masks the performance of the FWA. For ultra-white applications, concentrations of 5000 ppm are required. The selection of the best brightener for a given application or process can only be determined through testing in the given polyurethane formulation.

Polyolefins are sometimes "through-colored," as is the case with PE films used for paper coating. The light stability of FWA in polyethylene is generally low. A fundamental criterion for selecting suitable products is compatibility with the substrate. The compatibility of FWAs with polyolefins falls in the order PP > HDPE > LDPE.

Only a small group of FWAs is suitable for use in **PMMA**, with solubility in the monomer being a key requirement. Furthermore, the FWAs must not react with the polymerization initiators. Polymerization must be completely complete, otherwise interactions between FWAs and residual monomers can lead to discoloration. Various FWAs are highly effective in injection-molding PMMA applications. Both dry-blending and compounding are suitable methods for incorporating FWAs into PMMA.

The only possible way to incorporate FWA into **unsaturated polyesters** is to dissolve them in the polyester/styrene resin before it cures. Therefore, in addition to solubility, an additional basic property is necessary for the FWA: it must not negatively influence curing. Negative whiteness (*W*) values are an indication of incompatibility and the so-called "greening" effect. The greening effect occurs when FWAs are incompatible with unsaturated polyester resins or are not properly incorporated. This phenomenon is primarily caused by chemical interactions between the brightener and components of the resin or curing agents, which can alter the spectral properties of the FWA and shift the perceived color towards green. Poor solubility or inadequate dispersion of the FWA can also lead to uneven distribution within the polymer matrix, resulting in localized color shifts and an overall greenish tint. Additionally, overdosing the FWA may intensify this effect, as excessive fluorescence can cause the color balance to tip towards green rather than the desired blue–white.

The optical brightening of **PET** textile fibers is one of the most important applications of FWAs. The products can be incorporated into the spinning dope before the extrusion step. However, the introduction of FWAs during polyester polymerization is more important. To achieve this, the FWA must have optimal thermal and chemical stability throughout the entire production process, from the monomer to the finished yarn. For optimal performance, FWAs are additionally applied to the surface of PET fibers that already contain brighteners in the mass, e.g., during the dyeing or finishing steps.

While **polyamide** fabrics and yarns are usually brightened on the surface by FWAs, dope brightening of polyamide yarns is an important alternative to reduce the initial yellowing.

9.9.4 Testing of Fluorescent Whitening Agents

Examination for migration and exudation effects following incorporation of FWAs is best carried out in the light of a UV lamp. Migration and exudation effects result in random yellow and greenish spots, which fluoresce distinctly brighter than the surrounding area.

The brightening effect can be assessed visually (preferably under daylight) or using colorimetric measurements. The result of a visual assessment depends significantly on the intensity (luminous intensity) and spectral energy distribution of the incident light, as well as on the observer's physiology and color perception. Visual assessment is not entirely satisfactory due to numerous variables that are difficult to control.

For this reason, instrumental assessment methods for whiteness have been developed. "Whiteness", being a physiological impression, cannot be directly measured, and so it is calculated using colorimetric data. Based on these physical measurements, objective, characteristic values are therefore obtained that are free from subjective influences. A specified light source in the colorimeter is a prerequisite for reproducible results, independent of time, location, and the person performing the measurement. Many methods for measuring the whiteness of an object are described in the literature. They are mainly based on the color values (*X*, *Y*, *Z*) and the color space coordinates (*x*, *y*, *z*) that relate to the CIELAB color space (see Section 9.3).

Currently, there are no international standards for evaluating the light stability of FWAs in plastics. In practice, DIN 54004 or similar test methods are often used, which were originally developed for textiles. Manufacturers and testing laboratories typically use their own methods or adapt existing standards for related materials to assess the light stability of FWAs in plastics. International standards such as ISO 105-B06:2020 regulate the lightfastness of dyes and textiles, but either explicitly exclude samples containing FWAs or are not tailored to plastics. For other applications (e.g., inkjet prints), there are light stability standards (such as ASTM F1945-05), but these are not transferable to plastics and FWAs. The exposure of the samples is carried out in a device with filtered xenon irradiation, corresponding to daylight irradiation behind window glass (e.g. Xenotest®). An objective assessment can be made by measuring the change in whiteness as a function of irradiation time, as shown in Figure 9.6 for p-PVC.

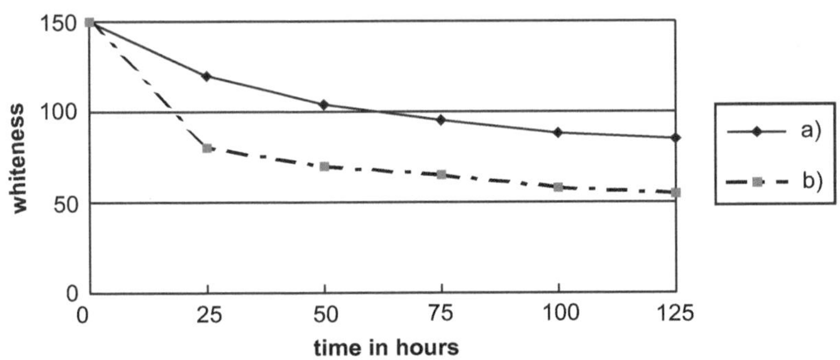

Figure 9.6 Lightfastness of the whitening effect (Ganz formula) of (a) a bis-benzoxazole type FWA and (b) a phenyl-coumarin-type FWA in plasticized PVC containing 5% anatase titanium dioxide, measured as a function of exposure time in a Xenotest® 150 with filtered radiation corresponding to daylight behind window glass

9.9.5 Technological Trends in Fluorescent Whitening Agents

No new FWAs used in plastics and fibers have been introduced in recent years. On the other hand, established producers of FWAs are faced with growing competition from producers in China and Korea. Latest technical developments are found in the field of synergistic blends of established FWAs and whitening boosters. Other developments include the use of small amounts of dyestuffs (blue, violet, red) for increased whitening effectiveness and modification of cast (shading colors).

9.10 Questions

1. What are the definitions for "dye" and "pigment"? What is the key difference between them?
2. What is meant by "dispersion" of pigments? Why is pigment dispersion critical in plastics?
3. How do organic and inorganic pigments differ in terms of properties and applications?
4. What are the typical challenges encountered when processing pigments in thermoplastic polymers?
5. What are the criteria for selecting a dye or a pigment for food packaging applications?
6. What is a fluorescent dye? Provide an example of its use in industry.
7. What is the mechanism by which optical brighteners enhance the perceived whiteness of plastics?
8. What test method should be used for determining the color strength of a pigment?
9. What is the metameric effect in plastics? How is it measured?
10. What is the migration of colorants? Why is minimizing it critical in plastics?
11. What are the environmental considerations for selecting and using colorants in plastics?

10 Flame Retardants and Smoke Suppressants

Content of this chapter was adapted from:

- M. M. Hirschler, "Flammability and fire performance" (Chapter 13), in: *PVC Handbook* (2nd edition), ed. R. P. Krock, M. Schiller, P. Frenkel, J. W. Summers, and C. A. Daniels, Hanser, 2025, pp. 559–648.
- P. F. Ranken, "Flame retardants" (Chapter 12), in: Plastics Additives Handbook (6th edition), ed. H. Zweifel, R. D. Maier, and M. Schiller, Hanser, 2009, pp. 701–717.
- G. Sarti, "Flame retardants and smoke suppressants" (Chapter 14), in: *PVC Handbook* (2nd edition), ed. R. P. Krock, M. Schiller, P. Frenkel, J. W. Summers, and C. A. Daniels, Hanser, 2025, pp. 649–702.

10.1 Key Terms and Definitions

The following definitions as based partly on those provided in ASTM E176.

Char	To form a carbonaceous residue during pyrolysis or incomplete combustion, and the residue itself.
Combustion	A chemical process of oxidation that occurs at a rate that is fast enough to produce a temperature rise, and also usually light, either as a glow or flame.
Fire hazard	The potential for a fire to cause harm to people or property.
Fire resistance	The ability of a material, product, or assembly to withstand fire or give protection from it for a period of time.

Fire-test response	A response characteristic of a material, product, or assembly to a prescribed source of heat or flame, under controlled fire conditions. Such response characteristics may include but are not limited to ease of ignition, flame spread, heat release, mass loss, smoke generation, fire endurance, and toxic potency of smoke.
Flame spread	Propagation of a flame front.
Flashover	The rapid transition to a state of total surface involvement in a fire of combustible materials within an enclosure.
Heat release rate	The thermal energy released per unit time by an item during combustion under specified conditions. The heat release rate is typically reported per unit of surface area.
Ignition	The initiation of combustion. The combustion may be evidenced by glow, flame, detonation, or explosion. The combustion may be sustained or transient.
Ignition temperature	The lowest temperature at which sustained combustion of a material can be initiated under specified test conditions. The ignition temperature is reached when combustion continues after the pilot source is removed.
Intumescent material	A coating or substance that expands and chars when exposed to heat, forming an insulating layer that protects the underlying structures.
Pyrolysis	Process of simultaneous phase change and chemical transformations in the absence of oxygen caused by heat.
Self-ignition	The minimum temperature at which the self-heating properties of the specimen lead to ignition or ignition occurs by itself, under specified test conditions in the absence of any additional flame ignition source.
Smoke	The airborne solid and liquid particulates and gases evolved when a material undergoes pyrolysis or combustion.
Smoke toxicity	The propensity of smoke to produce adverse biochemical or physiological effects.
Smoldering	Combustion of a solid without flame, often evidenced by visible smoke.

10.2 Thermal Degradation and Thermal Decomposition

All polymers are sensitive to excessive heat, which can lead to both physical and chemical changes. The term *thermal degradation* usually refers to alterations in electrical and physical (mechanical) properties, such as strength, and/or elasticity, while *thermal decomposition* also involves chemical composition changes. Only thermal decomposition is relevant in the context of fire, since the chemical decomposition of a solid material generates gaseous fuel vapors, which can burn above the hot solid material.

For combustion to be sustained, the gases released through thermal decomposition must burn in the presence of oxygen and produce enough heat to continue decomposing the underlying solid and generate gaseous fuel vapors or volatiles. The decomposition products react exothermically with oxygen in the surrounding atmosphere, releasing further heat, part of which transfers back to the solid and drives additional breakdown. The visible result of these gas-phase reactions is the formation of a flame, usually accompanied by light emission and soot formation.

Only a few polymers decompose without leaving any solid residue. Most materials, however, leave behind solid residues, which can be organic (carbonaceous char), inorganic (mineral matter derived from non-carbon atoms in the polymer), or a mixture of both. The formation of char layers is generally beneficial for fire resistance. Carbonaceous chars can swell and form an insulating foam (intumescent behavior), thereby slowing down further decomposition, while inorganic residues may form glass-like protective layers that hinder the release of volatiles and protect deeper material layers from heat exposure.

Many thermosetting polymers tend to produce a carbon-rich char upon heating. The physical characteristics of such char – typically low density and high porosity – make it an effective thermal insulator. This insulation reduces the heat flow from the combustion zone back into the polymer bulk, suppressing further decomposition and, consequently, lowering the material's flammability. As the char layer thickens, less heat reaches the undecomposed material, which further reduces the rate of thermal decomposition.[1)]

Polymers cannot burn if they do not break down.

10.3 Thermal Decomposition and Fire Performance

Studies of polymer thermal decomposition is largely motivated by its crucial role in understanding fire behavior and fire protection mechanisms. The way polymers break down under heat determines how readily they produce flammable gases and how these contribute to combustion. In general, flame reactions can be governed either by chemical reaction rates or by physical processes such as mixing, diffusion, or gas flow. In real systems, however, most flames exhibit characteristics that fall between purely kinetically controlled and purely diffusion-controlled regimes, representing a balance of both influences.

---

1 M. M. Hirschler, "Flammability and fire performance" (Chapter 13), in: *PVC Handbook* (2nd edition), ed. R. P. Krock, M. Schiller, P. Frenkel, J. W. Summers, and C. A. Daniels, Hanser, 2025, pp. 559–648.

As the condensed-phase fuel generation is closely coupled with gas-phase fuel consumption, the rates of these processes must be in equilibrium to maintain a steady combustion. A change in the rate of either process disturbs the equilibrium state and can ultimately lead to extinction of the flame.[1)]

Radicals are likely to be involved in the flames of all hydrocarbons, demonstrating that the chain-propagating moieties are largely independent of the initial fuel.

Most flame processes can therefore be described as high-temperature reactions of a small number of free radicals. The detailed reaction pathways in hydrocarbon flames, however, are strongly influenced by the fuel-to-oxidant ratio in the system. In the combustion of polymers, additional elements are often present in the flame gases. Particularly important are compounds containing nitrogen, sulfur, chlorine, and bromine. When chlorine or bromine present in organic polymers, these halogens are typically converted during combustion into the corresponding hydrogen halides, whereas sulfur is mainly oxidized to sulfur oxides (predominantly sulfur dioxide (SO_2)).[1)]

The nature of the fuel feeding the flame during combustion of an organic polymer is determined by the chemical composition of the parent polymer. In essence, a hydrocarbon flame can be considered as a system in which the primary reactive components are carbon monoxide, hydrogen, and oxygen.

Flames above burning organic polymers do not differ greatly, from the chemical point of view, from one polymer to another, although they are highly dependent on the nature of the gaseous oxidant.[1)]

Fire performance is characterized by a combination of several fire properties rather than a single parameter. These include ignitability, ease of extinction, flammability of the products generated, amount of heat released on burning, heat release rate, flame spread, smoke obscuration, and smoke toxicity.

Acceptance of materials in commerce usually requires compliance with one or more fire properties in accordance with codes and/or regulations. It was determined that the heat release rate (which determines the intensity of a fire) is a critical property that controls fire hazards. Typically, requirements in codes or regulations apply primarily to finished articles, but they may occasionally also apply to material components, such as polymers.

If the material does not ignite, then there is no fire. Therefore, low ignitability is the first line of defense in a fire.

All organic polymers ignite, but the higher the temperature a material has to reach before it ignites, the safer it is.

There are three kinds of measures of ignition:

- The minimum temperature of ignition (in the absence/presence of a pilot source of ignition)
- The time to ignition (TTI) under a certain applied heat flux
- The minimum heat flux required to reach ignition.

10.4 Main Test Methods

10.4.1 Self-ignition and Flash Ignition Temperatures

The traditional method of assessing ignition temperatures, particularly for plastics, has been the use test, such as the Setchkin chamber test (ASTM D1929[2)]). This has been employed to measure both the self-ignition temperature and the flash ignition temperature (with an external pilot flame) for a broad range of polymers (Table 10.1). ASTM D1929 typically uses pellets as specimens, placed inside an electrically heated vertical furnace tube under slow air flow.

The results of this test method are no longer considered to be very indicative of true material ignitability, because heat exposure according to this method does not accurately represent that in real fires. A more informative approach is to measure the time to ignition at a specific incident heat flux, which yields a quantitative criterion for whether that heat flux is sufficient to trigger the ignition. In practice, the critical heat flux for ignition is obtained by extrapolating the results from a series of experiments conducted over a range of incident heat flux levels.[1)]

Table 10.1 Flash Ignition and Self Ignition Temperatures of Polymeric Materials (ASTM D1929)[1)]

Material	Flash ignition (°C)	Material	Self-ignition (°C)
PA 6	490	PA 6.6	530
PA 6.6	490	PTFE	530
PC	467	PS	492
PVC	391	ABS	466
SAN	366	PMMA	456

[2] ASTM D1929 is broadly aligned in scope with ISO 871 "Plastics – Determination of ignition temperature using a hot-air furnace", although the apparatus and some procedural details differ.

Table 10.1 Flash Ignition and Self Ignition Temperatures of Polymeric Materials (ASTM D1929) (*continued*)

Material	Flash ignition (°C)	Material	Self-ignition (°C)
PS	352	PVC	454
PE	349	SAN	451
Polyether PU (rigid foam)	310	PA 6	450
PMMA	290	Polyether PU (rigid foam)	416

10.4.2 Ease of Extinction

Once a material is ignited, the easier it is to extinguish it, the lower the fire hazard associated with products manufactured from it.

ASTM D2863[3)] quantifies this behavior by determining the limiting oxygen index (LOI), which is the minimum oxygen concentration in a flowing oxygen–nitrogen mixture that sustains a downward, candle-like flaming combustion of the specimen.

The higher the LOI, the less likely it is that the material will continue burning in air.

The atmosphere is normally about 21% oxygen. In principle, a material with an LOI higher than about 25% should only sustain burning under relatively severe or highly oxygen-rich conditions. This test serves as a measure of the ease of flame extinction of the material.

In the test, a small specimen is mounted vertically in a glass column and ignited at the upper end with a small gas flame, while the oxygen–nitrogen ratio in the mixture is adjusted until just-sustained, candle-like burning is obtained. The method offers excellent repeatability and reproducibility, and provides numerical LOI values over a wide range of materials and fire performances (Table 10.2).

Results from the LOI test are primarily useful as an indicator of how readily a flame goes out under conditions of relatively low external heat exposure. As such, the method is not suitable for predicting the behavior of materials in full-scale or high-

3 The ISO counterpart to ASTM D2863 is ISO 4589-2 ("Plastics – Determination of burning behavior by oxygen index – Part 2: Ambient-temperature test"). ASTM D2863 and ISO 4589-2 are regarded as technically equivalent, provided that the gas measurement and control system specified in the ISO standard is used, as explicitly stated in ASTM D2863 itself.

heat fire scenarios. Despite these limitations, LOI values are widely reported in technical data sheets, and are often used as a convenient comparative parameter for material selection and formulation work.[1]

Table 10.2 LOI Values of Selected Polymeric Materials[1]

Material	LOI
Polyacetal	15.8
PE	17.0
PP	17.1
ABS	17.6
PS	17.7
PMMA	17.9
SAN	18.0
PET	20.0
PA 6.6	25.1
PC	26.2
PVC (rigid)	47.0
PVDC	60.0
CPVC	62.2
PTFE	95.0

10.4.3 Small-scale Flammability

The higher the flammability of a material, the greater are the hazards associated with it and articles made from it.

For plastics, a widely applied group of small-scale flammability tests is defined in UL 94.[4] In the vertical test (UL 94 V), a material specimen (127 mm × 13 mm, or 5″ × 0.5″) is exposed vertically to a Bunsen-burner-type flame, resulting in a rating (classification) ranging from V 0 (best, most resistant), through V 1 and V 2 to "B" (burns).[1]

[4] UL 94 methods are technically harmonized with several international standards, in particular ISO 9772 and ISO 9773 for small-flame tests on plastics, and with IEC/EN 60695-11-10 and 60695-11-20, which adopt essentially the same vertical and horizontal burning procedures.

10.4.4 Flame Spread

The tendency of a material to propagate flame beyond the initial ignition zone is an important indicator of its potential fire hazard. Flame spread tests can be carried out either on base polymers, or on finished articles, such as textiles or electrical insulation, as well as on composite assemblies, such as furniture.

Most flame spread methods can be categorized by the angle between the exposed surface and the horizontal plane. This angle can range from 0° for a horizontal specimen burning on its upper face, through 180° for a horizontal specimen burning on its lower surface, to 240° for a specimen burning on its lower surface but tilted a further 60 °C from the horizontal line, and up to 270° for a specimen burning vertically downward. These orientations strongly affect both the extent at which the gaseous by-products of combustion heat the surface before the flame front reaches it and the extent at which the hot gases penetrate the polymer.

For building materials in the USA, the "Steiner tunnel test" (ASTM E84)[5] has been widely adopted into building codes as the reference flame spread method for assessing lining and surface materials (Table 10.3).[1]

Table 10.3 Flame Spread Index (Steiner Tunnel Test – ASTM E84) of Selected Polymeric Materials[1]

Material	Flame spread index range	
	Low	High
ABS	200	275
Particleboard	135	180
PC	80	120
Flame-retardant ABS	10	100
PVC wallcovering on gypsum board	10	25
PVC rigid profile	15	20
PVC siding (1 mm)	10	15

5 A Steiner tunnel is a rectangular test duct used to measure flame spread and smoke development of building materials mounted on its ceiling. In the ASTM E84 test (also known as the "Steiner tunnel test"), a gas burner and controlled airflow create a standardized fire exposure, and the distance and rate at which flames travel along the material's surface are recorded as the flame spread index and smoke developed index. There is no single standard ISO test that is equivalent to ASTM E84, which is essentially a North American test method specific to surface burning characteristics of building materials. In the ISO/European framework, the same fire performance questions (reaction to fire, flame spread, smoke) are typically addressed using different test methods, such as EN 13823 (SBI test) together with EN ISO 11925-2 for Euroclass classification.

10.4.5 Heat Release

The maximum heat release rate is a very critical fire hazard characteristic. The heat release rate is generally regarded as a more critical parameter for describing fire hazards than ease of ignition, smoke toxicity, or flame spread. A bench-scale calorimetric test for measuring heat release from materials was developed by Edwin Smith at Ohio State University (OSU), later formalized as the OSU rate-of-heat-release (RHR) calorimeter and standardized in ASTM E906,[6] providing a controlled method for quantifying how rapidly energy is liberated during combustion. Table 10.4 shows peak heat release rates for a variety of polymeric materials, and Table 10.5 includes information about the heat release capacity of the selected polymers.[1]

Table 10.4 Ohio State University (ASTM E906) Peak Heat Release Rates of Selected Non-flame-retarded Polymers[a],[1]

Material	Peak heat release rate (kW/m^2)
PMMA	586.8
PE	476.9
PP	451.2
ABS	391.1
PS	376.7
PBT	316.0
PC	192.5
PVC	43.0

[a] Supplied heat flux 20 kW/m^2.

Table 10.5 Heat Release Capacity of Selected Polymeric Materials[1]

Material	Heat release capacity (J/g K)
HDPE	1450
PP	1106
PS	1088
HIPS	873

6 There is no single ISO standard equivalent to ASTM E906, but the closest widely used ISO counterpart for measuring heat release rate in bench-scale tests is ISO 5660 (cone calorimeter).

Table 10.5 Heat Release Capacity of Selected Polymeric Materials (*continued*)

Material	Heat release capacity (J/g K)
ABS	585
PC	578
PA 6.6	565
PMMA	480
PET	366
POM	200
PVC	157

In the early 1980s, a more advanced bench-scale method for determining heat release rate was introduced with the development of the cone calorimeter, now standardized as ASTM E1354.[7] This apparatus can simultaneously provide data on ignitability, heat release, mass loss, and smoke production for material specimens exposed to a controlled external heat flux. On a smaller scale, the intrinsic heat-release capacity of materials can be determined using micro-calorimetric techniques that measure the specific rate at which heat is evolved during controlled decomposition or combustion under well-defined conditions.[1]

10.4.6 Smoke Obscuration

The most common small scale test method for measuring smoke release from burning products is the traditional smoke chamber, which assesses the formation of visible smoke in a closed unit (ASTM E662).

Smoke obscuration, meaning the reduction of light transmission due to smoke, is a critical aspect of fire hazard because diminished visibility reduces available light and severely interferes with both evacuation and rescue operations. The loss of visibility in a real fire depends on two main factors: (a) the total amount of material that actually burns in the fire, and (b) the quantity of smoke produced per unit mass of burnt material. A widely used bench-scale method for evaluating smoke production from

7 The ISO equivalent to ASTM E1354 (cone calorimeter) is ISO 5660-1: "Reaction-to-fire tests – Heat release, smoke production and mass loss rate – Part 1: Heat release rate (cone calorimeter method) and smoke production rate (dynamic measurement)". Both ASTM E1354 and ISO 5660-1 describe the cone calorimeter method, and are considered technically aligned for determining heat release rate and related fire parameters such as smoke production and mass loss.

burning materials is the closed-chamber smoke test, commonly performed with a smoke chamber that measures the development of visible smoke within a sealed compartment according to ASTM E662.[8]

> Most materials with low flame spread (or low heat release) also exhibit low smoke release.

10.4.7 Smoke Toxicity

> Most fire fatalities are a result of inhalation of smoke and combustion products rather than the consequence of burns.

The following points are critical for understanding and assessing fire hazards associated with smoke toxicity.[1]

Most fatal fire incidents occur in scenarios where the fire grows very large in size. In such "flashover fires", the concentration of carbon monoxide (CO) in the atmosphere is influenced primarily by the available oxygen rather than by the detailed chemical composition of the fuels.

For most polymeric materials, the intrinsic toxic potency of their smoke is relatively low compared with that of typical poisons. Common construction and consumer materials tend to exhibit broadly similar smoke toxicity, so their principal fire hazards are generally not dominated by differences in toxic potency of the combustion gases.

Epidemiological and forensic data show a strong statistical correlation between carbon monoxide inhalation and fire-related fatalities, indicating that CO is a major contributor to lethal outcomes in fires.

The overall heat release and smoke generation from a burning material can be reduced by incorporating suitable flame retardants and smoke-suppressing systems, which in turn can limit fire growth and associated hazards.

8 The closest ISO counterpart to ASTM E662 is ISO 5659-2: "Plastics – Smoke generation – Part 2: Determination of optical density by a single-chamber test". Both ASTM E662 and ISO 5659-2 use a closed smoke chamber with a photometric system to measure optical smoke density, and they are widely recognized as analogous small-scale methods for assessing smoke production of materials, even though their heating configurations and some procedural details differ.

10.5 Mechanisms of Flame Formation and Flame Retardancy

Flame retardants generally impart their properties to plastics either in the condensed phase or gas phase. In the condensed phase, they can lower the effective thermal load on the polymer by acting as a heat sink or by promoting char formation that forms a barrier to heat and mass transfer, thereby protecting the underlying material from further decomposition and combustion. Additional condensed-phase effects include heat removal by conduction, cooling via evaporation, dilution of combustible gases, or participation in endothermic reactions that absorb energy. Many intumescent flame-retardant systems are built from three functional components: an acid source, a carbon-rich char former, and a gas-releasing agent. The gases produced by the endothermic decomposition of these systems expand and foam the char layer and, at the same time, dilute the surrounding oxygen concentration that sustains the flame.

A well-designed sequence of chemical reactions and physical changes is required for a good quality of intumescent coating to be formed. Timing of acid release, polymer degradation, and gas evolution must be aligned. Additionally, the viscosity of the polymer mass subjected to combustion must be suitable to trap the small gas bubbles produced in a char, which lead to its gelling and solidification.

Certain additives act as *condensed-phase flame retardants* without forming an intumescent layer but instead producing a protective surface coating that thermally insulates the polymer and reduces the release of additional fuel. For instance, some silicone-based flame retardants used in polyolefins promote the formation of a silicon dioxide layer on the material surface that limits further degradation of the underlying polymer. Another example is resorcinol diphenyl phosphate (RDP), an organophosphate that operates predominantly in the condensed phase by generating a non-volatile, non-combustible, phosphorus-containing surface barrier; this layer restricts the diffusion of flammable degradation products into the flame zone.

In contrast, *gas-phase flame retardants* act in the gas phase, and mainly disrupt the combustion chemistry of the fire, interfering with the radical chain reactions in which polymer-derived fragments react with oxygen and other reactive moieties to form oxygen, hydroxyl, and hydrogen radicals that sustain the flame (Figure 10.1).[9]

(1) $H + O_2 \rightarrow HO^{\cdot} + O^{\cdot}$
(2) $O + H_2 \rightarrow HO^{\cdot} + H^{\cdot}$
(3) $HO^{\cdot} + CO \rightarrow CO_2 + H^{\cdot}$

Figure 10.1 Key reactions during the combustion of hydrocarbon-containing polymers[9]

[9] P. F. Ranken, "Flame retardants" (Chapter 12), in: *Plastics Additives Handbook* (6th edition), ed. H. Zweifel, R. D. Maier, and M. Schiller, Hanser, 2009, pp. 701–717.

Certain plastic additives, particularly halogen- and phosphorus-containing types, can react with flame-propagating radicals to form less energetic moieties, thereby disrupting the radical chain reactions required for ignition and sustained combustion.

10.6 Selected Flame Retardant Types and Mechanisms of Action

10.6.1 Halogenated Flame Retardants

> In principle, any halogen-containing substance capable of releasing hydrogen halide (HX) at or just before the polymer's thermal decomposition/combustion stage can be considered as a flame retardant.

If the hydrogen–oxygen reactions above (Figure 10.1, steps 1 and 2) can be interrupted, the combustion of the fuel can be retarded (reduced). Hydrogen halides are efficient flame extinguishers (disruptors), and their effectiveness decreases in the order HI > HBr > HCl > HF (adjusted for molecular weights of HX). Certain commercially available bromine- and chlorine-containing compounds are the most suitable for generating hydrogen bromide and hydrogen chloride, respectively, during their decomposition within the flame zone of most polymers.

These hydrogen halides (HX) act primarily by reacting with energetic hydrogen radicals ($H^{\cdot}$) and hydroxyl radicals ($HO^{\cdot}$) and converting them into less reactive halogen radicals ($X^{\cdot}$). By reducing the concentration of these highly energetic radicals, the overall flame energy and the rate of radical chain propagation are markedly reduced, consistent with the mechanism illustrated in Figure 10.2.[9]

(1) $RX \rightarrow R^{\cdot} + X^{\cdot}$
(2) $X^{\cdot} + R'H \rightarrow R' + HX$
(3) $HX + H^{\cdot} \rightarrow H_2 + X^{\cdot}$
(4) $HX + HO^{\cdot} \rightarrow H_2O + X^{\cdot}$

Figure 10.2 Brominated or chlorinated flame retardants (RX) generate radicals ($X^{\cdot}$) (step 1) capable of abstracting hydrogen atoms from a polymer backbone (R'H) and yielding HX (step 2). This in turn suppresses combustion by neutralizing reactive radicals present in the flame (steps 3 and 4)[9, 10]

This action is a sort of "flame poisoning" that directly impacts the heat release rate, ignition, flammability, and flame spread. As a result of the action of halogenated flame-retardants, the less energetic flame gives more carbon particles (soot) and smoke. It also changes the chemistry of the fuels in the flame, yielding the increased CO/CO_2 ratio in the flame.

Organobromine compounds that function as flame retardants as a result of their thermal decomposition while they are in contact with the thermally decomposing polymers are also called brominated flame retardants. They can be either non-polymeric or polymeric. The term “polymeric” in this case means that they are bonded to the main polymer backbone. Since HBr is heavier (less volatile) than HCl, HBr can remain in the flame longer, enabling it to more effectively scavenge the hot (energetic) radicals that fuel the flame. The efficiency of these flame retardants is influenced by the bromine content and how well their thermal decomposition temperature overlaps with that of the polymer in which they are incorporated (i.e., how much HBr they can make available during the thermal decomposition/combustion of the polymer).[10)]

Halogenated compounds impart flame retardancy when used either by themselves or in combination with antimony trioxide (ATO) (Section 10.6.2) and/or organophosphorus flame retardants (Section 10.6.6). Their dosages range from 0.8–1.0% in less demanding applications to 20–24% in the most critical uses (e.g., cables), particularly for highly combustible polymers with inadequate flame retardancy, such as polyolefins. Often, the use of combinations of organobromines with ATO is necessary for improving flame retardancy further at a reduced loading of the total flame-retardant system. Typically, but with few exceptions, organobromines are predominantly used in non-halogenated polymers.

> In contrast to many polymers, PVC possesses a certain degree of inherent flame retardancy due to its chlorine content. This significantly reduces the need for including brominated flame retardants in the PVC compounds.[10)]

10.6.2 Antimony Trioxide (ATO)

One of the most efficient flame-retardant synergists is antimony trioxide (ATO, Sb_2O_3). It does not act as a flame retardant by itself but requires the presence of chlorine- or bromine-containing components to function effectively. The synergism of ATO is attributed to the formation of volatile antimony halides, such as antimony trihalides, in the flame zone. The mechanism explaining the flame-inhibition functionality of the combined halogen–antimony trioxide system takes place in the gas phase, and is described in Figure 10.3.[9)]

The first step is reaction of the formed HX with antimony trioxide to yield volatile antimony halide (step 1). This is capable of interrupting the combustion process by reacting with energetic hydrogen radicals in the flame according to a multi-step pro-

[10] G. Sarti, “Flame retardants and smoke suppressants” (Chapter 14), in: *PVC Handbook* (2nd edition), ed. R. P. Krock, M. Schiller, P. Frenkel, J. W. Summers, and C. A. Daniels, Hanser, 2025, pp. 649–702.

cess (steps 2–7). The antimony oxide-catalyzed recombination of hydrogen radicals is the second mode of inhibiting combustion. In either case, scavenging the $H^{\cdot}$ radicals in the gas phase lowers flame energy. Moreover, because it acts as a catalytic precursor, ATO becomes effective at very low dosages. Generally, ATO is added to polymers at a wide concentration range of 1–20%, depending on the application and the type of organobromine additive.[10)]

(1) $Sb_2O_3 + 6HX \rightarrow 2SbX_3 + 3H_2O$
(2) $SbX_3 + H^{\cdot} \rightarrow HX + {}^{\cdot}SbX_2$
(3) ${}^{\cdot}SbX_2 + H^{\cdot} \rightarrow HX + SbX$
(4) $SbX + H^{\cdot} \rightarrow HX + {}^{\cdot}Sb$
(5) $Sb + O^{\cdot} \rightarrow {}^{\cdot}SbO$
(6) $SbOH + H^{\cdot} \rightarrow {}^{\cdot}SbO + H_2$
(7) ${}^{\cdot}SbO + H^{\cdot} \rightarrow SbOH$

Figure 10.3
Synergistic action of ATO in combination with halogenated flame retardants[9)]

ATO is also used as a synergist with other flame retardants, such as zinc borates (ZnBO), organophosphates, magnesium dihydroxide (MDH), and aluminium trihydroxide (ATH).

ATO does not promote char formation, because it acts only in the gaseous phase.

ATO is a white pigment, and therefore cannot be used in transparent articles.

However, regulatory pressure on ATO and its increasing cost are driving significant R&D efforts focused on identifying cost-effective and non-toxic alternatives to it.

10.6.3 Borates

Boric acid, sodium borates, zinc borates, and barium borate have been used for a long time as flame retardants and smoke suppressants in applications across various industries, including polymers.

Zinc borates (ZnBO) are broadly used as flame retardants and smoke suppressants in PVC, PO, EVA, elastomers, and certain engineering thermoplastics like PA. In non-halogenated polymers, the mechanism of action of the ZnBO hydrated forms is physical, and it proceeds in both the gas and condensed phases. Endothermic dehydration and water release make the borates function as a heat sink, diluting the flame's fuels and reducing the flame energy. These actions are also combined with the formation of a low-melting, glassy protective char that serves as a physical barrier against oxygen penetration and heat transfer. In contrast, the anhydrous form of zinc borates acts essentially in the condensed phase. To be effective in halogen-free polymers, the dosage of zinc borates must be relatively high, or they need to be combined with organophosphorus compounds, organobromines, ATO, ATH, or MDH in flame-retardant packages.

In PVC, zinc borates work the same way as in non-halogenated polymers. In vinyl compounds, however, they act as incipient Lewis acids (precursors of actual Lewis acids such as $ZnCl_2$, which are formed by the reaction of ZBO with HCl from the combusting PVC). This incipient catalytic reactivity promotes crosslinking of the polymer matrix, which in the second stage of the PVC decomposition results in the formation of a char that enhances flame retardancy and reduces smoke generation. Therefore, zinc borates also function as char modifiers, i.e., they enhance the barrier layer's consistency more than other incipient Lewis acids commonly used in PVC as components of mixed-metal carboxylate heat stabilizers. On the other hand, they may promote PVC degradation due to the generation of $ZnCl_2$ during processing. The zinc borate dosages in PVC are substantially lower than those in halogen-free systems.

Zinc borates are particularly beneficial as partial replacements for ATO because they exibit synergy with it in terms of flame retardancy, decreasing the formulation cost and smoke production, and suppressing smoldering combustion. They also show a synergistic effect with ATH, as well as MDH and $MgCO_3$.

The only zinc borates registered in the EU are $2ZnO \cdot 3B_2O_3 \cdot 3.5H_2O$ (a fully hydrated form) and $4ZnO \cdot B_2O_3 \cdot H_2O$ (a partially hydrated form). However, both zinc borate grades are under regulatory pressure from the authorities.[10)]

10.6.4 Incipient Lewis Acids

Zinc stannate (ZS, $ZnSnO_3$) and **zinc hydroxystannate** (ZHS, $ZnSn(OH)_6$) are known flame retardants and smoke suppressants. In PVC, ZHS shows a slightly lower LOI than ZnBO, but it imparts a longer time to ignition (TTI), lower peak heat release rate (pHRR), and lower average smoke release rate. Usually, ZS and ZHS are not used alone but in combination with synergistic substances (such as calcium carbonate, magnesium hydroxycarbonate, and aluminium trihydroxide) to increase fire performance. ZS combined with ZnBO, ammonium octamolybdate, and organophosphate plasticizers resulted in a more consistent char.

ZS and ZHS are utilized in halogenated plastics, such as PVC and polychloroprene, and function there as incipient Lewis acids capable of promoting crosslinking of PVC during its combustion and forming a char layer. In PVC, the main mechanism for flame retardancy occurs in the condensed phase, where $ZnCl_2$ (a Lewis acid derived from ZS) (a) promotes crosslinking of PVC matrix, reducing the heat release, and (b) inhibits benzene formation and therefore reduces smoke generation. It should also be noted that these stannates may reduce the heat stability of PVC compounds containing them, in a similar way to the action of zinc borates.

Molybdenum compounds are a class of flame retardants and smoke suppressants used in PVC for reducing heat release and smoke generation. An example of those compounds includes the above-mentioned **ammonium octamolybdate** (AOM, $(NH_4)_4(Mo_8O_{26})$). This is an established flame retardant and smoke suppressant, primarily used in applications where keeping the smoke production at a minimum level is critical. An example of this kind of application is PVC plenum cables. In these cases, AOM is used at dosages reaching up to 30 phr. AOM functions primarily as a flame retardant and smoke suppressant in the condensed phase, where it decomposes, and its decomposition product reacts with HCl that is released during the combustion of PVC compounds. This reaction also leads to the formation of various chlorides, including MoO_2Cl_2, a key intermediate in the flame-retardant action of AOM in PVC articles (Figure 10.4).[10)]

$$(1)\ (NH_4)_4(Mo_8O_{26}) \rightarrow 8MoO_3 + 4NH_3 + 2H_2O(200\text{–}370\,^\circ C)$$
$$(2)\ MoO_3 + 2HCl \rightarrow MoO_2Cl_2 + H_2O$$
$$(3)\ HCl + NH_3 \rightarrow NH_4Cl$$

Figure 10.4 Transformation of AOM in PVC[10)]

10.6.5 Flame-Retardant Functional Fillers

Regardless of the mechanism, functional fillers help to reduce smoke production.

The smaller the particle size of functional fillers, the lower the decomposition temperature.[10)]

The thermal decomposition of **aluminium trihydroxide (ATH)** starts at 180–200 °C and ends at 320 °C (Figure 10.5). It proceeds through two endothermic steps with the formation of boehmite (AlOOH) as an intermediate. Each kilogram of ATH absorbs 1051 kJ of energy, yielding 0.654 kg of aluminium oxide (Al_2O_3) and releasing 0.346 kg water. On the other hand, **magnesium dihydroxide (MDH)** has an onset decomposition temperature of about 300–320 °C, with the process ending at about 450 °C (Figure 10.5). Each kilogram of MDH releases 0.309 kg of water in the flame, leaving 0.691 kg of MgO on the surface, which protects the polymer.

$$2Al(OH)_3 \rightarrow Al_2O_3 + 3H_2O$$
$$2Mg(OH)_2 \rightarrow 2MgO + 2H_2O$$

Figure 10.5 Thermal decomposition of ATH and MDH[10)]

Thermal decomposition of ATH and MDH in the condensed phase does not lead to the formation of char or a protective layer on the burnt polymer surface. They function as heat sinks and also release non-combustible water that dilutes flammable gases near the flame. Additionally, Al_2O_3 and MgO form a layer on the unburnt polymer that protects it from heat and oxygen permeation. They are also capable of altering the viscosity of the melted polymer during its combustion. This action increases flame retardancy and the anti-dripping effect. Furthermore, water release impedes the aggregation of carbon particles formed in the flame, reducing smoke production and soot formation.

In the so-called halogen-free flame-retardant (HFFR) polymers, ATH is frequently used as the principal flame retardant, particularly in the polymer formulations with processing temperatures below 180 °C. In general, substantial dosages of these additives are needed to impart flame retardancy to a resin. Resin formulations containing 40–60% ATH are not unusual, and in specific HFFR cable compounds, ATH loading levels can be as high as 70%. In other applications, such as foamed PVC-U sheets, lower dosages of ATH are used (typically less than 20 phr).

ATH is frequently combined with other flame retardants and/or synergists to enhance its effectiveness. Common synergistic agents include ATO, zinc borates, and incipient Lewis acids.

In PVC-P, every 20 phr of ATH added raises the LOI by one point, but when synergistic agents are present, this increase can be up to two points.

This synergistic interaction underlines the importance of formulating flame-retardant systems that maximize the flame performance of the compound, while minimizing the total quantity of flame-retardant additives and also minimizing effect of ATH on the mechanical properties of the polymer system. However, ATH is not suitable for use in polymers with higher processing temperatures, such as PA and PC.

MDH is used sometimes as a partial replacement for ATH. It exhibits somewhat lower flame performance but at lower cost.

Magnesium carbonate ($MgCO_3$) is also a flame-retardant filler that absorbs heat and decomposes at about 550 °C, acting as a heat sink and lowering the temperature of the combustion zone, while releasing CO_2, diluting flammable gases and reducing the availability of oxygen in the flame. It forms the MgO protective barrier that inhibits further flame spread by limiting heat and mass transfer. However, efficiency of magnesium carbonate as a flame retardant is generally lower compared to ATH or even MDH when used in PVC. Magnesium carbonate decomposes with absorbing significantly less energy than ATH or MDH. Additionally, if used in PVC, a portion of magnesium carbonated can be used up on scavenging HCl that is released during thermal decomposition of the polymer. It results in the formation of magnesium chloride, carbon dioxide, and water.

Hydromagnesite ($Mg_5(CO_3)_4(OH)_2(H_2O)_4$ or $4MgCO_3{\cdot}Mg(OH)_2{\cdot}4H_2O$) and **huntite** ($Mg_3Ca(CO_3)_4$) are minerals often used in combination with dolomite and magnesite, and have been studied as alternatives to ATH and MDH. Hydromagnesite decomposes gradually over a wide temperature range (200–700 °C), releasing both water and carbon dioxide. Huntite decomposes via two high-temperature endothermic steps, releasing CO_2. Therefore, the decomposition temperature range of the huntite/hydromagnesite-containing combinations may vary depending on the ratio of the components. Huntite and hydromagnesite are recognized as non-toxic flame-retardant fillers.[10)]

10.6.6 Phosphorus-containing Flame Retardants

Phosphorus compounds are widely used as flame retardants in various applications, such as coatings, adhesives, textiles, and plastics. In polymers, they are mainly employed in rigid and flexible PU, TPU, polyesters, PO, PC, and its blends with ABS, TPE, PBT, PET, PA, epoxy and phenolic resins, HIPS, rubber, and PVC.

Phosphorus (P) compounds can be organic and inorganic. Specific classes of P-containing substances include phosphonates (P oxidation state +3), hypophosphites (+1), phosphine oxides (–1), red phosphorus (0), and phosphates (+5). Many P-containing compounds are used to impart flame-retarding characteristics on plastic articles.

The action of these compounds occurs in both the gas and condensed phases. During thermal decomposition in the flame, components with a lower P oxidation state can produce P-containing radicals ($^{\bullet}PO$ and $^{\bullet}HPO$), which catalyze the recombination of the high-energy radicals $H^{\bullet}$ and $HO^{\bullet}$ present in the flame into hydrogen (H_2) and water (H_2O), respectively, effectively diminishing flame propagation. Components with a higher P oxidation state typically generate phosphoric acid, which yields pyrophosphoric acid at about 210 °C and metaphosphoric and polyphosphoric acids upon condensation about 300 °C. These endothermic reactions result in P-containing compounds acting as heat sinks, while their by-products dilute the fuel. However, their main effect is in the condensed phase, where phosphoric acid catalyzes the elimination of terminal OH groups, promoting crosslinking of polyene sequences, yielding the char, and insulating the polymer from heat and oxygen.[10)] In polymers lacking oxygen, adding polyols can facilitate char-formation during combustion. These polyols often work synergistically with blowing agents. These additives can release gases, such as nitrogen or carbon dioxide, to expand and thicken the protective char layer and enhance its barrier properties against heat and oxygen. Flame-retardant packages of this type, which promote the formation of an insulating layer, are called *intumescent* systems. One example of an intumescent system is a mixture consisting of melamine polyphosphate, ammonium polyphosphate, and polyols, which can be used in a variety of non-halogenated polymers.

Inorganic hypophosphites or phosphinates (such as salts of aluminium or calcium) are used in non-halogenated polymers like PO, PA, PC, PBT, PC/PBT, ABS, PU, and TPU. Red phosphorus (RP) is used in ABS, PU, PA, PMMA, HIPS, PS, and PET. It functions in the condensed phase in oxygen-containing polymers and in the gas phase in polymers with no oxygen.[10)]

Phosphate esters can be aryl phosphates, alkyl phosphates, or alkyl–aryl phosphates. Alkyl phosphates are used mainly in polyesters, rubber, and cellulosics. Tricresyl phosphate (TCP), being developed in the 1930s, was one of the first alkyl phosphate flame retardant; more recently developed and commercialized alkyl–aryl phosphate esters include 2-ethylhexyl diphenyl phosphate and isodecyl diphenyl phosphate. Triphenyl phosphate (TTP) is employed mainly in PO, EPDM, PC/ABS blends, HIPS, TPU, flexible and rigid PU, epoxy and phenolic resins, and rubber. Organophosphates (phosphate esters) are broadly used as flame-retarding plasticizers in PVC-P. These are less combustible than the conventional ester plasticizers and are used for replacing them partially or completely in manufacturing films, conveyer belts, tarpaulins, flooring, coated textiles, technical membranes, and cables, where they are applied in combination with AOM and/or halogenated phthalates.[10)]

10.6.7 High-temperature Acid Scavengers

An acid scavenger is a substance designed to trap acidic substances through either a chemical reaction or physical absorption. The acid scavengers discussed in this section are specific to HCl. Acid scavengers that also act as co-stabilizers in processing PVC compounds can be both organic and inorganic. However, for high-temperature applications, the scavenger must be inorganic and remain thermally stable even at the maximum temperature of the test that is used for assessing its acidity, corrosivity, or halogen content.

HCl scavengers operate in both the gas phase and the condensed phase. Scavenging HCl in the condensed phase, by forming solid or liquid reaction products with it, reduces the concentration of HCl available in the gas phase.

Calcium-containing substances are the most effective acid scavengers. Upon trapping HCl, they form $CaCl_2$, which is thermally stable up to 950 °C. Calcium carbonates representative acid scavengers capable of trapping HCl at high temperatures.

> The efficiency of calcium carbonate as an acid scavenger is increased by decreasing the particle size, making the ground additive effective at a particle size of about 0.1 μm. A further increase in efficiency is achieved with the use of an ultrafine precipitated calcium carbonate (uPCC) grade.

10.7 Properties and Applications of Organic Flame Retardants in Polymers

The properties and typical applications of selected organic flame retardants in polymer systems are listed in Table 10.6.

Table 10.6 Selected Organic Flame Retardants, their Physical Properties, and Typical Uses in Plastics[9)]

Common name	Chemical structure	Physical properties	Typical use
Bisphenol A diphenylphosphate	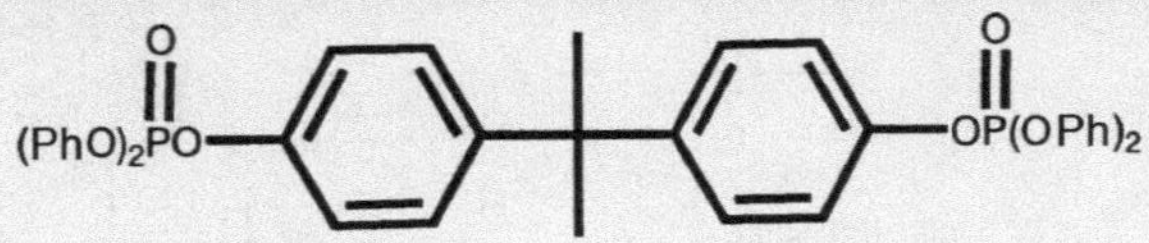	Liquid, mixture	PC/ABS, PPO/HIPS
1,2-Bis(tribromophenoxy) ethane	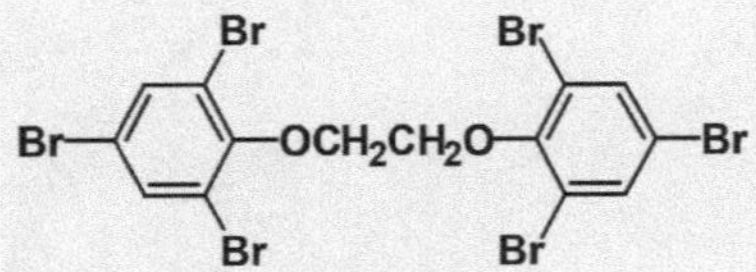	M.p. 223–228 °C	ABS
Brominated epoxy oligomers	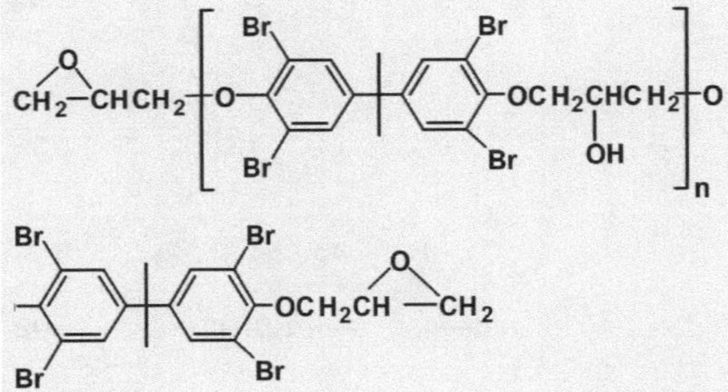	Softening range 130–158 °C	ABS, PBT, HIPS, PA
Brominated polystyrene	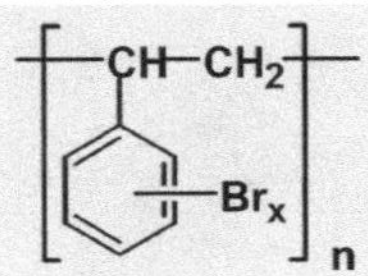 , x = 2-3	T_g 175–195 °C	PBT, PET, PA
Chlorendic anhydride	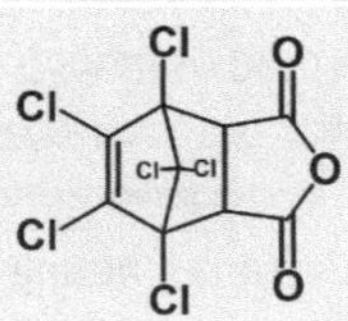	M.p. 235–239 °C	Unsaturated polyesters

Table 10.6 Selected Organic Flame Retardants, their Physical Properties, and Typical Uses in Plastics (*continued*)

Common name	Chemical structure	Physical properties	Typical use
Chlorinated paraffins	Contains 40–70% Cl	Liquid, mixture	PVC
Decabromobiphenyl	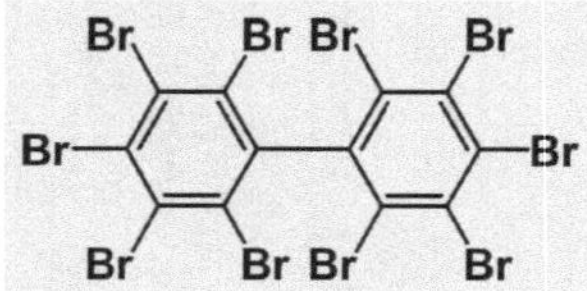	M.p. 385–389 °C	PBT
Decabromodiphenylethane	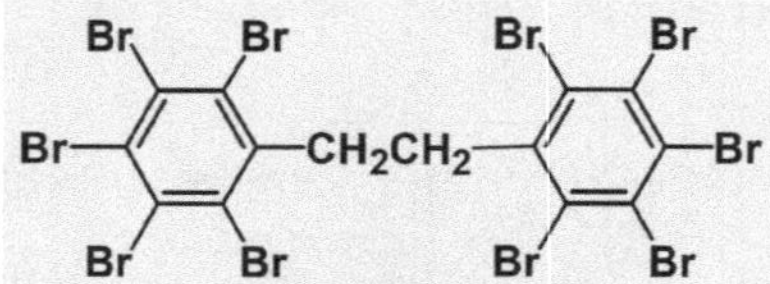	M.p. 340–344 °C	HIPS, PBT, PE, PP
Dibromoneopentylglycol	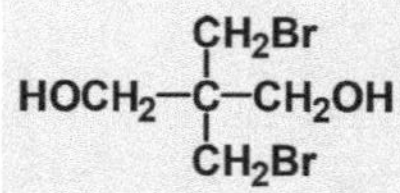	M.p. 109–110 °C B.p. 134 °C (1 mmHg)	PUR, unsaturated polyesters
Ethylene bis(5,6-dibromo-norbornane-2,3-dicarboximide)	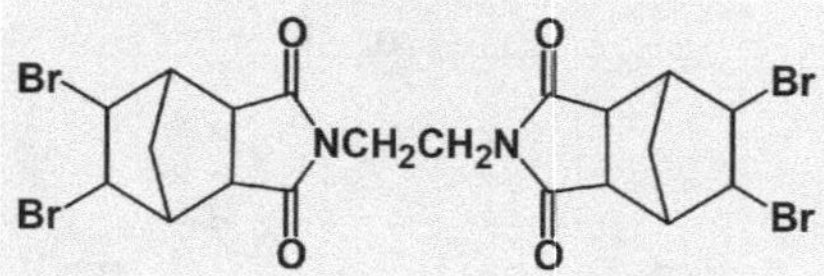	M.p. 294 °C	PP
Ethylene bis(tetrabromophthalimide)	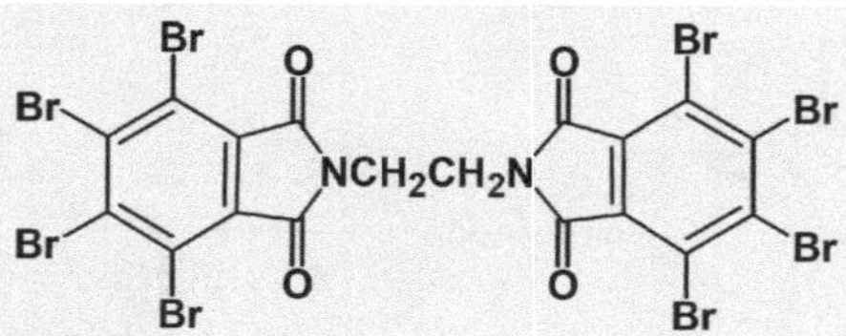	M.p. 450 °C	HIPS, PBT PE, PP
Halogenated polyetherpolyols	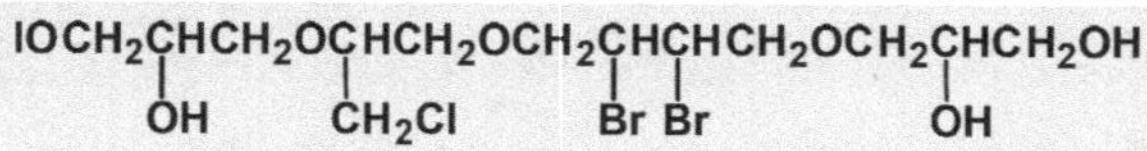	Liquid, mixture	PUR

Common name	Chemical structure	Physical properties	Typical use
Hexabromocyclododecane	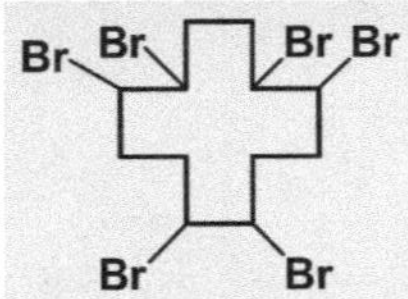	M.p. 175–195 °C (mixture)	PS foam
Octabromotrimethylphenylindane	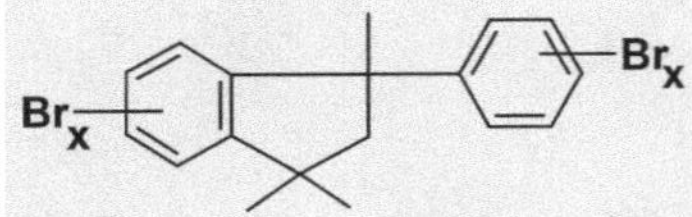 , x=7-8	M.p. 230–250 °C	HIPS, ABS
Poly(dibromostyrene)	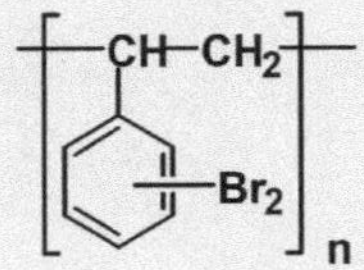	M.p. 210–230 °C	PBT, PET, PA
Poly(pentabromobenzylacrylate)	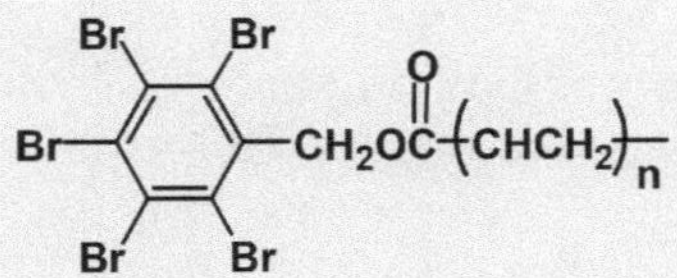	M.p. 190–210 °C	PBT
Resorcinol diphenylphosphate	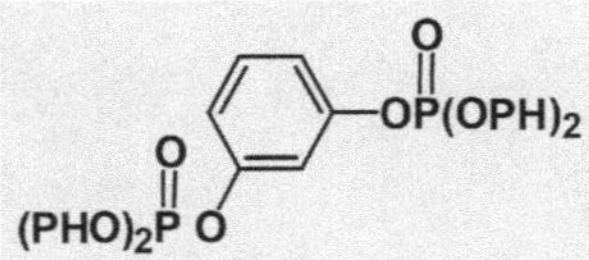	Liquid, mixture	PC/ABS, PPO/PS
Tetrabromobisphenol A	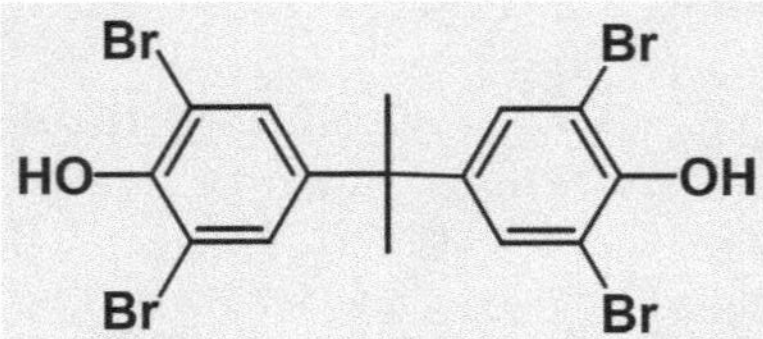	M.p. 179–181 °C	Epoxy resins, ABS

Table 10.6 Selected Organic Flame Retardants, their Physical Properties, and Typical Uses in Plastics (*continued*)

Common name	Chemical structure	Physical properties	Typical use
Tetrabromobisphenol A, bis(2,3-dibromopropyl ether)	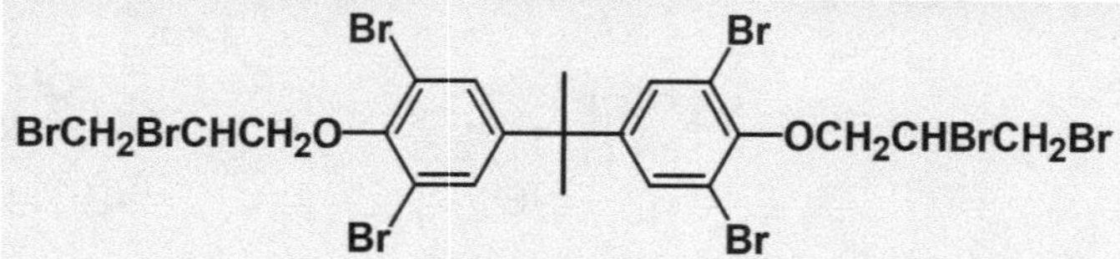	M.p. 90–100 °C	PP
Tetrabromobisphenol A, carbonate oligomers	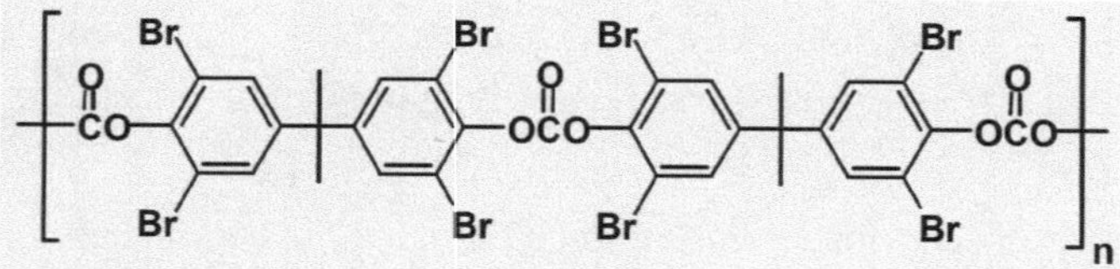	M.p. 210–260 °C	PBT
Tetrabromophthalate diols	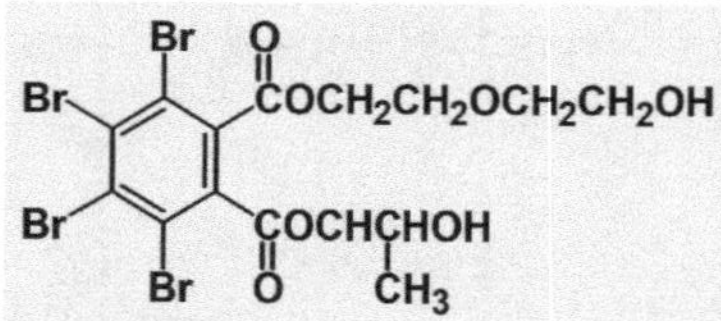	Liquid, mixture	PUR
Tetrabromophthalic anhydride	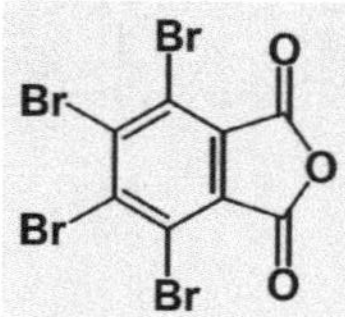	M.p. 270–276 °C	Unsaturated polyesters
Tetradecabromodiphenoxybenzene	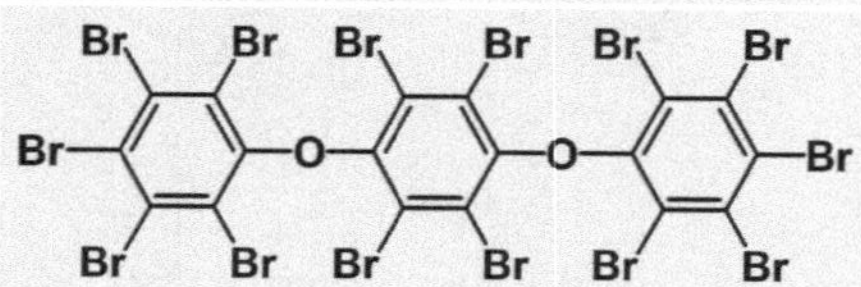	M.p. >350 °C (decomp.)	PET, PA, PBT
Triphenyl phosphate	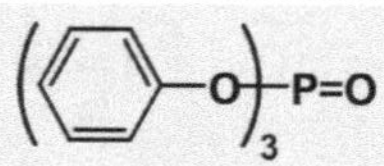	M.p. 50–52 °CB.p. 244 °C (10 mmHg)	PC/ABS, PPO, PS

10.8 Questions

1. What are the differences between thermal degradation and thermal decomposition of polymers?
2. Can a polymer burn without breaking down?
3. What serves as fuel that is generated from a burning polymer?
4. What are the consequences of polymer burning?
5. What are the main characteristics of char formed as a result of polymer combustion?
6. To maintain combustion, what should be the rate of fuel generation in comparison with the rate of fuel consumption?
7. What are the most critical high-energy free radicals formed in burning materials?
8. What are the main fire performance characteristics?
9. What is LOI and how is it measured?
10. What are the hazards associated with smoke obscuration?
11. How can the comparative smoke toxicity of all burning polymers be characterized in general?
12. What are the material changes in the condensed and gas phases during polymer combustion?
13. What components are needed to form intumescent flame retardant systems? Give an example.
14. What is the mechanism of action of halogenated flame retardants?
15. What are the advantages of HBr over HCl in the flame?
16. How does antimony trioxide work as a flame-retardant synergist?
17. What are the flame retardant effects of zinc borates?
18. What is an incipient Lewis acid? Give an example of this kind of flame retardant.
19. Give examples of the flame-retardant functional fillers. What is their mode of action?
20. What is the role of phosphate esters in PVC?
21. Give an example of a high-temperature acid scavenger, and explain its flame-retardant action mode.

11 Antistatic Agents

Content of this chapter was adapted from:

- F. Wylin, "Antistatic Additives" (Chapter 10), in: *Plastics Additives Handbook* (6th edition), ed. H. Zweifel, R. D. Maier, and M. Schiller, Hanser, 2009, pp. 655–673.
- E. Richter, "Antistatika" (Section 5.4), in: *Handbuch der Kunststoff-Additive* (4th edition), ed. R. D. Maier and M. Schiller, Hanser, 2016, pp. 642–659.

11.1 Terms and Definitions

Surface resistivity	Electrical resistance measured across the surface of an insulating material.
Static electricity	Electric charge accumulation on a surface due to friction or contact separation.
Ionic molecules	Molecules containing charged atoms or groups of atoms, resulting in ionic bonding.
Non-ionic molecules	Molecules without a net electric charge, typically covalently bonded.
Antistatic effect	Reduction or prevention of static charge accumulation on a material's surface.
Quaternary ammonium salts	Nitrogen compounds with four alkyl or aryl groups, used for antistatic and antimicrobial purposes.

Component migration within polymers	Movement of additives or components through the polymer matrix over time.
Crystallinity of plastics	Degree of ordered arrangement of polymer chains in a solid plastic material.
Adsorption	Adhesion of atoms, ions, or molecules to a surface from a gas or liquid phase.
Masterbatch	Concentrated blend of additives on a carrier resin for easy incorporation into polymers.

11.2 Introduction

Amber has a property of attracting lightweight objects after rubbing, and Thales of Miletus was already familiar with this property in ancient times (~500 B.C.). This phenomenon arises due to a build-up of electrons on the surface of the amber, and is known as static electricity. Investigations were carried out by Peregrinus in the 13th century, and these were continued by Gilbert in the 16th century, who found that other materials can also be electrified by friction.

The branch of physics that deals with static electrical charges, charge distributions, and the electrical fields of charged bodies is known as electrostatics, and the following relationship is central to this branch of physics:

Coulomb's Law: The magnitude of the electrostatic force between two point charges is directly proportional to the product of the magnitudes of their charges, and inversely proportional to the square of the distance between them.

When two materials are rubbed against each other, an electrostatic charge is created. If one material releases electrons easily, and the other accepts electrons easily, the electrons pass from one material to the other. This then results then in the two materials becoming positively and negatively charged, respectively, and voltages of 30–40 kV can be built up.

Plastics are usually very good insulators, and the built-up charges cannot flow away because plastic materials do not conduct electricity. As a result, static electricity charges on plastic surfaces cannot easily be dissipated, and the electrical voltage can only be reduced when the charged material comes in contact with another object that has a significantly different potential. If this happens, then a spark (discharge) will pass between them, and an electrical current will flow until the potentials are equilibrated.

The potential of the formed electric charge depends on several parameters:

- Type of contact
- Strength of the friction
- Humidity of the environment
- Dielectric constant of the materials
- Specific electrical resistance of the materials.

In most cases, electrostatic charges arise unintentionally – for example, when filling plastic bags with powders, or when pumping liquids or powders through funnels and pipes (e.g., pneumatic conveying). This not only causes problems and significantly disrupts production processes, but can also present a considerable potential risk and hazard. Some challenges associated with electrostatic charging of plastics are:

- Films or fibers sticking together during production
- Damage to packaged goods (e.g., electronic components)
- Attraction of dust to a considerable extent during processing and service life
- Sparking increasing the risk of explosion in the presence of dust, fine powders, or flammable vapors
- Difficulties arising with painting and printability.

Most of these problems can be addressed by lowering the surface resistivity of plastics from the order of 10^{15}–10^{17} ohms to 10^{9}–10^{12} ohms. This reduction in surface resistance (SR) is achieved through the use of antistatic agents (Figure 11.1).

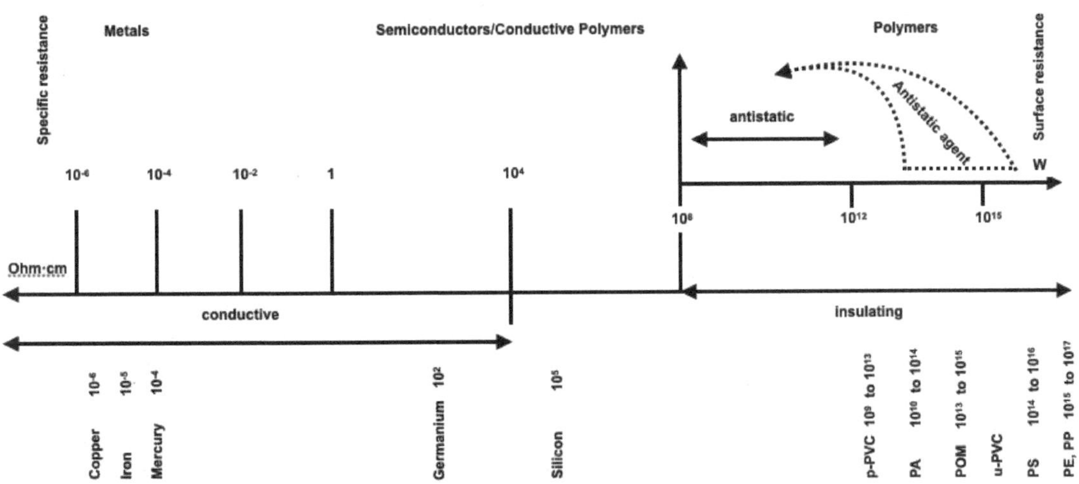

Figure 11.1 Specific electrical resistance of different materials and the effect of antistatic agents

11.3 Chemistry of Antistatics

There are three categories of antistatic agents: (a) external, (b) internal, and (c) permanent antistatic agents. They differ not only in their chemistry, but also in their mechanisms of action and the concentration used. External antistatic agents are not considered here, since they are dissolved in a solvent and applied to the plastic surface, e.g., in the form of sprays.

A major factor affecting the surface resistivity is the relative humidity of air, as shown in Figure 11.2 for PE. The SR reduces from 10^{15} ohm·cm to $<10^{12}$ ohm·cm, when the humidity increases from 50% to 100%.

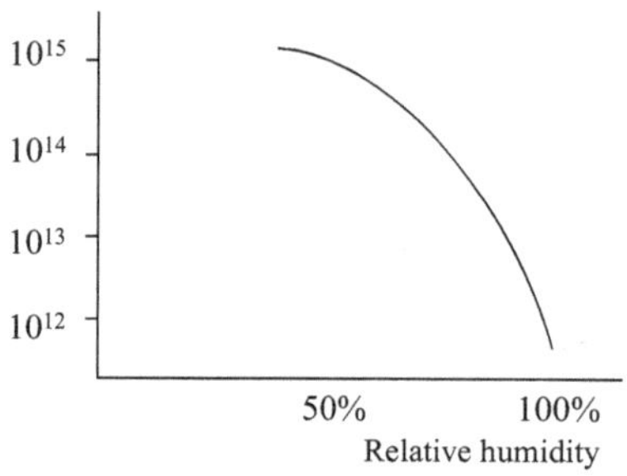

Figure 11.2
Influence of atmospheric humidity on the surface resistivity (in ohm·cm) of the surface of PE

Ideally, antistatic agents should have limited compatibility with plastics, so that after processing plastic compounds with the antistatic agents, they migrate to the surface of an article and form a film on the surface (Figure 11.3). In this film, their non-polar functional groups protrude into the polymer matrix, while the polar groups have a certain affinity to moisture in the air. This water forms a thin film on the surface in which the electrostatic charge can dissipate. However, this water film on the surface can only be formed when the air humidity is relatively high.

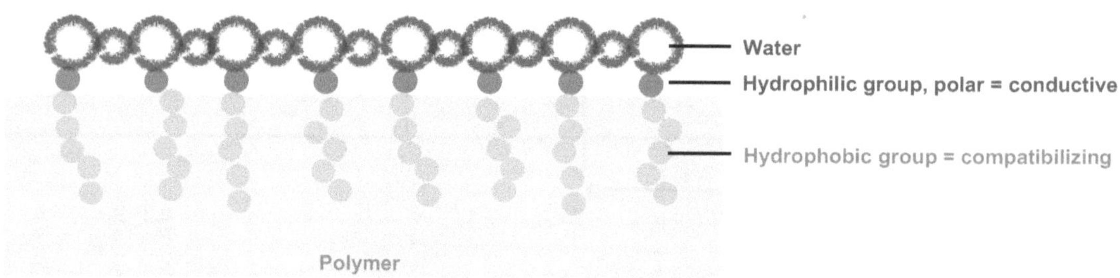

Figure 11.3 Schematic mechanism of action of antistatics

Amongst antistatic agents, the internal antistatic agents are the most dominant and commercially viable class. All of them have a polar group as a required structural element, which is responsible for the affinity of the antistatic agent to moisture. Four groups of internal antistatic agents can be distinguished, and are described in the following sections.

11.3.1 Non-ionic Antistatics

Ionic molecules contain charged functional groups (ions), either positive (cations) or negative (anions), which form charged particles in water and conduct electricity. Non-ionic molecules do not carry an electrical charge within the molecule. They usually contain neutral but polar groups (like hydroxyl or ether groups) and do not form ions in water.

Non-ionic antistatics (Table 11.1) include:

- Fatty acid esters:
  - Glycerol monoesters with fatty acids
  - Glycerol diesters with fatty acids
  - Partial esters of other polyols than glycerol like pentaerythritol
  - Polyenthylene glycol esters, such as polyethylene glycol monolaurate.
- Sorbitan partial esters:
  - Ethoxylated esters of fatty acids (~20 ethylene oxide units, and stearic acid or oleic acid)
  - Sorbitan monolaurate (Figure 11.4)
  - Ethoxylated sorbitan monolaurate (Figure 11.4).
- Ethoxylated alkyl amines:
  - *N,N*-Bis(2-hydroxyethyl)tallow amine.
- Diethanol amides:
  - Lauric diethanol amide.

Table 11.1 Chemical Structures of Non-ionic Antistatics

Chemical structures	Substituents		Chemical class
HO-CH_2-CH(OH)-C(O)OR	R: C_nH_{2n-1}	*n*: 11–17	Saturated fatty acid monoesters
HO-CH_2-CH(OH)-C(O)OR	R: C_nH_{2n-3}	*n*: 11–17	Unsaturated fatty acid monoesters
RO-CH_2-CH(OH)-C(O)OR	R: C_nH_{2n-1}	*n*: 11–17	Saturated fatty acid diesters

Table 11.1 Chemical Structures of Non-ionic Antistatics (*continued*)

Chemical structures	Substituents		Chemical class
RO-CH_2-CH(OH)-C(O)OR	R: C_nH_{2n-3}	*n*: 11–17	Unsaturated fatty acid diesters
R-N(CH_2-CH_2-OH)$_2$	R: C_nH_{2n-1}	*n*: 12–18	Ethoxylated alkylamines
R-C(O)-N(CH_2-CH_2-OH)$_2$	R: C_nH_{2n-1}	*n*: 11–17	Diethanolamides
R-O(CH_2-CH_2-O)$_m$-R	R: C_nH_{2n-1}	*n*: 12–18	Ethoxylated alcohols

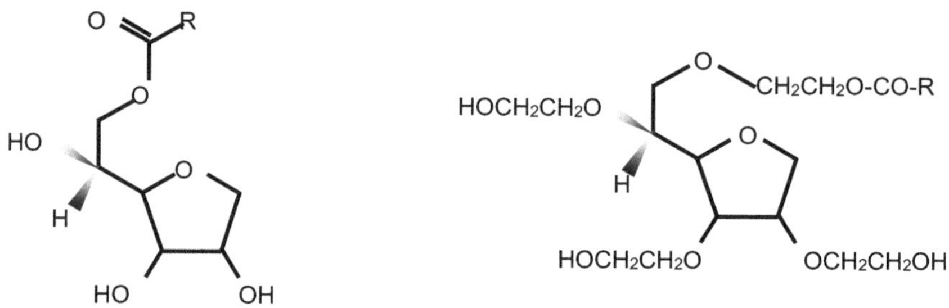

R = $C_{11}H_{23}$: Sorbitan monolaurate
R = $C_{17}H_{33}$: Sorbitan monooleate

R = $C_{17}H_{33}$: Sorbitan monooleate, ethoxylated

Figure 11.4 Chemical structures of sorbitan esters

Glycerol monostearate (GMS) and **glycerol monooleate** (GMO) are the main antistatic agents of the group of fatty acid esters. They act as lubricants too (see Chapter 7). These agents are always mixtures of the mono-, di-, and triesters of glycerin, and they vary in the concentration of the monoester from 40% to over 90%. Both GMS and GMO impart an antistatic effect very quickly, since after compounding they easily migrate to the surface of the article due to their limited compatibility with polymers. Most of the GMS grades are suitable for food contact applications, and the mono- and diglycerides of fatty acids are registered in the EU as food additives under number E471. Their disadvantage is that they only show a short-term effect. The antistatic effect is strongly dependent on the monoester content and the dosage of GMS (Figure 11.5). At 0.25% GMS the initial antistatic performance is not sufficient enough, and the optimum performance is reached at 0.50% GMS. At 0.75–1.00% it exhibits the same performance as at 0.50%. However, over time, at higher dosages the article loses its antistatic performance (see Section 11.5).

Ethoxylated alkylamines usually have an aliphatic substituent in their structure, with a typical chain length of C_{12}–C_{18}. In contrast to GMS and GMO, these antistatics do not have a strong short-term effect, because they migrate to the surface more slowly and have a pronounced long-term effect. As a result, they are often used as in synergistic mixtures in combinations with GMS or GMO (Figure 11.6). The synergism is based on the fact that the monoglycerolates provide a fast antistatic effect, and the alkylamines a long-lasting antistatic effect. As seen in Figure 11.6, in the case of using GMS alone, the surface resistivity (SR) drops to <10^{12} ohm·cm after one day. However, after 2 weeks the SR increases and reaches values just below the blank. In the case of just ethoxylated alkylamine being used, the effect on SR is below 13.5 × 10^{13} ohm·cm, and it takes about 3 months to reach a minimum of <10^{11} ohm·cm. GMS combined with ethoxylated alkylamine reduces SR below 10^{11} ohm·cm after one day, and the SR remains below <10^{11} ohm·cm even after about a year. However, ethoxylated alkylamines have limited food contact approvals depending on their dosages. This needs to be taken into consideration in the case of food packaging applications.

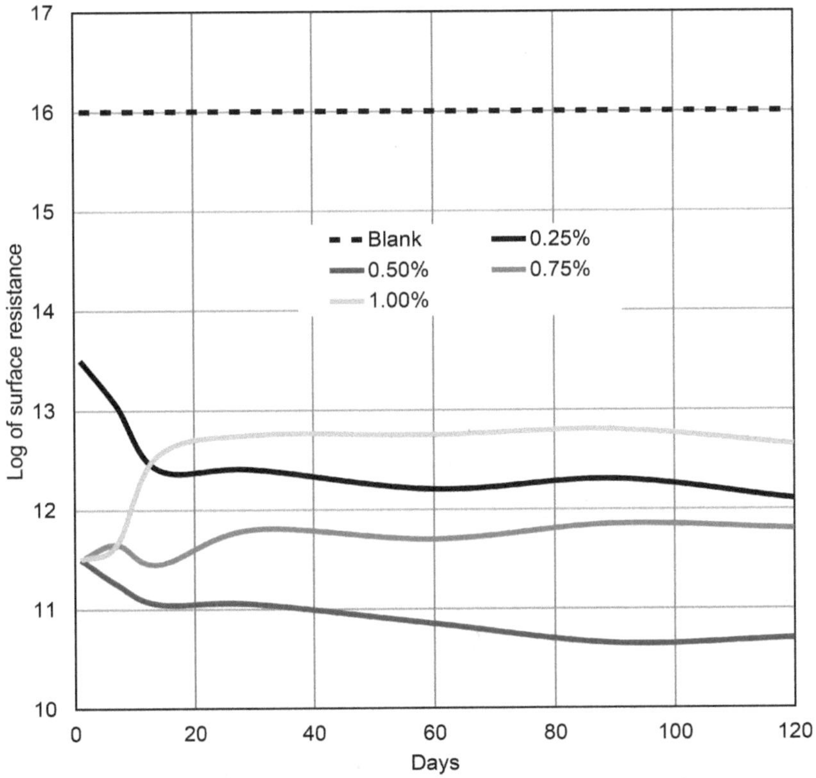

Figure 11.5 Antistatic performance of GMS at varied dosages in PP

Diethanol amide derivatives are very similar in their chemistry to ethoxylated alkylamines. They are not suitable for food contact, but their advantage is their good performance even under low-humidity conditions.

They migrate much slower to the surface of the polymer and need time to reach their optimal performance. As a result, they provide long-term performance as antistatic agents. In order to impart a more immediate antistatic effect on plastic articles, they are often combined with GMS.

Ethoxylated alcohols are especially useful in p-PVC. In addition to their antistatic effect, they also contribute a strong viscosity-reducing effect. These agents differ considerably in the degree of ethoxylation and thus in the hydrophilic–lipophilic balance (HLB) value.

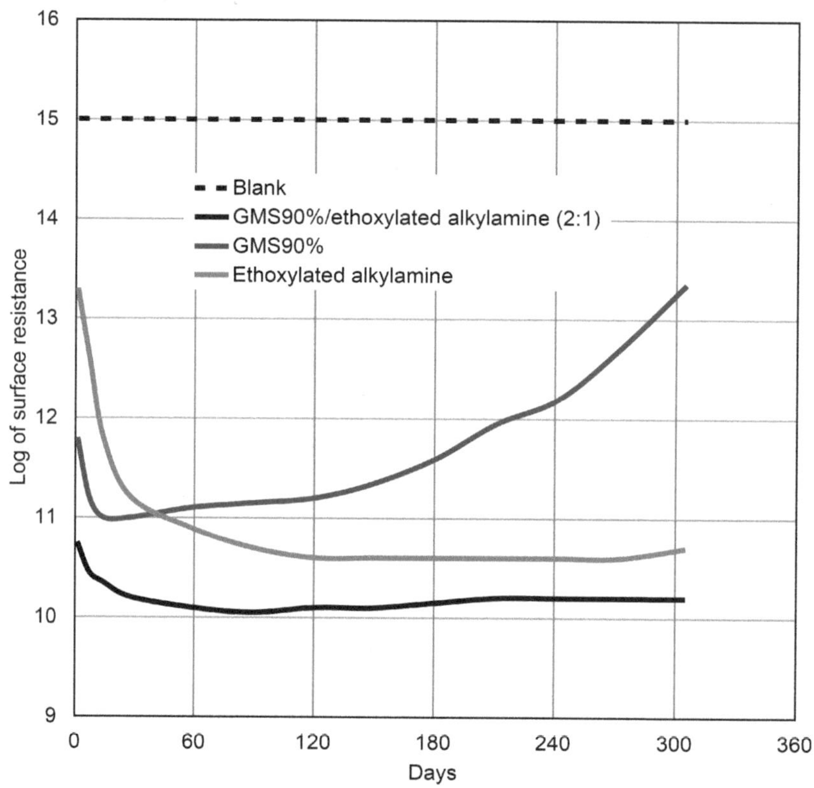

Figure 11.6 Antistatic effect of GMS, ethoxylated alkylamine, and the combination of both in injection-molded PP

11.3.2 Ionic Antistatics

Anionic antistatic agents are shown in Figure 11.7. The anion is usually an alkyl sulfonate or an alkyl phosphate and the cation is an alkali metal; individual products differ in the chain length of the alkyl substituent.

Sulfate or sulfonated antistatic agents, such as alkane sulfonates, provide effective dissipation of static electricity in plastics. Alkane sulfonates are particularly useful in thermoplastic compositions, such as polycarbonate blends, imparting a good antistatic effect at relatively low concentrations (e.g., 2–4 wt%), while maintaining the article's mechanical properties. However, sulfate-based antistatic agents are generally considered less effective compared to other types, such as quaternary ammonium salts.

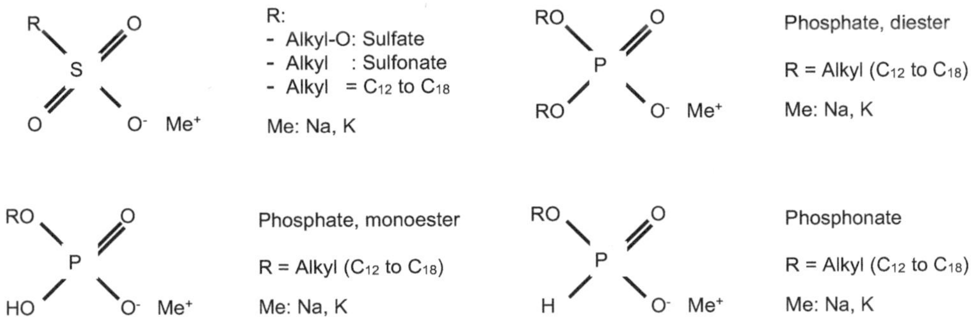

Figure 11.7 Chemical structures of anionic antistatic agents

Phosphate-based antistatic agents are widely used in various industries due to their unique performance attributes. Their advantages are:

- They are extremely effective in reducing static electricity buildup, making them ideal for a variety of applications
- A wide range of phosphoric acid esters is commercially available, allowing for tailored solutions depending on the requirements
- Potassium salts are particularly favored due to their superior performance in many formulations
- Low molecular weight (MW) phosphate salts are more effective antistatic agents compared to their higher MW counterparts, as they demonstrate better solubility.

Their disadvantages are:

- Solid phosphates are difficult to disperse and incorporate into formulations of plastics

> As the molecular weight of the phosphate esters increases, their effectiveness as antistatic agents decreases.

- Un-neutralized acid esters exhibit a reduced antistatic effect, requiring additional processing steps
- Many phosphate salts fail to comply with EPA and REACH regulations, limiting their use in certain regions or applications
- Low-molecular-weight antistatic agents are more likely to be absorbed into polymers, such as polyamides and polyether–polyurea, potentially reducing their surface effectiveness
- Phosphate esters prepared using phosphorus oxychloride ($POCl_3$) can leave behind corrosive halide ion residues, which may affect the durability and safety of the final article.

Cationic antistatic agents with bulky cations are represented in most cases by quaternary tetra-alkyl or alkyl aryl ammonium salts (Figure 11.8).

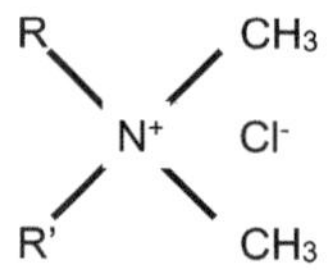

R: Alkyl (C_{12} to C_{18})

R': CH_3-, C_6H_5-CH_2-

Figure 11.8
Chemical structures of cationic antistatic agents

11.3.3 Amphoteric Antistatics

Amphoteric antistatic agents are a kind of internal salt, containing both cationic and anionic hydrophilic groups in one molecule.

Alkyl betaines (Figure 11.9) belong to this type of antistatics, although they are considered to be substances with a limited thermal stability.They are amphoteric (zwitterionic) surfactants with a long-chain alkyl group and a betaine headgroup. Chemically, they are 2-(alkyldimethylammonio)acetates, consisting of a quaternary ammonium center with two methyl groups, a long-chain alkyl group (typically C_{12}–C_{18}), and a carboxymethyl group that is internally neutralized, giving a zwitterionic structure. The betaine head simultaneously carries a positive (N^+) and a negative (COO^-) charge, which makes alkyl betaines amphoteric and allows them to interact, depending on pH, with both anionic and cationic surfactants.

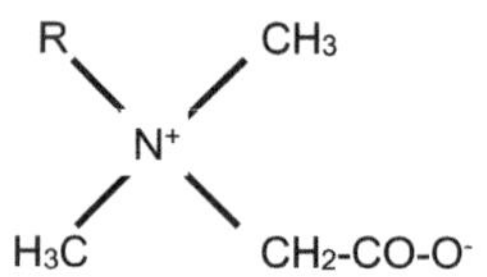

R: Alkyl (C_{12} to C_{18})

Figure 11.9
Chemical structure of alkyl betaines, as an example of an amphoteric antistatic agent

Ammonium-based antistatic agents are widely used in the plastics industry due to their ability to dissipate or prevent buildup of electrostatic charges on plastic surfaces. When applied specifically to plastics, their advantages are:

- A variety of these antistatic systems are commercially available, and can be selectively chosen for different plastic formulations and applications
- They can be readily incorporated into lubricant systems used in plastic processing, simplifying the manufacturing process
- They are commonly used in cosmetic-grade plastics due to their very low toxicity
- Compared to phosphate-based antistatic agents, they exhibit lower deposition (plate-out) issues, ensuring a cleaner surface finish
- They maintain moderate antistatic capabilities even under low relative humidity, which is critical for certain regional environments.

Their disadvantages are:

- For cationic antistatic agents based on quaternary ammonium salts ($NR_4^+ X^-$), the choice of counterion X^- (e.g. chloride, methyl sulfate) is constrained by EPA and EU REACH requirements, which may restrict their use in certain regions or applications. Zwitterionic alkyl betaines, in contrast, do not contain a separate counterion; they are regulated as individual organic substances under REACH/EPA, but their use is not limited by counterion selection
- Their effectiveness is more dependent on concentration than the phosphate-based systems, often requiring 1.5–2 times higher dosages to achieve similar results
- Certain anions, such as methyl sulfate, raise health and environmental concerns that need careful consideration
- They exhibit more variability in their partition coefficient compared to the phosphate systems, affecting consistency across applications
- Low-molecular-weight ammonium antistatic agents may migrate within polymer matrixes potentially reducing their surface effectiveness.

11.3.4 Permanent Antistatics

Permanent antistatic agents are a relatively new addition to this group of plastic additives, and unlike those mentioned so far, they are not based on forming a surface film that is responsible for discharging static electricity.

Inherently conductive polymers (ICPs) are one group in this category, and are positioned on the boundary between antistatic agents and conductive fillers (e.g., conductive carbon black, graphite (for both see Section 12.5.5), carbon nanotubes, and graphene). These are polymers with conjugated double-bond systems (Figure 11.10), which being chromophores means that these antistatics have a strong intrinsic color, greatly limiting their applicability.

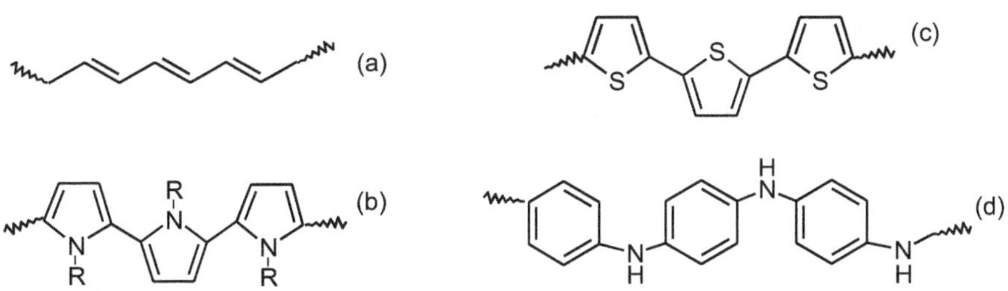

Figure 11.10 Chemical structures of inherently conductive polymers used as permanent antistatics: (a) polyacetylene; (b) polypyrrole; (c) polythiophene; (d) polyaniline

Block copolymers are another group of permanent antistatic agents, which have both hydrophilic blocks (as conductive units) and hydrophobic blocks in their chemical structures. These structures result in unique polarity and ensure compatibility with specific plastic materials. They usually consist of polymethyl methacrylate or polyamide segments. The conductive, hydrophilic blocks consist of polyethylene glycol, polyether, or epichlorohydrin molecular segments (Figure 11.11).

Figure 11.11 Block copolymer antistatics

11.4 Mechanism of Action of Antistatics

The basic mechanism of action of antistatics is already explained in Figure 11.3. However, the efficiency of internal antistatic agents depends on their incompatibility with the plastic to be treated and, consequently, on their tendency to migrate within the polymer. Furthermore, the effectiveness of internal antistatic agents is directly related to the humidity level in the air. Internal antistatic agents for plastics are surface-active substances. They must be sufficiently incompatible with the plastic to migrate to its surface. Once there, they form a continuous hydrophilic film. The non-polar structural groups of these antistatic agents extend into the polymer matrix, while the polar residues protrude from the surface into the air. Due to their polar properties, these polar groups exhibit a certain affinity to water vapor in the air. Water forms a second thin film, through which charges can dissipate via ionic conduction. This continuous water film can only develop under sufficiently humid conditions. Most internal antistatic agents become effective sometime after the production of plastic articles. The surface-active film must first be formed through migration from within. The internal migration speed depends on several factors:

The compatibility of the antistatic agent with the polymer must not be too high; otherwise, it will not migrate out of the matrix to the surface.

As the concentration of the antistatic increases, its migration speed also increases due to greater incompatibility at higher concentrations, and a larger number of migrating molecules available to form the active film.

Higher crystallinity of plastics inhibits their tendency to migrate. The antistatic agent molecules can move much more easily within the amorphous phase and reach the surface faster.

Within a chemical class of antistatic agents, the chain length of the alkyl groups also significantly affects the rate of migration. As the chain length increases, mobility of the molecules decreases. This effect may be accompanied by better compatibility with the polymer matrix as the chain length increases.

Double bonds in the alkyl group reduce the migration tendency because the alkyl groups exhibit restricted mobility and therefore cannot move as easily within the polymer matrix.

Fillers, such as chalk, or pigments tend to adsorb antistatic agents due to their relatively large surface area. This adsorbed portion of the active substance is no longer available for migration.

Other formulation components, such as slip additives or anti-fogging agents that also migrate to the surface, enhance migration of antistatic agents.

At elevated temperatures, migration speed of the antistatic agents increases due to the enhanced micro-Brownian motion of polymer chains and as well as mobility of low molecular weight molecules.

Combinations of certain different antistatic agents can exhibit synergistic effects.

For both product groups, it is necessary to include a sufficient concentration of conductive substances so that they can form a kind of network. In both cases, concentrations of 5 to 15% are required. These antistatic agents have the advantage of being fully effective immediately after processing, as they do not need time to migrate to the surface. Their effectiveness is not reduced over time and or as a result of treating the polymer surface mechanically or by water. These additives provide not only surface conductivity but also through-conductivity. Their effectiveness is also independent of humidity and ambient temperature.

11.5 Failure of Antistatic Agents

Antistatic agents can fail under certain circumstances, causing the antistatic effect to be lost. There are numerous reasons for it. A common issue is insufficient humidity. Without enough moisture in the air, a conductive film cannot be formed on the surface. To a limited extent, increasing the dosage of the antistatic agent can help in such cases. However, at excessively high concentrations, antistatic agents may form double layers or recrystallize. As a result, no continuous conductive film of the active substance remains on the surface of the plastic.

Depending on climate conditions, antistatic agents may also evaporate, leading to a depletion of the active substance on the surface. The antistatic effect is permanently lost, if the antistatic agent was applied externally or if the matrix is already depleted of the active substance. Migration of the antistatic agent under the polymer surface occurs when it was externally applied and exhibits excessive compatibility with the polymer.

A significant increase in surface area, through biaxial stretching, for example, can tear apart the conductive film. Continuous dissipation of the surface charges would no longer be possible. If the concentration of antistatic agents in the matrix remains sufficient, migration of the active substance can replenish the film on the surface.

Additionally, adsorption of antistatic agents by goods that come in contact with plastic articles is possible. Furthermore, they can be washed off by water or removed from the surface through abrasion caused by solid materials.

Antistatic agents accumulating on surfaces also function as a lubricating film. These substances reduce sliding friction and thus decrease friction of plastic parts during conveying. This also results in fewer static charges being generated.

11.6 Usage of Antistatic Agents

Antistatic agents are generally used at concentrations in the range 0.05–3 wt%. Particularly at lower dosages, it is often very difficult to maintain consistent performance over time. For effective and uniform antistatic action across finished parts, it is crucial that the antistatic agents are evenly incorporated into the polymer. This is often achieved by using masterbatches.

General recommendations regarding the suitability of antistatic agents for certain plastics and applications are provided in Table 11.2. The optimal choice of an antistatic agent for a specific application depends on many factors. For instance, it is important to determine what level of surface resistivity is desired in the final article, whether the article needs to be transparent, the time interval after processing when the antistatic effect should become pronounced, and how long the effect must last. Additionally, climatic conditions play a significant role, and dimensions (surface-to-volume ratio) of the manufactured molded parts also influence the dosage requirements.

In dry environments, higher dosages of antistatic agents are required compared to those under humid conditions, such as those in tropical climates.

A thick molded part has a large internal reservoir from which the active antistatic ingredient can gradually migrate to the surface. Therefore, the required dosage in such cases is lower than for thin-walled applications.

Table 11.2 Recommended Dosages and Suitability of Selected Antistatic Classes for Applications in Various Plastics

Plastic	Recommended dosages (wt%)			Suitability[a] [1]					
	Fatty acid ester	Alkyl sulfonate	Ethoxylated Amine	Phosphates			Ammonium salts		
				Liquid	Semi-solid	Solid	Liquid	Semi-solid	Solid
PP	0.3–2	—	0.1–0.2	5	3	2	3	0	–1
LDPE	—	—	0.05–0.2 (C_{12})						
LLDPE	—	—	0.05–0.2 (C_{12})						
HDPE	—	—	0.05–0.2						
PVC	1–2	1–2	—	—	—	—	—	—	—
Styrene polymers	(1–2)	1–2	—	—	—	—	—	—	—
POM	—	2	—	—	—	—	—	—	—
PA	—	2–3	—	4	5	3	4	3	2
PET, PBT	—	2–3	—	5	5	5	4	4	3
PC	—	2	—	—	—	—	—	—	—
TPU	—	1	—	–1	1	3	–1	1	3
Aramides	—	—	—	5	5	5	4	4	4
Polyketones	—	—	—	5	5	5	4	4	3
Fluoropolymers	—	—	—	5	5	5	4	4	4
Carbon fiber	—	—	—	5	5	5	4	3	3

[a] 5 = Excellent, –5 = Very poor.

[1] P. Seemuth, "Anti-static agents: How to select the right grade for plastics?", ***SpecialChem***, *https://polymer-additives.specialchem.com/selection-guide/antistatic-agents-for-polymers#Inorganic* [accessed on 07 April 2025]

11.7 Methods for Characterizing Antistatic Agents

Due to the chemical variety of antistatic agents, it is not possible to define a list of analytical methods that would be applicable to the characterization of all groups of antistatic agents.

However, the most critical test method for assessing the efficiency of antistatic agents and quantifying the antistatic effect is to measure the imparted surface resistivity. This is often confused with electrical resistance, and indeed in the German language, there is only one term for both resistivity and resistance. However, the difference between the two in English should be clearly understood,[2] as this could also be the case in other languages.

- **Electrical resistance** (R) is a measure of an object's opposition to the flow of electric current. It depends upon the object's length, area of cross-section, and nature of the material, and is expressed in ohms (Ω)
- **Electrical resistivity** (ρ) is the characteristic property of a material by which it resists the amount of current through it. It depends upon the nature of the material and physical conditions like temperature, and is expressed in ohm-metres (Ω·m).

Surface resistivity is a measure of a material's resistance to electric current flowing across its surface. It is defined as the resistance between two opposite sides of a square array of electrodes on the surface and, because this resistance is independent of the size of the square, the quantity is expressed in ohms (Ω), commonly referred to as "ohms per square".

According to the test method for measuring surface resistivity (ASTM D257), a specimen (usually stored at 20 °C and 65% relative humidity, with a thickness of 2 mm) is placed beneath two electrodes. A voltage (usually 100–500 V) is applied for 60 s, and the resistance is measured. The procedure is repeated several times to obtain a representative average value for the single measurement. Surface resistivity is calculated as K_sR, where K_s is a calculated constant based on electrode dimensions, and R is the measured resistance reading in ohms (Ω).

[2] M. Singh, "What is the difference between resistance and resistivity?", *Teachoo*, 2024, *https://www.teachoo.com/10647/3084/What-is-the-difference-between-resistance-and-resistivity-/category/Extra-Questions/* [accessed on 19 March 2026].

11.8 Questions

1. What is the mechanism of static electricity build-up when two plastic materials are rubbed against each other?
2. What problem can be caused by electrostatic charging of plastics in industrial processes?
3. What is the difference between internal and external antistatic agents in terms of their application methods and the mechanism of action?
4. How does relative humidity affect the effectiveness of antistatic agents and why?
5. What is the mechanism of action of internal antistatic agents? What are the essential structural elements these molecules must possess to act as antistatic agents?
6. What are the advantages and disadvantages of using glycerol monostearate (GMS) and ethoxylated alkylamines as antistatic agents?
7. Why are GMS and ethoxylated alkylamines often used in combination? How do their mechanisms of action complement each other?
8. What are the essential chemical differences between anionic and cationic antistatic agents? Give specific examples of each and their respective applications in plastics.
9. What are the advantages and disadvantages of phosphate-based antistatic agents in terms of their effectiveness, environmental impact, and application limitations?
10. How do the permanently acting antistatic agents (inherently conductive polymers) work? How are they different from traditional antistatic agents? What are the challenges of using the permanently acting antistatic agents in plastics?

12 Functional Fillers

Content of this chapter was adapted from:

- W. Hohenberger, "Fillers and Reinforcements/Coupling Agents" (Chapter 17), in: *Plastics Additives Handbook* (6th edition), ed. H. Zweifel, R. D. Maier, and M. Schiller, Hanser, 2009, pp. 919–966.
- M. Knerr and E. Hersche, "Füllstoffe und Verstärkungsmittel" (Chapter 12), in: *Handbuch der Kunststoff-Additive* (4th edition), ed. R. D. Maier and M. Schiller, Hanser, 2016, pp. 1069–1138.

12.1 Terms and Definitions

Functional filler	An additive introduced into a polymer to impart specific functional properties beyond mere cost reduction.
Performance additive	A substance added to improve or tailor the performance characteristics of a polymer material.
Surface modification	Techniques used to alter the surface chemistry or structure of filler particles to enhance compatibility or functionality in composites.
Particle morphology	The shape, size, and structure of individual filler particles (factors that significantly influence composite properties).
Dispersion technology	Methods and processes to achieve uniform distribution of fillers within a polymer matrix.
Mechanical reinforcement	The enhancement of a material's mechanical strength, stiffness, or durability through the incorporation of fillers.

Coupling agent	A chemical that improves adhesion and interfacial bonding between a filler and the polymer matrix.
Thermal conductivity	The ability of a filled material to conduct heat, typically modified by introducing thermally conductive fillers.
Electrical conductivity	The property describing how well a filler-enhanced polymer can conduct electricity.
Nanofiller	A filler with nanometre-sized particles, used to significantly modify material properties at low loading levels.
Compatibility	The degree to which a filler is physically and chemically compatible with a specific polymer matrix, affecting dispersion and final properties.
Composite	A material consisting of a polymer matrix with embedded reinforcing fibers or particles, designed to combine and tailor mechanical and physical properties to make them superior to those of the pure polymer.

12.2 Introduction

It is generally accepted that fillers in plastics are used to partially replace the base material and thereby reduce material costs. Actually, this type of definition is too simple and misleading because, in addition to reducing costs, fillers exhibit other desired effects on plastics, enabling improvement or modification of certain properties of the end product. They are solid additives available in the form of powders, fibers, or granules that significantly differ from the plastic matrix in terms of their composition and structure. These fillers are usually inorganic materials, and more rarely, organic materials. There are two main categories of fillers:

- **Functional fillers** are used to enhance specific properties of the plastic, such as mechanical strength, stiffness, impact resistance, thermal conductivity, and electrical properties. They can also reduce polymer shrinkage and deformation during processing.
- **Inactive fillers or extenders** primarily serve to increase the volume of the polymer and reduce material costs, without significantly altering the physical properties of the end material.

By carefully selecting and combining fillers, properties of plastics can be tailored to meet the specific requirements of various applications.

Functional fillers may influence the following characteristics: processing parameters like torque and pressure in extrusion, output, cooling rate, density of finished articles, transparency, shrinkage, heat transmission, coefficient of thermal expansion, mechanical properties (such as impact strength, stiffness, elongation, surface quality, and gloss), electrical properties, fire retardancy, weathering, and the price of the final material.

While they were originally incorporated to reduce the cost of plastic compounds, the focus is currently on specifically improving the properties of these compounds. In certain cases, this can justify the cost of the filled plastic, which may well be higher than that of the matrix polymer.

12.3 Functional Fillers and Their Applications

Table 12.1 shows the main plastics and the functional fillers used with them.

Table 12.1 Common Applications of Mineral Fillers in Plastics[1]

Mineral filler	PE	PP	PVC	PET	ABS	Fluoro polymers	PS	PC	TPE	PEU[a]	EP[a]	PUR	Phenolic resin	Thermosets
Calcium carbonate	Y	Y	Y		Y	Y	Y		Y		Y	Y	Y	
Talc	Y	Y		Y			Y					Y	Y	
Kaolin	Y		Y	Y					Y	Y		Y		
Aluminium trioxide	Y		Y		Y				Y		Y	Y	Y	
Silica	Y		Y		Y		Y		Y		Y	Y		
Wollastonite	Y	Y	Y	Y			Y	Y	Y					Y
Barium sulfate			Y							Y		Y		
Feldspar	Y	Y	Y				Y		Y		Y	Y	Y	
Mica	Y	Y		Y	Y	Y		Y	Y					Y

[a] PEU: Polyether urethane; EP: Epoxy/epoxide.

1 J. W. Leonhardt and M. Knerr, "Functional fillers, mineral additives and reinforcing agents in PVC" (Chapter 7), *PVC Handbook* (2nd edition), ed. R. P. Krock, M. Schiller, P. Frenkel, J. W. Summers, and C. A. Daniels, Hanser, 2025, p. 328.

12.4 Physicochemical Properties of Functional Fillers

Different functional fillers affect the properties of thermoplastics in different ways. In addition to the chemical composition, especially in the case of mineral additives, their particle geometric shape, particle size, and the particle size distribution, as well as surface energy and the specific surface area of the particles, are filler parameters that allow assessment of their effect on the plastic.

12.4.1 Particle Shape and Size

Geometric particle shapes can be described by the aspect ratio, which represents the ratio of length to width/thickness of the particle. It is usually determined using sophisticated microscopic methods, but it is often sufficient to know the particle shape qualitatively. Figure 12.1 shows the five most common particle shapes of fillers.

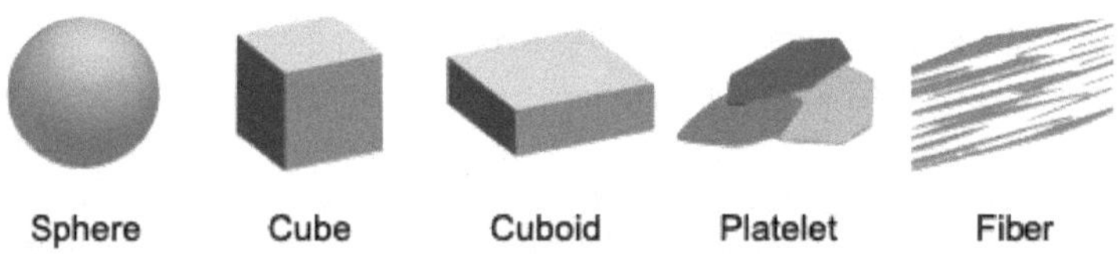

Figure 12.1 Particle forms of additives used in plastics

There are a number of methods for determining the particle size distribution. The most common method for coarser powders is sieving. For finer powders, laser scattering or sedimentation analysis is often used. A single figure – the mean particle diameter of the particle size distribution curve (D50%) – can be used for characterizing the particle size, but this parameter does not take into account the width of the particle size distribution.

Both the measurement process and interpretation of the results are influenced by several factors, one of which is the choice of dispersing medium. Options include liquids (such as water and organic solvents) to air. It is essential to ensure that the filler neither dissolves in nor reacts with the chosen medium. Additionally, the refractive index of the dispersing medium plays a significant role.

The dispersing medium should preferably have a refractive index close to that of the filler.

Refining the approach to selecting the dispersing medium can greatly enhance the accuracy and reliability of particle size distribution measurements; as demonstrated in Figure 12.2 to Figure 12.6. Figure 12.2 shows a filler with a hexagonal platelet structure. These particles are significantly smaller than 1 mm in diameter and have a thickness of about 100 µm. This is not reflected in the evaluation in volume mode in Figure 12.3, where water used as the dispersing medium, and where supposedly there is a large number of particles with >5 µm. In Figure 12.4, where isopropanol was used as the medium, the expected grain size distribution was also not found. Changing the evaluation in Figure 12.4 to the surface mode (Figure 12.5) gave results that were closer to the reality, although a false picture is still suggested. After changing the evaluation to the number mode (Figure 12.6), the results become similar what is expected from Figure 12.2.

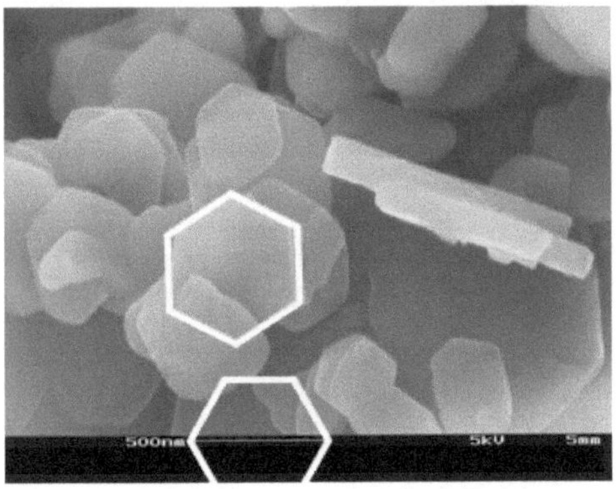

Figure 12.2 Scanning electron microphotograph of a plate-form particle

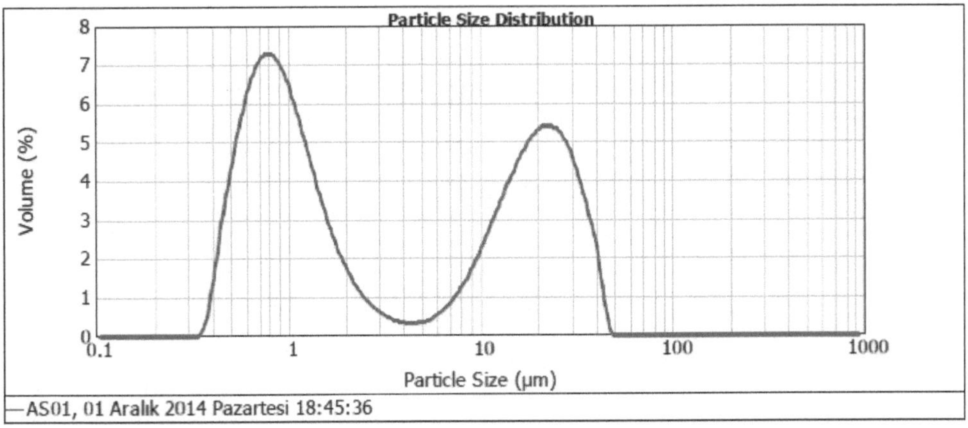

Figure 12.3 Particle size distribution of the particles shown in Figure 12.2 in water (volume mode)

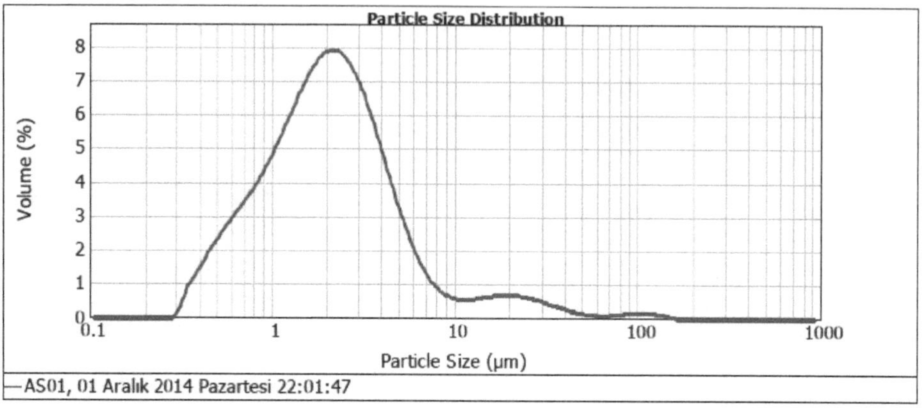

Figure 12.4 Particle size distribution of the particles shown in Figure 12.2 in isopropanol (volume mode)

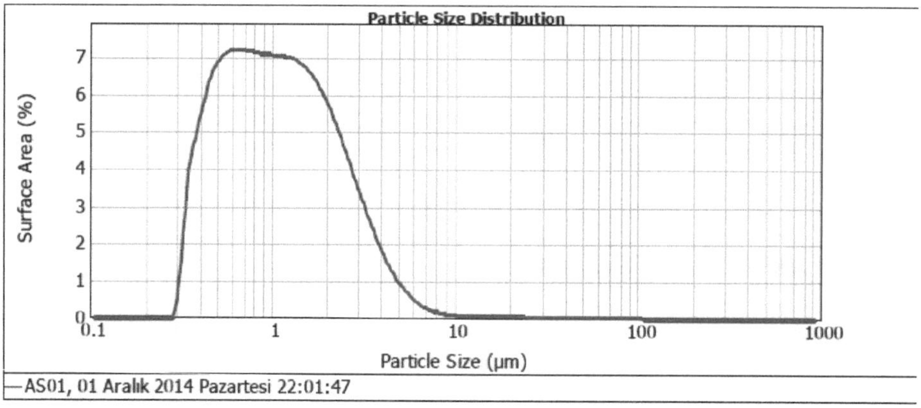

Figure 12.5 Particle size distribution of the particles shown in Figure 12.2 in isopropanol (surface mode)

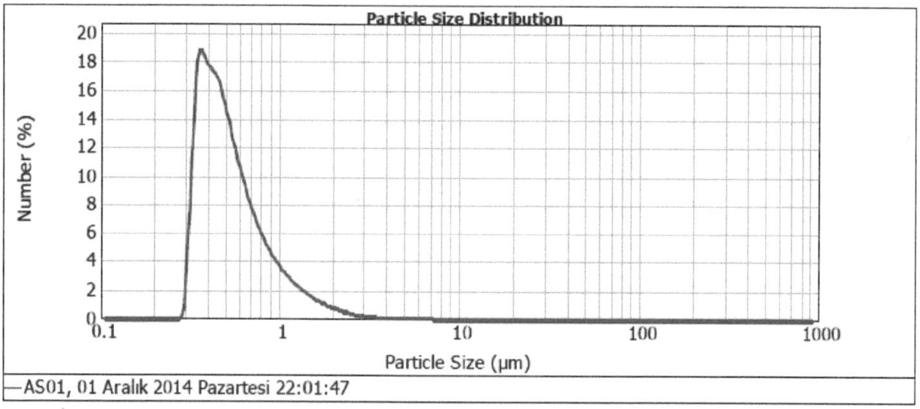

Figure 12.6 Particle size distribution of the particles shown in Figure 12.2 in isopropanol (number mode)

When analyzing particle size distributions using laser diffraction, there are four primary modes of evaluation: volume, surface, length, and number. Each mode represents the distribution in a different way, depending on the specific requirements of the analysis (Table 12.2). Choosing the right dispersing medium and the correct mode ensures that the analysis provides accurate results.

Table 12.2 Modes for Evaluating Particle Size Distributions

Mode	Description	Application	Comments
Volume	Based on the volume of the particles. Larger particles contribute disproportionately to the distribution, because their volume is proportional to the cube of their diameter.	This is the most commonly used mode and is suitable for applications where particle volume is critical, such as in the filler industry or when analyzing bulk materials.	Use when volume or mass is critical.
Surface	Calculated based on the surface area of the particles. Particles with larger surface areas have greater weight in the results.	Ideal for materials, where specific surface area is important, such as in catalysts or adsorption materials.	Use when surface area is central.
Length	Calculated based on the length or diameter of the particles, providing a linear average value.	Less commonly used but can be useful when the longest dimension of a particle is crucial for the application.	Suitable for specialized applications focusing on linear dimensions.
Number	Calculated based on the number of particles. Small particles dominate this representation, because they are often more numerous in a sample.	Suitable for applications, where the number of small particles is important, such as in nanotechnology or suspensions.	Ideal for analyses where small particles are dominant.

The D50% value, also known as the median particle diameter, divides the mass distribution curve into two equal areas, with 50 wt% having a diameter greater than D50% and 50 wt% having a diameter less than D50%. When the particle diameter is averaged by particle number rather than weight, significantly lower median values result, depending on the width of the distribution. Although the D50% is a highly informative value, further analysis of the upper and lower ends of the particle size distribution curve is beneficial for more comprehensive characterization of a mineral additive's particle distribution. Modern measurement technology enables the determination of a precise particle size upper limit, often called the "top cut" of the particle size distribution curve. The top cut is determined at different points depending on the application and production requirements. For example, the top cut D98% represents the particle diameter that is exceeded by only 2% by weight of the particle ensemble, meaning 98 wt% of the particles have a diameter smaller than this value.

12.4.2 Specific Surface Area

The specific surface area of a particle ensemble, expressed in m^2/g, is primarily determined by the content of "fines" for a given mean particle diameter (D50%) (see Figure 12.7), and is defined as the total surface area of a filler per unit weight. This parameter is typically measured using nitrogen adsorption according to the BET method (ISO 9277), named after Brunauer, Emmett, and Teller, who published the fundamental aspects of this theory in 1938. The BET method offers the advantage of measuring the surface area of fine particles and textures on the particles.

As particle size decreases, the specific surface area increases, as does the surface area of porous particles.

The specific surface area of a mineral additive is a critical parameter, as it determines the number of potential adhesion points of the filler with polymer chains. Consequently, it significantly influences the mechanical properties of the filled plastic material.

The higher the specific surface area of the filler, the greater the viscosity of the polymer melt containing it, and the higher the torque during extrusion.

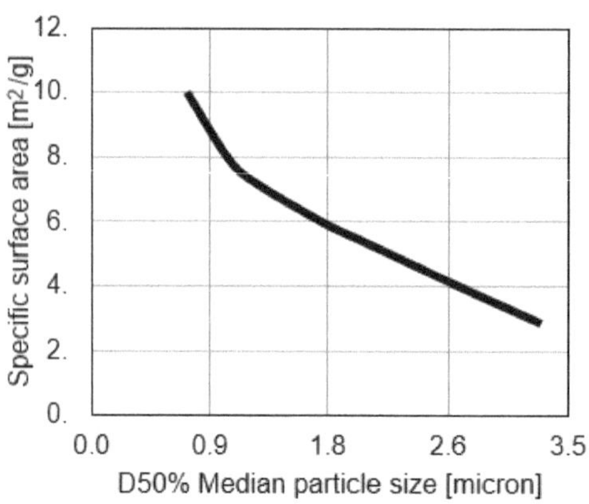

Figure 12.7 Specific surface area as a function of the average particle size of commercially available calcium carbonates[2)]

An increase in the specific surface area of the filler generally leads to the following changes in the properties of the filled plastic:

- Increased tensile and tear strength
- Higher surface gloss

[2] M. Schiller, *PVC Additives* (2nd edition), Hanser, 2022, pp. 156–186.

- Increased impact and notched impact strength
- Higher rigidity (higher modulus of elasticity)
- Reduced output.

The specific surface area of the additive must be tailored to the respective processing conditions. Particle behavior during processing is influenced by both van der Waals forces acting between particles and dispersive shear forces introduced by processing equipment, such as mixers and extruders.

> The behavior of relatively large particles with diameters of 10 µm or greater is primarily determined by shear forces. In contrast, attractive forces dominate for small particles with diameters of 1 µm or less.

The above distinction between large and small particles is significant because below a critical particle size, undesirable agglomerates can form. These agglomerates can negatively impact the gloss, surface quality, and notched impact strength of finished parts.

Therefore, relationships between particle size and dominant forces can be summarized as follows:

- Particles >10 µm: Shear forces dominate
- Particles <1 µm: Van der Waals forces dominate
- Particles 1–10 µm: Transition zone, where both forces play significant roles.

Proper dispersion techniques and careful selection of particle size distribution can help mitigate agglomeration problems and enhance the overall quality of the composite material.

12.4.3 Surface Energy

The surface energy of a substance, expressed in mJ/m^2, is the energy required to increase the surface area of a solid material by one square metre. For liquids, surface energy is equivalent to the more commonly known surface tension, measured in mN/m.

Surface treatment can dramatically alter the surface energy of a mineral additive. For instance, treating calcium carbonate with stearic acid can reduce its surface energy from 200 to 40 mJ/m^2. This treatment can transform an originally hydrophilic surface into a hydrophobic one, significantly changing the material's interaction with its environment. The modification of surface energy has profound implications for the behavior of mineral additives in various applications, particularly in polymer composites. It affects properties, such as wettability, dispersion, and adhesion, which are critical performance attributes of the final product. Understanding and controlling surface energy is essential in industries ranging from coatings and adhesives to composite materials and nanotechnology, where surface interactions play a pivotal role in product performance and functionality.

12.4.4 Filler Purity

The quality of fillers is significantly influenced by their purity. For mineral additives of natural origin, only high-purity deposits are suitable for extraction and use in plastics. Contamination, particularly by heavy metals, can have detrimental effects, including:

- Drastically reduced weathering stability and increased discoloration
- Organoleptic (odor) impairments
- Interference with the cycle of autoxidation (as shown in Figure 1.19).

These impurities can compromise the performance and longevity of the final plastic article, potentially leading to premature degradation, discoloration, or unintended changes in physical properties. Therefore, rigorous quality control and careful selection of mineral deposits are essential in the production of high-quality fillers for the plastics industry.

When selecting fillers, their hygroscopic properties must be considered. Moisture absorption of fillers can vary significantly, ranging from a few parts per million (ppm) to as high as 30% in the case of wood flour. The moisture content of substances and material components, not only fillers, can be measured relatively easily using "loss on drying" (LoD).

Excessive moisture content typically has detrimental effects on the properties of the compound, necessitating careful attention during their storage and processing. Key options for addressing it include:

- Storage conditions: Fillers should be stored in a controlled environment to minimize moisture absorption
- Pre-drying: It may be necessary to pre-dry the compound before processing to reduce moisture content to acceptable levels
- Degassing: During processing, degassing techniques may be required to remove residual moisture and prevent defects in the final product.

12.4.5 Particle Hardness

Particle hardness is a crucial factor in the processing of plastics, with lower hardness generally being advantageous. The hardnesss of minerals is typically quantified using the Mohs scale, which is a qualitative ordinal scale characterizing the scratch-resistance of various minerals through the ability of a harder material to scratch a softer material.[3] While greatly facilitating the identification of minerals in the field, the Mohs scale does not show how well materials perform in an industrial setting. However, there are other hardness scales, including those developed by Smith and Sandland (Vickers hardness, H_V) and by Tröger (Table 12.3). It is interesting to note that the Tröger values quite well fit to the geometric series H_M* in Table 12.3.

Table 12.3 Hardness Scales

Mineral	Formula	Mohs hardness	Rosiwal hardness		Vickers hardness	Tröger hardness	Geometric series H_M* [a]
			Ref. [4]	Ref. [5]	Ref. [5]	Ref. [4]	
Talc	$Mg_3Si_4O_{10}(OH)_2$	1	0.03	0.03	2.4	1.08	1.56
Gypsum	$CaSO_4{\cdot}2H_2O$	2	1.04	1.25	36	2.36	3.12
Calcite	$CaCO_3$	3	3.75	4.5	109	6.99	6.25
Fluorite	CaF_2	4	4.2	5.0	189	12.1	12.5
Apatite	$Ca_5(PO_4)_3(OH,Cl,F)$	5	5.4	6.5	536	25.7	25
Orthoclase feldspar	$KAlSi_3O_8$	6	30.8	37	795	49.5	50
Quartz	SiO_2	7	100	120	1120	100	100
Topaz	$Al_2SiO_4(OH,F)_2$	8	146	175	1427	143	200
Corundum	Al_2O_3	9	833	1000	2060	342	400
Diamond	C	10	117,000	140,000	10,060	850	800

[a] $H_M^* = (5/4)^2 \cdot (2^{\text{Mohs}-1})$, where "Mohs" is the Mohs hardness

[3] Created in 1812 by German geologist and mineralogist F. Mohs, this is one of several definitions of hardness in materials science, some of which are more quantitative. The method of comparing hardness by determining which minerals can scratch others is of great antiquity, having been mentioned by Theophrastus in his treatise *On Stones* (ca. 300 B.C.) and by Pliny the Elder in his *Naturalis Historia* (ca. 77 A.D.).

[4] P. Drucker, "Abrasivity of granular soil and tool wear in ground engineering" [Doctoral Thesis], Vienna University of Technology, 2013, *https://repositum.tuwien.at/bitstream/20.500.12708/10992/2/Drucker%20Petra%20-%202013%20-%20Abrasivity%20of%20granular%20soil%20and%20tool%20wear%20in%20ground...pdf.*

[5] "Mineralienatlas – Fossilienatlas: Härte", Geolitho Stiftung gemeinnützige GmbH, *https://www.mineralienatlas.de/lexikon/index.php/H%C3%A4rte?lang=de* [accessed on 03 March 2025].

12.4.6 Density

Since the density of mineral fillers is usually significantly higher than the density of plastics, adding them to a plastic leads to an increase in the density.

Exercise 12.1

If a plastic has a density of 1.0 g/mL and a filler has a density of 2.5 g/mL, and both are mixed 1:1 by weight, what is the density of the mixture?

In order to solve this correctly, the calculation must use the principle of weighted harmonic mean, which is appropriate for calculating the density of mixtures in which the components are mixed by weight. Namely:

$$1/\rho_{\text{mix}} = (w_1/\rho_1) + (w_2/\rho_2)$$

where:

ρ_1,ρ_2 and w_1,w_2 = densities and weight fractions of the two components, respectively

So given the density of plastic (ρ_1) = 1.0 g/mL, and density of filler (ρ_2) = 2.5 g/mL, and $w_1 = w_2 = 0.5$, we have:

$$1/\rho_{\text{mix}} = (0.5/1) + (0.5/2.5)$$
$$1/\rho_{\text{mix}} = 0.5 + 0.2$$
$$1/\rho_{\text{mix}} = 0.7$$

And therefore:

$$\rho_{\text{mix}} = 1/0.7 \approx 1.429 \text{ g/mL}$$

In plastics processing, the weight-based output, expressed in kg/h, has an economic significance, and the volume-based output has a technological significance. Volume-based thinking is illustrated in the following exercise.

Exercise 12.2

A mixture consisting of 40 wt% PP (density 0.91 g/mL) and 60 wt% magnesium hydroxide (density 2.381 g/mL) as a flame retardant is extruded. What is the density of the mixture, and what volume percent of PP and magnesium hydroxide does it contain?

Using the same formula as in Exercise 12.1 gives us $\rho_{\text{mix}} \approx 1.446$ g/mL. The volume percentages are obtained by taking the weight percentages and dividing by the corresponding densities to obtain the relative volumes, and then dividing each figure into the sum of both to obtain the percentages. This gives us 63.6 vol% PP and 36.4 vol% magnesium hydroxide.

12.4.7 Ash Content and Loss on Ignition

The ash content ("%ash") is a portion of the material that *remains* as a residue after the complete combustion of the material at high temperatures. This residue primarily consists of non-volatile inorganic components, such as minerals and metals.

The loss on ignition ("%LoI") is a portion of the material that is *lost* during heating and combustion at high temperatures. This loss mainly includes organic components, volatile substances, and water.

Both values are connected by the following equation:

$$100\% = \%\text{LoI} + \%\ \text{ash}$$

The ash content and loss on ignition are important characteristics of fillers used in plastics. Accurate determination of ash content is essential for quality control and optimization of filled plastic formulations, for the following reasons:

- **Filler purity:** The ash content provides insight into the purity of the filler. A high ash content can indicate impurities or unwanted components that may negatively affect properties of the final product
- **Validation of the composition:** The ash content helps determine the ratio of organic to inorganic components in the filler. This is particularly important for fillers that contain both organic and inorganic components
- **Processability:** Variations in ash content might affect the processing properties of the filled plastics, as well as undesirable abrasion and wear on processing equipment, altering the melt flow characteristics
- **Mechanical properties (such as strength and stiffness):** The ash content can influence the mechanical properties of the final product. The optimal ash content can help improve the strength and stiffness of the plastic, while excessive ash content can impair these properties.
- **Heat resistance:** The ash content of fillers can also affect the thermal stability of plastics. Inorganic components in the filler can enhance the thermal conductivity and heat resistance of the final product
- **Filler quality:** Determining the ash content is critical for ensuring consistency of filler quality. Consistent filler quality is crucial for producing plastic products with uniform properties.

Below, we summarize the chemical processes involved in ash content determination for calcium carbonate ($CaCO_3$). Its decomposition involves its conversion to calcium oxide (CaO) and carbon dioxide (CO_2):

$$CaCO_3 \rightarrow CaO + CO_2$$

However, the exact decomposition temperature can vary. Complete thermal decomposition and calcination of calcium carbonate takes place at temperatures above 800 °C. However, from about 540 °C, calcium carbonate begins to decompose, and the resulting CaO is able to react with carbon dioxide below 800 °C.[6]

$$CaO + CO_2 \rightarrow CaCO_3$$

This results in an equilibrium with calcium carbonate on the one side and CaO and CO_2 on the other side. This means the ash produced by heating calcium carbonate can be higher than the theoretical amount, when the temperature of the oven is below 800 °C and when enough CO_2 is available in the atmosphere inside the oven (coming from, for example, the oxidation of organic substances). As the partial pressure of CO_2 increases, the decomposition temperature of calcium carbonate increases, and as the partial pressure of CO_2 becomes lower, the decomposition of calcium carbonate begins at lower temperatures.

Furthermore, this equilibrium in the decomposition of calcium carbonate is indirectly influenced by the Boudouard equilibrium:[7]

$$C + CO_2 \leftrightharpoons 2\,CO$$

This reaction is endothermic, and the equilibrium shifts to the carbon monoxide (CO) side at higher temperatures. At lower temperatures, however, the equilibrium is on the side of carbon (C) and CO_2. These considerations are of course also valid for other carbonate-containing functional fillers, such as dolomite. They also lead to the recommendation that determination of the ash content or the LoI should be carried out at higher temperatures (>950 °C), as specified by some standards.

Everything that has been said up to this point is valid for pure fillers, but not for plastic articles that contain the fillers.

There is another method that is often used for measuring ash content of plastic articles with fillers. This is the determination of sulfated ash (IS 8769:1978 and ISO 1270). It is used, among other things, for quality control of U-PVC pipes, and is carried out in several steps:

1. A first annealing step is used to burn the sample in order to remove organic components
2. The addition of sulfuric acid is used to convert the residue into sulfates:

   $$CaO + H_2SO_4 \rightarrow CaSO_4 + H_2O$$

3. The second annealing step, at 775–850 °C, is necessary to remove excess acid.

---

6 J. A. H. Oates, *Lime and Limestone*, Wiley-VCH, 1998, p. 142.

7 The Boudouard equilibrium, named after Octave Leopold Boudouard, is the redox reaction of a chemical equilibrium mixture of carbon monoxide and carbon dioxide at a given temperature; or, to put it another way, it is the disproportionation of carbon monoxide into carbon dioxide and graphite, or its reverse.

While the first method for pure calcium carbonate gives a value of 56% ash, the sulfated ash value is 136%. Therefore, when discussing the ash content, it is advisable to check what method was used for determining it.

In addition to mechanical, thermal, and optical properties, fillers can also influence coloration of the plastic composites containing them. Two key considerations are:

The more the filler's inherent color differs from that of the plastic, the greater the change in the final material's color.

A higher filler dosage leads to a more significant color difference in the filled plastic article.

Functional fillers positively affect (increase) the thermal conductivity of polymers, in addition to reducing the coefficient of thermal expansion and increasing heat resistance. This results in shorter cooling times for hot finished articles (Figure 12.8).

The higher the thermal conductivity of the added filler and the higher its dosage, the shorter the resulting cooling time.

These improvements can lead to increased production efficiency and potentially superior product quality due to more uniform cooling.

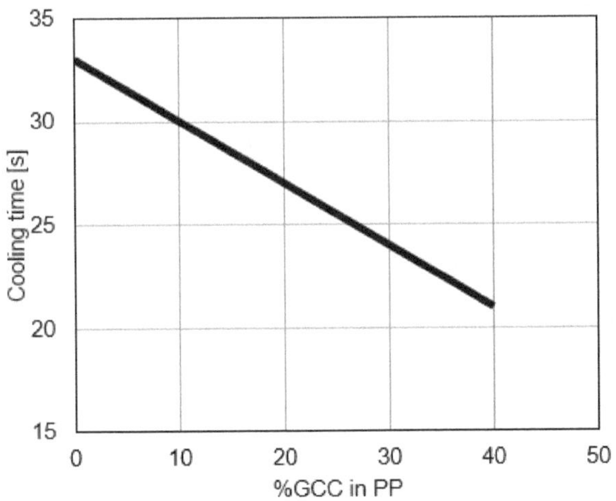

Figure 12.8
Change in the cooling time of an injection-molded PP part (4 mm thick) with increased levels of calcium carbonate filler

12.5 Compositions of Functional Fillers

12.5.1 Carbonates and Hydroxides

Natural calcium carbonates are often sedimentary rocks, usually of marine origin, which are mined, purified, and ground, after which they are known as "ground calcium carbonate" (GCC). A rough distinction is made between the following types:

- Chalk: Barely solidified, consisting of fine calcite crystals and fragments of shells with a supporting matrix derived from nanofossils
- Limestone: More solidified, consisting of macrofossils of shells, and generally older (Jurassic formations)
- Marble: Formed by metamorphism of a sedimentary rock, i.e. recrystallization in the earth's interior at high pressures and temperatures around 600 °C. The resulting stone is relatively hard with a dense, coarse-grained structure.

Furthermore, a distinction is made between the different crystal forms of calcium carbonate, with the main forms being:

- **Calcite:** The most stable and common form, which crystallizes in the trigonal system, and which is characterized by double refraction and being transparent to visible light
- **Aragonite:** A form that is stable at high pressure and temperature, and which crystallizes in the orthorhombic system
- **Vaterite:** The least stable polymorph of calcium carbonate ($CaCO_3$), typically forming fine-grained, often spherical or cauliflower-like aggregates, and crystallizing in a hexagonal or pseudo-hexagonal system; under ambient conditions it is metastable and gradually transforms into the more stable forms calcite or aragonite.

Calcium carbonate hemihydrate (CCHH) is a crystalline form discovered in 2022, and forms when amorphous calcium carbonate crystallizes in the presence of magnesium. In addition to these crystalline forms, calcium carbonate can also exist in amorphous states, which play an important role in the formation of calcareous shells in living organisms. However, only aragonite and calcite are of interest for their use as fillers. They can be distinguished in plastic compounds containing them, using FTIR spectroscopy (Table 12.4):

Precipitated calcium carbonate (PCC) is a synthetic form of calcium carbonate with unique properties and applications in various industries, particularly in thermoplastics, such as PVC. PCC is produced through a carbonation process, which involves introducing carbon dioxide into an approximately 20% suspension of calcium hydroxide ("milk of lime"). Thisreaction causes calcium carbonate particles to precipitate out of the solution; the process is conducted under controlled conditions of temperature, pressure, and agitation to ensure optimal conversion and particle size distribution.

Table 12.4 FTIR Wavelengths of Calcium Carbonate Crystal Forms

Crystal form	Band wavelength (cm^{-1})[a]			
	ν_4	δ	ν_1	ν_3
Calcite	713	875	—	1420–1480
Aragonite	700	855	—	1460–1490
Vaterite	870	745	1080	1400–1480

[a] ν_4: In-plane bending of the CO_3^{2-} group. δ: Out-of-plane bending of the CO_3^{2-} group. This band is characteristic of each of the three crystal forms. ν_1: Symmetric stretching (ν_1) of the CO_3^{2-} group. ν_3: Asymmetric stretching of the CO_3^{2-} group.

PCC can be produced in various crystal modifications, including calcite and aragonite. The morphology can be varied widely, with rhombohedral shapes being most common. The average particle diameter can be in the range 0.03–0.40 µm, which is significantly smaller than that of GCC. PCC typically has higher purity than natural calcium carbonates. Surface treatment of PCC enhances its dispersibility within polymer matrices, leading to improved properties in thermoplastics like PVC, namely:

- Increased impact strength, especially in u-PVC up to certain dosage levels
- High surface gloss of final products
- Good elongation at break, tear propagation, and tensile strength
- Increased modulus of elasticity
- Strong influence on the rheology of PVC plastisols.

When requirements can be met with natural GCC, it is often preferred due to lower cost. Some markets offer coarser PCC grades comparable to GCC with similar target applications and properties.

Dolomite is a double carbonate of calcium and magnesium formed by metamorphic processes. The $CaMg(CO_3)_2$ content is 98–99.5%, and high-quality dolomite has a pronounced crystalline structure.

Similar to dolomite chemically is **huntite** (magnesium–calcium carbonate, $Mg_3Ca(CO_3)_4$, CAS no. 19569-21-2), while lacking the calcium is **hydromagnesite** (magnesium hydroxy carbonate, $Mg_5(CO_3)_4(OH)_2(H_2O)_4$ or $4MgCO_3Mg(OH)_2{\cdot}4H_2O$, CAS no. 12125-28-9). They are often found combined with dolomite and magnesite. They function as “exotic” non-toxic flame retardants in plastics, but are less high-performing than aluminium hydroxide (ATH) and magnesium dihydroxide (MDH), and are used predominantly as a partial or total flame retardant replacement in less demanding applications. Mineral sources of huntite and hydromagnesite can differ in composition and efficiency in flame retardant actions. They are mined, ground, and, if necessary, coated to produce the final product. Some scientists have studied this class of flame retardants extensively (see Section 10.6.5).

In terms of their performance at high temperatures, hydromagnesite decomposes gradually over a wide temperature range (200–700 °C), releasing both water and carbon dioxide. Huntite has two endothermic decomposition stages releasing CO_2 – the first between 400 °C and 630 °C, and the second between 630 °C and 750 °C. Therefore, the whole decomposition range of huntite/hydromagnesite minerals changes depending on the ratio of huntite to hydromagnesite. More information on these materials (and also about MDH and ATH) can be found in Section 10.6.

It should also be noted that calcium hydroxide is used as a major component of heat stabilizers applied in manufacturing certain PVC cable compounds. Keeping its moisture content very low requires significant efforts on the part of quality control personnel.

12.5.2 Sulfates

Although it is chemically related to calcium carbonate, calcium sulfate exhibits different properties and therefore is used in different applications. Calcium sulfates, commonly known as **gypsum** ($CaSO_4{\cdot}nH_2O$), are utilized in the polymer industry in three distinct forms:

- Calcium sulfate dihydrate: Primarily used in foamed polymers
- Calcium sulfate hemihydrate: Crystallizes as microfibers
- Calcium sulfate anhydrate: Also known as anhydrous calcium sulfate, this can be treated to minimize moisture absorption. If the technical challenges associated with controlling moisture absorption can be overcome, it represents a promising option as a filler. The anhydrate form offers the advantageous properties of very white appearance, acid resistance, low surface energy (facilitating easy incorporation), and a minimal wear effect on processing equipment.

Similar to gypsum but containing barium rather than calcium is another functional filler known as **barite**, baryte, barium sulfate, or “blanc fixe” ($BaSO_4$). It is chemically similar to calcium sulfate, but significantly less so to calcium carbonate. Similar to the distinction between GCC and PCC, barium sulfate is available in both natural and synthetic (precipitated) forms. It has the highest density among mineral fillers in the plastics industry (4.3–4.6 g/mL).

Natural, ground barium sulfate has an average particle size between 2–15 µm, but the precipitated types are generally much finer, with particles in submicrometer and nanometer ranges. The high density of barium sulfate makes it effective for noise reduction in foamed materials, floor coverings, plastisols, and sewage pipes. Due to its high electron density, barium sulfate can effectively attenuate high-energy radiation, providing high X-ray contrast. Due to this property, it is used in medical applications.

12.5.3 Silicates and Silicas

Talc, also known as talcum, is a natural, hydrated magnesium silicate with the formula $2MgO{\cdot}4SiO_2{\cdot}H_2O$. In terms of the volumes used in the plastics industry, this mineral is second only to calcium carbonate. Talc occurs in four forms: fibrous, platelet-like, acicular, and modular (steatitic). The platelet-like form is predominantly used in plastic applications, and exhibits inherent hydrophobicity, high chemical inertness, and good sliding properties due to its planar lamellae and low surface tension.

Talc improves various properties in thermoplastics:

- Increased stiffness, flexural and torsional modulus
- Enhanced flexural strength
- Reduced creep tendency even at higher temperatures
- Improved heat resistance and dimensional stability
- Better shrinkage and warpage control
- Enhanced thermal conductivity
- Reduced water content and oxygen permeability
- Increased surface hardness of molded parts
- Nucleating effect in semi-crystalline polymers (e.g., PP).

Fine-particle talc grades exhibit good nucleation properties for smaller-cell structures in foamed u-PVC. However, if not selected correctly, talc can lead to:

- Decreased impact strength and elongation at break
- Reduced tensile strength (with coarse grades)
- Deteriorated seam strength
- Color changes in the composite material
- Negative influence on thermal resistance (with unfavorable chemical composition)
- Abrasive effects (in materials containing magnesium carbonate or dolomite)
- Higher costs compared to calcium carbonate.

Kaolin, also known as kaolinite or China clay, is a mineral formed from the weathering of granite and feldspar. The typical chemical composition of hard kaolin is described with the molecular formula $Al_2O_3{\cdot}2SiO_2{\cdot}2H_2O$. Kaolin imparts a high degree of whiteness, it is not electrically conductive, and it is chemically resistant even to strong acids. Its pH is slightly acidic (pH 5.0–5.6 for hard kaolin). It improves various properties when used in filled polymers:

- Enhanced electrical properties (e.g., in cable insulation)
- Improved chemical resistance

- Reduced water absorption
- Decreased cracking tendency in end materials
- Increased impact resistance
- Enhanced surface quality and hardness.

Calcined kaolin is a specially processed form of kaolin that offers:

- Higher hardness than the natural kaolin
- Significantly superior electrical properties in polymers used to increase dielectric strength in cable insulation (p-PVC and EPDM)
- Ion scavenging properties

Silane-coated kaolins are also available to facilitate better dispersion in plastics.

Mica is a mineral filler and reinforcing agent derived from two primary sources, muscovite ($K_2Al_4(Al_2Si_6O_{20})(OH,F)_4$) and phlogopite ($K_2(MgFe)_6(Al_2Si_6O_{20})(OH,F)_4$). The platelet-shaped structure of mica is crucial to its reinforcing properties. Mica enhances the performance of thermoplastics, imparting:

- High rigidity
- Excellent dimensional stability
- Good electrical properties
- Improved heat resistance.

Silanization of the mica surface facilitates its incorporation into plastics, and in particular, treatment with aminosilanes leads to improved mechanical properties in the resulting composites. Mica coated with inorganic interphase layers (e.g., silicon dioxide and titanium dioxide) exhibits a "pearlescent effect", adding aesthetic value to plastic materials.

Feldspar and **nepheline** are anhydrous alkali–aluminium silicates that are used as naturally occurring mineral fillers. They are highly chemically resistant, have a slightly alkaline pH of 8–9, a density of ~2.6 g/mL, and a refractive index of 1.53, which results in good reflectivity with low coloring strength. Furthermore, they are coarser compared to kaolin or talc with an oil absorption number in the range 13–24 g/100 g. Feldspar and nepheline are broadly used in PVC and thermosets due to their low oil absorption, and are available in silanized forms for improved compatibility with polymers.

Wollastonite ($CaSiO_3$) is a mineral filler that is notable for being the only commercially available pure white mineral with needle-shaped crystals, with a length/diameter ratio up to 15:1. β-Wollastonite, the most common form, competes with talc and mica in thermoplastic applications. It has high chemical resistance, imparts a stiffening effect due to its needle-shaped crystalline structure, and enhances both the electrical properties and the thermal properties. Silanized grades of wollastonite are available for improved compatibility with polymer matrixes, including scratch-resistant PP, PA, and TPE.

Silica-based minerals are commonly used as fillers in the plastics industry. These include sand, quartz, quartzite, novaculite, perlite, tripoli, and diatomaceous earth (also known as diatomite). These silica-based fillers differ in several key aspects like particle size, degree of crystallinity and hardness. β-Quartz stands out among these as the hardest of the common minerals used in plastics, with density of 2.65 g/mL. The diverse range of silica-based fillers allows solutions to be tailored to the needs of various plastic applications, depending on the specific requirements for hardness, particle size, and crystallinity.

Synthetic silicas (SiO_2) are also commercially available as functional fillers. They are characterized by fine, mostly spherical primary particles that aggregate into secondary particles (also called aggregates and agglomerates). Their specific surface area can range from 50 to 800 m^2/g due to their amorphous structure (in contrast to crystalline natural quartz). They do not impose any risk of silicosis due to their amorphous nature, and are also available in silane-coated grades for enhanced compatibility.

Synthetic silicas, even if added in small quantities, can impart the following performance attributes in thermoplastics:

- Reduced shrinkage and increased crack resistance
- Film blocking (see Chapter 7)
- Enhanced heat dimensional stability
- Reduced coefficient of thermal expansion
- Increased volume resistivity
- Increased hardness
- Reduced strand expansion during extrusion
- Thickening and rheology control
- Anti-plate-out action, especially in u-PVC.

12.5.4 Glass and Other Fibers

Glass beads and other spherical silicate fillers are available for use in plastics formulations/composites. For example, fly-ash-derived grades (cenospheres and fillite) or aluminium silicate-based microspheres, which are available in compact and hollow forms, are used in PVC plastisols and thermosets, and offer the following advantageous properties:

- Low density (with hollow spheres)
- Uniform particle size
- Low oil absorption number
- High chemical resistance.

However, hollow spherical fillers are sensitive to pressure and shear forces during processing due to their thin walls. These fillers are also available in specialized surface modifications, including silanized versions for improved compatibility with polymers, and metallized versions for enhanced properties.

Increasing demand for lightweight materials, particularly in aerospace and automotive industries, has led to replacing certain metal parts by plastics. In other technical applications, plastics are also increasingly used for cost reasons. When their strength and stiffness are insufficient, plastic composite materials containing reinforcing fibers are applied.

Glass fibers (chopped strand) increase the rigidity, tensile strength, HDT, creep resistance, and impact strength of plastic composites. They also provide excellent dimensional stability and a low coefficient of linear thermal expansion (CLTE). Their reinforcing effect depends on several factors, including the type of glass, coupling agent (sizing), fiber length, and polymer matrix. The minimal fiber length in most plastics is in the range 300–600 µm. Disadvantages of glass fibers include abrasiveness, warpage, low surface quality, weld lines, and cost. Typical applications are aviation, automotive, marine, weapons, and household equipment.

Aramid fibers are high-performance synthetic fibers with an outstanding strength-to-weight ratio and heat resistance. The name derives from "aromatic polyamide", introduced in 1972 and defined as a distinct fiber category by the U.S. Federal Trade Commission in 1974. Chemically, they are polyamides based on terephthalic acid and diamines. In plastics reinforcement, they offer:

- High tensile strength (they are up to 6 times stronger than steel wire while having a much lower density)
- High modulus of elasticity (2–3 times that of steel wire and glass fiber)
- Excellent thermal stability (maintain strength at temperatures up to 200 °C)
- Good chemical resistance (especially to organic solvents, fuels, and seawater)
- Low flammability.

They are also used as replacements for asbestos and in lightweight parts. Aramid fibers significantly enhance tensile, bending, and interlaminar shear strength of plastics, making them suitable for high-strength, low-weight, durable applications. Their drawbacks include high cost, notable water absorption, and poor dispersion in plastics.

Carbon fibers (Figure 12.9) are high-performance synthetic reinforcements known for their exceptional strength-to-weight ratio and unique properties. As reinforcing agents in plastics, they result in lightweight yet strong composite materials. They are made starting with polyacrylonitrile (PAN) fiber (Table 1.1) in several steps:

1. Flame proofing furnace (200–300 °C)
2. Carbonization furnace (up to 1700 °C)

3. A graphitizing furnace (up to 2800 °C)
4. Final sizing surface treatment.

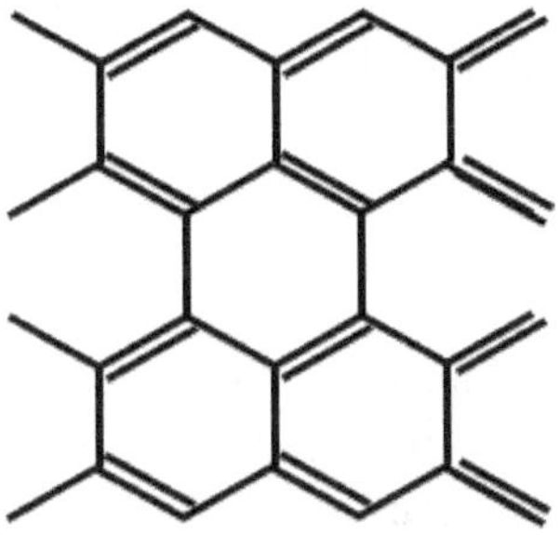

Figure 12.9
Chemical structure of carbon fibers

These fibers, which when chopped are 0.5–6.0 mm long, show excellent tensile strength and chemical resistance, and good thermal and electrical conductivity. Their high linear coefficient of thermal expansion minimizes shrinkage of the plastic composites containing them. Applications of carbon fiber reinforced plastics (CFRP) include:

- **Aerospace:** Reducing weight and improving fuel efficiency through use in aircraft fuselages, wings, and tail assemblies
- **Automotive:** Reducing weight and improving fuel economy in cars. They are particularly useful in hydrogen-powered vehicles for manufacturing lightweight but durable fuel tanks
- **Sports equipment:** Achieving high strength and low weight in high-performance rackets, fishing rods, and protective gear
- **Industrial applications:** Enhancing the properties (e.g. durability) of conveyor belts, hoses, and gaskets
- **Electronics:** Providing protective barriers against electrical and high-frequency interference in printed circuit boards
- **Medical applications:** The low X-ray absorption of carbon fibers makes them suitable for use in medical imaging equipment.

12.5.5 Carbon Black and Graphite

Although **carbon black** (CB) and **graphite** are both forms of carbon, they have somewhat different properties. CB is composed of relatively small, amorphous carbon particles, whereas graphite has a hexagonal crystalline lattice and a layered structure, giving it highly anisotropic properties, and most importantly much higher electrical and thermal conductivity than graphite.

Therefore, graphite is more suitable for applications requiring high electrical and thermal conductivity, while carbon black is used primarily as a pigment and mechanical reinforcing filler in plastics and rubber. Globally, only a small single-digit percentage of the total carbon black production is used in plastics (e.g., PE and ABS), whereas in PVC carbon black is mainly applied as a black pigment.

12.5.6 Wood Flour and Rice Husks

Wood flour is a commonly used organic filler for plastics, which consists mainly of α-cellulose (up to 55%), hemicellulose (up to 25%), lignin (up to 30%), and water (up to 6%), as well as minerals, waxes and fats. It is obtained from various types of wood, including hard and soft wood. Wood flour is often combined with PE, PVC, or PP to produce wood–plastic composite (WPC). The wood flour content in WPC can be 40–90%, with proportions of 50–90% being common in Germany and Central Europe. In North America, WPCs with 50% PE and 60% PVC content dominate.

Wood flour improves the mechanical properties of the composites, such as rigidity and dimensional stability when exposed to heat, compared to pure plastics or talc-filled variants. Wood flour enables a wood-like appearance, while being processable with the use of plastic processing techniques, such as extrusion and injection molding. It can be partially made from recycled materials. The processing temperatures for WPC are below 200 °C to avoid thermal decomposition of the wood. WPC is often used for decking, flooring and other outdoor applications. It is also applied in the automotive industry and other areas, where high rigidity and low shrinkage are required.

Like wood flour, **rice husks** increase the mechanical strength and rigidity of plastics. They contain a high proportion of silicon dioxide, which improves abrasion and chemical resistance when used in PE, PP, and PVC.

Stone powder can also be used an additive, to produce stone–plastic composites (SPCs). These typically comprise a mixture of limestone powder and PE or PVC. The core layer of SPCs usually contains 80% calcium carbonate and 20% HDPE or PVC.

12.5.7 Other Fillers

High electrical conductivity in polymers can be achieved through various methods: coating the polymers with conductive paint; metallizing finished plastic parts; or incorporating electrically **conductive fillers**. Besides conductive graphite, carbon black, and carbon fibers, ground copper, bronze powder and stainless steel fibers are also used. Carbon nanofibers (or the more expensive carbon nanotubes) enable meaningful electrical conductivity to be achieved at very low filler contents.

Nanofillers are materials with a primary particle diameter of well below 1 µm. Carbon black, precipitated silica, and precipitated calcium carbonate consist of such small particles. However, their primary particles form solid aggregates or loose agglomerates, which usually cannot be completely dispersed when incorporated into thermoplastics. In recent years, the use of numerous nanofillers in thermoplastics has been investigated. Particularly promising results have been achieved with clays (montmorillonite and smectite) that have a layered structure. Special additives can be intercalated between the layers and then exfoliated, resulting in superior mechanical properties. Materials reinforced in this way are gradually finding their way into industrial applications.

Some fillers exhibit low to no affinity or compatibility with the polymer matrix. To exploit their reinforcing effect, a strong bond between the matrix and fillers must be formed through applying coupling agents or adhesion promoters to the filler surface. These agents typically consist of an organic component that reacts with or is highly compatible with the polymer matrix and the filler surface.

The specific surface area of the filler determines the maximum amount of coupling agent that can be bonded to it. Excess coupling agent does not contribute to improved bonding between the filler and polymer matrix. The coupling agent can be applied to the filler as a pre-treatment or added *in situ* during compounding. For glass fibers, this pre-treatment is called sizing.

12.5.8 Coupling Agents and Adhesion Promotors

Common coupling agents include silanes, titanates, zirconates, chromium complexes, maleic anhydride grafted polymers, and acrylic acid oligomers.

Silanes (Table 12.5) have organofunctional groups (amine, vinyl, epoxy, and methacrylate) for polymer bonding and a methoxy- or ethoxy-group for filler surface bonding, and are used on glass fibers, silicates, and silica. **Titanates** and **zirconates** (Table 12.6) bond to fillers with monoalkoxy- or neoalkoxy- groups and with polymers that contain reactive groups. They can also be used on carbonates and carbon black. **Maleic anhydride grafted polymers** and **acrylic acid oligomers** bond to minerals like calcium carbonate and magnesium hydroxide and are highly compatible with polyolefins.

Coupling agents improve tensile strength and reduce water absorption due to the enhanced matrix bonding. However, they may potentially decrease the impact strength, and are rarely used in PVC.

Table 12.5 Functional Silanes and Their Applications in Plastics and Rubbers[8)]

Functional group	Thermosets	Thermoplastics	Rubbers
Amino	MF, PF, PU, EP	PVC, PE, PP, PS, PA, PC, PBT, PPO, PPS, PSU	Silicone rubber
Epoxy	EP, PU, MF, PF	ABS, PBT, PS, PVAC	NBR, EPDM, BR
Vinyl	UP	PS, EVA, PE, PP	EPDM
Methacryl	UP, DAP, acrylics	PS, ABS, PE, PP	BR

Table 12.6 Neoalkoxy Titanates, Zirconates and Their Applications in Plastics and Rubbers[9)]

Functional group	Thermosets	Thermoplastics	Rubbers
(Pyro-) phosphato	PF, UP, EP, acrylic	PE, PP, PVC, PS, PA, PET	XPE
Benzene-sulfonyl	EP	PE, PP, PSU	TPR
Ethylene-diamino	PU	TPU, EVA, CAB	Polysulfide
Amino-phenyl	PU	PC, PPO, PPS, ABS, POM	EPDM, SBR

12.6 Effect of Functional Fillers on Processing of Plastic Articles

Before using fillers, it is essential for the processor to clearly define the desired properties of the final plastic material. Functional fillers significantly affect a wide range of polymer properties. They expand the application boundaries of pure polymers and, in some cases, enable the use of specific plastics that would otherwise be unusable in their unmodified form.

When incorporating fillers and reinforcing agents, several factors must be considered, including:

- The desired property profile of the compound
- Dispersibility and compatibility with the plastic matrix
- Ease of processing and handling
- Optimal particle size distribution of the filler

[8] See Figure 1.1 for abbreviations.

- The wear effect of the filler on processing machinery
- Catalytic activity of the filler surface
- Occupational safety concerns related to dust emissions
- Legal and regulatory requirements
- Costs.

Particles with a small shape factor, such as spherical, cubic, or cuboid particles, generally do not increase tensile and tear strength. However, they typically improve the modulus of elasticity and, in brittle thermoplastics, often enhance toughness properties. Examples of this effect can be observed in u-PVC and PP homopolymer.

When strong adhesive forces occur between the mineral particles and the polymer matrix, even particles with a small shape factor can contribute to increased strength values of the composite material. A notable example is polyamide, where the addition of uncoated calcium carbonate increases both tensile and tear strength.

Spherical or cubic additives can result in the following property improvements in thermoplastics:

- Increased density
- Enhanced modulus of elasticity, compressive strength, and flexural strength (stiffening)
- Reduced shrinkage
- Improved hardness and surface quality
- Enhanced heat resistance and heat deflection temperature (HDT)
- Reduced temperature dependence of mechanical properties
- Reduced cost.

These additives can significantly influence the performance characteristics of thermoplastics, making them suitable for a wide range of applications.

Particles with a large aspect ratio, such as fibers or elongated platelets, significantly enhance the tensile and tear strength, as well as stiffness in plastics.

Figure 12.10 illustrates how the stiffness of PP (expressed as the flexural modulus) depends on the amount and aspect ratio of the selected filler. A comparable trend is observed for u-PVC composites.

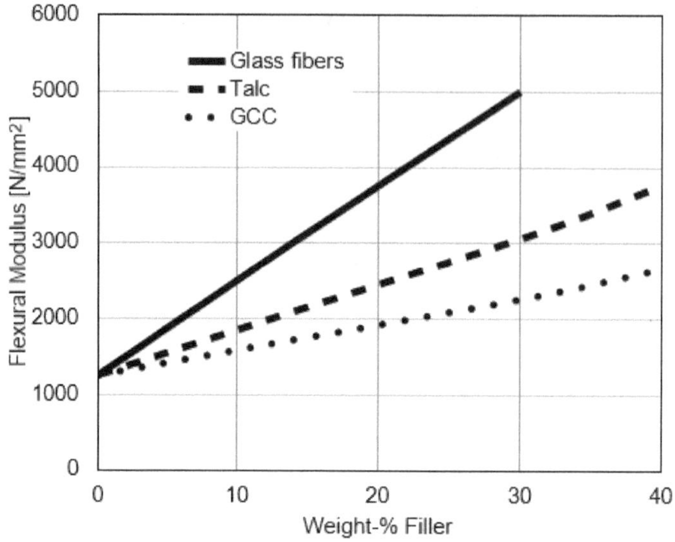

Figure 12.10 Flexural modulus of PP composites as a function of the filler content

All fillers, including spherical types like calcium carbonate or glass beads, increase the modulus of elasticity and stiffness of plastic materials. However, their tensile strength generally cannot be improved with round particles, and similarly, their HDT and stiffness at elevated temperatures cannot be increased by spherical additives to the same extent as it is with fiber reinforcements. Platelet-shaped fillers such as talc or mica, however, result in significant increase in these characteristics.

It should be noted that adding particles with a large aspect ratio is not without drawbacks. In many cases of this sort, simultaneous embrittlement of the composite material may be observed, and the increased stiffness and strength are often accompanied by reduced impact strength.

Compared to spherical or cube-shaped additives, platelets and fibrous fillers in thermoplastics offer additional advantages:

- Increased tensile, tear, compressive strength, and shear strength
- Reduced creep even at elevated temperatures
- Reduced viscoelastic yield under load
- Increased impact strength (with fibrous fillers).

The disadvantages of reinforcing fillers are mainly due to their anisotropic (oriented) reinforcing effect.

Reinforced plastics always consist of two discrete phases. The discontinuous filler phase should have higher tensile strength and modulus of elasticity than the polymer matrix, while the continuous plastic phase must have a higher elongation at break than the filler. For this reason, fibers are particularly suitable for use as reinforcing materials.

The higher the modulus of elasticity of the polymer matrix, the shorter the minimum length of the reinforcing fiber can be. Adhesion can be significantly strengthened by coupling agents between the filler and the plastic.

When a plastic material with a functional filler is subjected to mechanical stress, any cracks or fractures are formed first in the immediate vicinity of the coarse particles, due to stress concentrations.

The following properties can therefore be improved by reducing the amount of the large particle fraction ("top cut"):

- Impact and notched impact strength: Figure 12.11 shows a 2.5-fold increase in notched impact strength for u-PVC when reducing the top cut from 25 µm to 5 µm, consistent with the general trend that smaller particles result in improved mechanical properties
- Tensile strength: The tensile strength of composites can be increased by decreasing the particle size of the filler, although this relationship is not always linear
- Surface gloss: Figure 12.12 shows the difference in gloss levels of ground calcium carbonate (GCC) at different dosages in a PVC window profile compound
- Abrasion resistance: This improvement is consistent with the general trend that smaller particles result in improved mechanical properties.

It's worth noting that particle size effects can be complex. While smaller particles generally strengthen the composite, they also weaken particle cracking under high load. The optimal particle size depends on the specific application and desired properties of the composite.

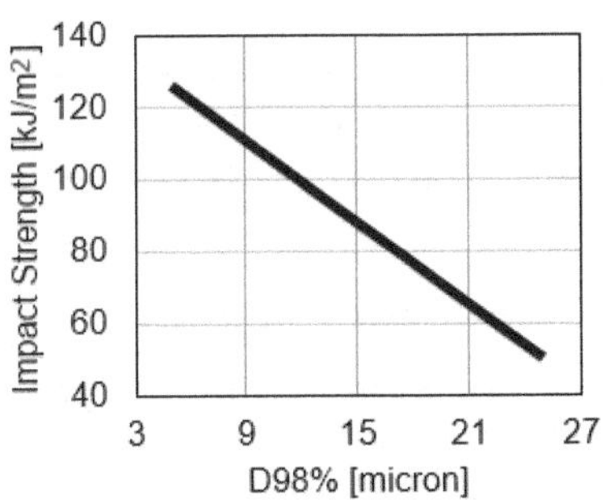

Figure 12.11
Izod notched impact strength of u-PVC (23 °C) as a function of the "top cut" D98% of calcium carbonate

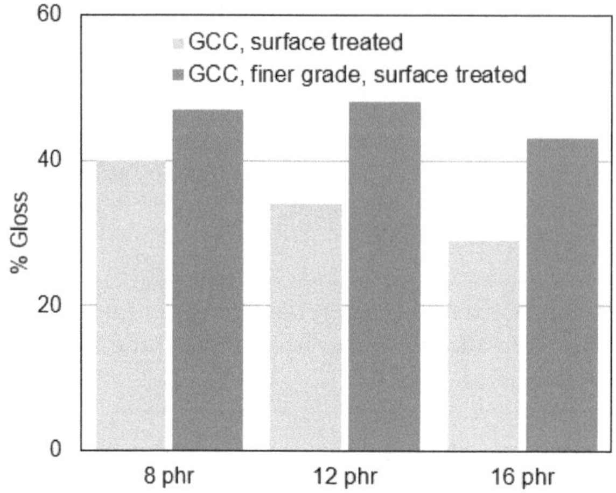

Figure 12.12 Gloss of window profile compounds with increasing levels and different particle sizes of calcium carbonate fillers

Properties of filled plastics required for a specific application, and the characteristics of fillers or reinforcing agents needed for it, are summarized in Table 12.7. This table serves as general guidance for selecting the appropriate fillers.

Table 12.7 Filler or Reinforcing Agent Properties Needed to Achieve Certain Characteristics of Filled Polymer Compounds/Composites[9]

Required polymer characteristic	Required filler properties
Mechanical properties	
High tensile strength and elongation	High length/diameter ratio; good filler/matrix adhesion; high strength compared to the matrix; good distribution in the matrix
High flexural strength	Same requirements as for high tensile strength; important is a flawless surface of the finished article
High compressive strength	Low compressibility; small, round filler particles (for semi-crystalline polymers)
High modulus of elasticity (high rigidity)	Large length/diameter ratio of the fiber or platelet-shaped reinforcing agents; high modulus of elasticity compared to the matrix; high orientation; good bonding with the matrix
Optimal impact strength	Long fibers; no perfect adhesion with the polymer matrix
Good long-term behavior (fatigue)	Permanent particle/matrix adhesion; no corrosive impurities in the filler; small, round particles that result in a smooth article surface

[9] R. D. Maier, *Handbuch Kunststoff-Additive* (4th edition), Hanser, 2016, pp. 1084–1085.

Required polymer characteristic	Required filler properties
High hardness	High hardness; grain size distribution suitable for high packing density; nucleation properties (for semi-crystalline polymers)
Low abrasion and good sliding properties	Low hardness; good sliding properties (formation of a sliding film)
High density	High density of the filler; good packing behavior
Thermal properties	
High heat resistance	High modulus of elasticity; good thermal stability
High resistance to thermal shock	High thermal conductivity; good filler/matrix contact
Low thermal expansion	Small coefficient of thermal expansion; anisotropic, hard particles; good adhesion to the matrix
Good flame and annealing resistance	Suitable chemical composition; good thermal conductivity; high thermal stability
Electrical properties	
Favorable dielectric loss factor, favorable dielectric constant, good insulation properties	Low moisture and plasticizer absorption; high chemical purity; good adhesion to the matrix, suitable electrical properties; insulating properties; nucleating effect with semi-crystalline polymers
Chemical properties	
High chemical resistance	Suitable chemical composition
Good weathering properties	High light stability; good chemical resistance; water resistance; permanent bonding to the matrix
Low water absorption	Surface with low affinity to water or suitable surface coating; low porosity; permanent bonding to the matrix
Optical properties	
Good transparency	Low refractive index difference to the matrix; particles significantly smaller than light wavelength
Good color	Suitable chemical composition
Good surface quality	Small, round or platelet-shaped filler particles
Costs	
Highest possible filling level	Particle shape, sufficiently narrow particle size distribution and competitive volume-based filler cost

Table 12.8 to Table 12.11 provide an overview of different filler types in several thermoplastic polymers.

Table 12.8 Effect of Fillers on PVC Compounds[a]

Polymer	Filler	Proportion (wt%)	Stiffness	Toughness	HDT	Processing	Remarks
PVC-P	Calcium carbonate	20–80	+	–	~	+	Cables
	Dolomite	20–80	+	–	~	+	Flame retardancy
	Kaolin	10–30	+	–	+	–	
	Silica	1–15	~	~	~		
	ATH	1–40	+	–	~	–	
PVC-U	Calcium carbonate	10–50	+	+	+	+	
	PCC	10–30	+	– –	+	–	
	Talc	10–20	++	–	++	~	
	Glass fiber	10–30	++	++	++	–	
PVC plastisols	Calcium carbonate	20–80	~	~	~	+	
	Synthetic silica	1–15	~	~	~	++	

[a] ++ Very good, + Good, ~ No effect, – Negative effect, – – Very negative effect.

Table 12.9 Effect of Fillers on Polyolefins[a]

Polymer	Filler	Proportion (wt%)	Stiffness	Toughness	HDT	Processing	Remarks
LD-PE	Calcium carbonate	20–40	+	+	~	++	
	Talc, mica	20–40	++	– –	++	+	
	Glass fiber	20–40	++	++	++	–	Creep resistance
	Silica	1–10	~	+	~	–	Antiblocking
	Carbon black	1–25	~	+	~	–	Conductive
HD-PE	Calcium carbonate	20–75	+	+	~	++	
	Talc	20–40	++	– –	+	+	
	Kaolin	20–40	+	–	+	~	
	Glass fiber	20–30	++	++	++	–	Creep resistance

Polymer	Filler	Proportion (wt%)	Stiffness	Toughness	HDT	Processing	Remarks
PP	Glass fiber	20–30	++	+	++	–	
	Talc	15–40	++	– –	+	++	Nucleating agent
	Mica	10–40	++	–	++	–	
	Calcium carbonate	20–50	+	+	~	++	
	Wollastonite	10–40	++	– –	++	–	
EVA	Calcium carbonate	20–50	+	~	~	++	
	Barium sulfate	10–75	+	–	~	+	Acoustic dampening
	Talc	10–40	++	– –	+	–	
	ATH	30–70	+	– –	+	– –	Flame retardancy

[a] ++ Very good, + Good, ~ No effect, – Negative effect, – – Very negative effect.

Table 12.10 Effect of Fillers on Engineering Thermoplastics[a]

Polymer	Filler	Proportion (wt%)	Stiffness	Toughness	HDT	Processing	Remarks
PA	Glass fibers	20–30	++	++	++	–	
	Glass beads	20–40	~	~	~	++	
	Carbon fibers	20–30	++	–	++	–	
	Silica	10–20	~	~	~	~	
	Talc	20–40	+	–	+	–	
	Calcium carbonate	20–40	~	~	~	+	
	Kaolin	20–40	+	+	+	+	
	Wollastonite	20–40	++	–	+	–	
	Mica	20–30	++	~	+	–	
	MoS_2	10–20	~	–	~	~	

Table 12.10 Effect of Fillers on Engineering Thermoplastics[a] (*continued*)

Polymer	Filler	Proportion (wt%)	Stiffness	Toughness	HDT	Processing	Remarks
PET, PBT	Glass fibers	20–30	++	++	++	–	
	Aramid fibers	10–30	++	+	++	–	
	Carbon fibers	20–30	++	–	++	–	
	Wollastonite	10–50	++	–	+	–	
	Mica	20–30	+	–	+	–	
	Talc	20–40	+	–	+	+	
PC	Glass fibers	10–30	++	–	++	–	
	Wollastonite	20–40	+	–	+	–	
POM	Glass fibers	20–30	+	+	++	–	
	Glass beads	10–40	~	~	~	++	
	Talc	20–40	+	–	+	~	
	Calcium carbonate	20–40	~	~	~	~	
PPO	Glass fibers	10–40	++	–	++	–	
	Carbon fibers	10–20	++	–	++	–	
PPS	Glass fibers	10–40	++	+	++	–	
	Carbon fibers	10–30	++	–	++	–	
PSU, PES	Glass fibers	10–40	++	+	++	–	
	Carbon fibers	10–40	++	–	++	–	
PEEK	Glass fibers	10–30	++	++	++	–	
	Carbon fibers	10–30	++	~	++	–	
PAI	Glass fibers	10–30	++	+	++	–	
	Carbon fibers	10–30	++	–	++	–	
	Graphite	10–20	+	–	+	+	Friction reduction
PEI	Glass fibers	10–40	++	+	++	–	
PMMA	Glass fibers	10–40	++	–	++	–	

[a] ++ Very good, + Good, ~ No effect, – Negative effect, – – Very negative effect.

Table 12.11 Effect of Fillers on Styrenic Polymers[a]

Polymer	Filler	Proportion (wt%)	Stiffness	Toughness	HDT	Processing	Remarks
PS	Glass fibers	20–40	++	– –	++	++	
	Glass beads	20–40	+	–	+	+	
	Calcium carbonate	20–30	+	–	~	–	
	ATH	20–40	+	–	~	–	Flame retardancy
EPS	Calcium carbonate	1–15	~	~	~	–	
	Talc	1–3	~	~	~	++	Nucleator
ABS	Glass fiber	20–40	++	–	++	+	
	Glass beads	20–40	+	~	~	~	
	Calcium carbonate	20-30	~	~	~	–	
SAN	Glass fiber	20–30	++	–	++	–	
	Glass beads	15–40	+	~	~	++	

[a] ++ Very good, + Good, ~ No effect, – Negative effect, – – Very negative effect.

12.7 Trends and Outlook

The market for polymer fillers is steadily growing, with an expected annual growth rate of approximately 3.1% from 2024 to 2032. Several trends are currently shaping the development and application of fillers in plastics:

- Rising demand for lightweight materials across industries, such as automotive, aerospace, and packaging. Filler-like glass microspheres, carbon nanotubes, and hollow ceramic spheres are increasingly used to reduce weight and enhance the weight/performance ratio of polymers
- A focus on sustainability and regulations are driving the development and use of bio-based (biogenic) and recycled fillers. Manufacturers are integrating environmentally sound filler options into their polymer formulations
- Innovative nanofillers, such as nanoclays, graphene, and carbon nanotubes, are being increasingly used to improve the mechanical, thermal, and barrier properties of polymers

- Use of functional fillers is gaining momentum. Filler-like flame retardants, conductive fillers, heat stabilizers, and antimicrobial agents are being adopted to meet specific requirements in sectors like electronics, construction, and healthcare
- Fillers are also becoming deployed to improve thermal conductivity of plastics, particularly using new manufacturing processes, such as fused filament fabrication (FFF).

These trends reflect ongoing innovation in the field of plastic fillers and highlight the potential for future advancements in polymer technology.

12.8 Questions

1. What is the definition of functional fillers? How are they different from traditional fillers used in polymer composites?
2. What are the methods for surface modification? How do they affect filler compatibility with plastics?
3. How does particle morphology influence the mechanical properties of a plastic composite?
4. Why is proper dispersion of fillers crucial for manufacturing high-quality filled polymer systems?
5. What are the mechanisms that contribute to the mechanical reinforcement of plastics by fillers?
6. How do coupling agents enhance the interfacial adhesion between fillers and polymers?
7. What are the factors responsible for electrical conductivity of filled polymers?
8. What is the definition of nanofillers? What are their advantages over conventional micro-scale fillers?
9. How does surface modification of fillers improve their dispersion in hydrophobic polymers?
10. How does the use of nanofillers affect both mechanical and thermal properties of polymers?

13 Blowing Agents

Content of this chapter was adapted from:

- H. Hurnik, "Chemical blowing agents" (Chapter 13), in: *Plastics Additives Handbook* (6th edition), ed. H. Zweifel, R. D. Maier, and M. Schiller, Hanser, 2009, pp. 719–743.

13.1 Terms and Definitions

Cellular plastics structure	Plastics containing numerous gas-filled cells that reduce density and weight.
Foamed materials	Plastics expanded with gas or blowing agents to create a foam structure.
Onset decomposition temperature	Temperature at which noticeable thermal decomposition of a material begins.
Maximum decomposition temperature	Temperature at which decomposition rate reaches its peak.
Enthalpy of decomposition	Total heat absorbed or released during a material's decomposition process.
Rate of decomposition	Speed at which a material chemically breaks down under specific conditions.
DSC (Differential scanning calorimetry)	A technique that measures heat flow in/out of a sample as the temperature changes.

13.2 Functions and Types of Blowing Agents

Physical and chemical blowing agents (also known as foaming agents) are additives used in the manufacturing of foamed plastics. The main purpose of manufacturing foamed plastics is their greatly reduced density, which translates into material savings, weight reduction of the finished article, and decreased costs. Additional considerations include improved heat and sound insulation characteristics, enhanced resilience and shock absorption, superior dielectric properties, and decorative and aesthetic effects.

Disadvantages of foamed materials may include reduced strength, a longer cooling time (because more cooling is required), a narrower processing window, and lower melt strength (because of the need to achieve a balance between melt strength and density reduction).

Various blowing agents used to foam plastics, and can be classified according to their mode of liberating the blowing gas, i.e., they can be divided into physical and chemical blowing agents.

Physical blowing agents do not undergo chemical transformations during the foaming process, and the blowing gas is liberated by physical processes, namely by vaporization of low-boiling liquids or by pressure release of compressed gases. Most physical blowing agents are liquids. The foaming process with physical blowing agents is irreversible and endothermic due to the volatilization of the liquid.

Physical blowing agents include:

- Chlorofluorocarbons (CFCs; ozone depletants, banned by the 1987 Montreal Protocol)
- Hydrochlorofluorocarbons (HCFCs; ozone depletants, therefore being phased out)
- Hydrofluorocarbons (HFCs)
- Hydrofluoro-olefins (HFOs; actively in use)
- Hydrocarbons (e.g., pentane, isopentane, cyclopentane)
- Liquid carbon dioxide
- Water.

Examples of foamed plastics manufactured with physical blowing agents like water and HFO include rigid and flexible polyurethane (PU) foams, which are used among other applications for building insulation and furniture cushions, respectively.

Chemical blowing agents (CBAs) are mostly organic chemicals, although some inorganic compounds are in use. These additives generate gas as a result of their thermal decomposition. The gas then foams the plastic matrix. CBAs may be technically pure substances, blends of blowing agents, or preparations containing both the CBA and functional additives (e.g., kickers, inhibitors, co-reagents). During the foaming process at elevated temperatures, CBAs undergo thermal decomposition that liberates the blowing gas. Most CBAs are solids.

CBAs are classified into three groups, according to their mode of gas generation:

- Compounds that split off the gaseous products as the result of irreversible thermal decomposition. The organic CBAs belong to this group, and include ADC, TSH, OBSH, TSSC, 5-PT, and DNPT, which are described in this chapter (for full chemical names see).

  The mode of decomposition is: AB → C + gas ↑. These are typically first-order reactions, and the gaseous components may be N_2 (nitrogen), CO (carbon monoxide), CO_2 (carbon dioxide), HNCO (isocyanuric acid), H_2O (water, at temperatures >100 °C), HCHO (formaldehyde), and NH_3 (ammonia).

- Compounds that split off the gaseous products as the result of an equilibrium reaction. The main representatives are inorganic bicarbonates and carbonates of the alkali metals and alkaline earth metals, such as sodium bicarbonate that liberate CO_2.

  The mode of decomposition is: AB → C + gas ↑

- Combinations of compounds that liberate gaseous products as a result of chemical reactions with other components.

  The mode of decomposition is: A + BG → AB + gas ↑. Typical examples of this group of blowing agents are blends containing sodium bicarbonate and citric acid.

13.3 Chemical Blowing Agents

13.3.1 General requirements

A CBA used to foam polymers has to fulfill several requirements:

- The decomposition temperature of the blowing agent must be comparable to the processing temperature of the polymer
- The liberation of the blowing gas must occur within a temperature range of about 10 °C and be controllable during the process
- The decomposition must not be autocatalyzed, to avoid overheating and damage to the polymer
- The blowing gas must be chemically inert (such as nitrogen and carbon dioxide) or be chemically compatible with the polymer.

It is desirable that neither the blowing agent nor its decomposition products present health hazards, adversely affect the thermal stability or mechanical strength of the plastic, or have a corrosive effect. The decomposition residues should also be compatible with the plastics, not bloom out to the surface, or cause discoloration. The blowing agent should have a high gas yield and be economical to use.

The following thermal decomposition characteristics of chemical blowing agents are also critical:

- Onset decomposition temperature
- Maximum decomposition temperature
- Decomposition temperature interval from onset to the end of gas yield
- The gas yield under the given temperature conditions (mL of gas per g of blowing agent)
- Exothermal/endothermal effects (enthalpy) of decomposition.

Characterization of CBAs is conducted using thermal analysis, using differential thermal analysis (DTA), differential scanning calorimetry (DSC), or thermogravimetric analysis (TGA). Preparatory methods for studying the gas generation and more closely simulating actual industrial conditions include the decomposition of a blowing agent in a plasticizer, any other liquid carrier (e.g., paraffin oil), or a polymer. Typically, 1 g of blowing agent is dispersed in 15 mL of a plasticizer in a 50 mL flask connected to a gas burette. The decomposition is conducted isothermally, and the gas evolution is recorded over time.

13.3.2 Classes and Mechanisms of Action

Commercially viable CBA classes, specific substances, their performance attributes, and applicability for use in plastics and elastomers are listed in Table 13.1.

The economic importance of an individual CBA varies greatly. Azodicarbonamide and its modifications account for about 85% of the total volume of all chemical blowing agents, followed by the sulfonylhydrazides. Endothermic blowing agent formulations based on citric acid and sodium carbonate or sodium bicarbonate are ranked third. TSSC and 5-PT are small-segment specialties for thermoplastics with high processing temperatures.

Table 13.1 Selected Chemical Blowing Agents Used for Plastics and Elastomers

Chemical class	Blowing agent	Abbreviation	Molecular weight	Decomposition temperature range (°C)	Gas yield (mL/g)	Plastic and elastomer application[1]	Enthalpy of decomposition (cal/g)	Components of blowing gas
Azo-compounds	Azodicarbonamide	ADC, AZO	116.1	205–215	220	PVC, PE, PP, ABS, PS, PA, PPO, TPE	86	N_2, CO, (NH_3, CO_2)
	ADC, activated	ADCA	—	155–215	150–220	PVC, PE, PP, EVA, PS, ABS, EPDM, NBR, SBR, NBR/PVC	86	N_2, CO, (NH_3, CO_2)
Hydrazine derivatives	*p*-Toluenesulfonylhydrazide	TSH	186.2	105–110	115	EVA, EPDM, SBR, NBR, NR	—	N_2, H_2O
	4,4′-Oxybis (benzenesulfonylhydrazide)	OBSH	358.4	155–165	120–125	LDPE, EVA, NBR/PVC, EPDM/CR, SBR, CR, NBR, NR, PUR	123	N_2, H_2O
Semicarbazides	*p*-Toluenesulfonyl semicarbazide	TSSC	229.2	226–235	120–140	PP, ABS, HDPE, U-PVC, PPO, HIPS, PA	—	N_2, CO_2, NH_3
Tetrazoles	5-Phenyltetrazole	5-PT	146.1	240–250	190–210	ABS, PPE, PC, PA, PBT	—	N_2
Nitroso-compounds	*N,N′*-Dinitroso-pentamethylenetetramine	DNPT	186.2	195	190–200	NR, SBR, HI-PS, CR, BR, P-PVC	537	N_2, (NH_3), HCHO
Inorganic carbonates	Sodium bicarbonate/Citric acid	SBC/Citric acid	—	150–230	185–240	PS, ABS, PA, PE, PP, U-PVC	—	CO_2, H_2O
	Sodium bicarbonate	SBC	84.0	125–175	125	U-PVC	—	CO_2, H_2O

[1] See Figure 1.1 for abbreviations.

13.4 Azodicarbonamide (ADC)

Azodicarbonamide (ADC, AZO) (Figure 13.1) is one of the most dominant commercial blowing agents used in the plastics industry, by virtue of its inherent decomposition temperature range, ability to be activated to lower process temperatures, and its lower cost compared to other chemical blowing agents. ADC was first suggested as a potential blowing agent for plastics during World War II, and was introduced in the USA in the early 1950s.

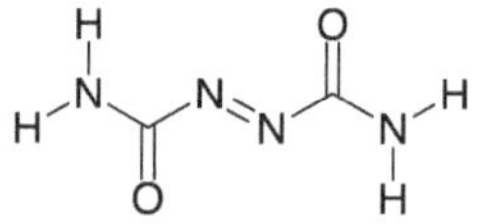

Figure 13.1
Chemical structure of ADC

Although ADC is a yellow-to-orange-yellow powder in its pure state, it generates white- to off-white-colored residues if completely decomposed. Some traditional approaches to addressing this problem have included the use of white pigments, such as titanium dioxide and/or optical brighteners, to mask the yellow discoloration.

ADC has excellent storage stability, unlike a number of other CBAs. Additionally, it is self-extinguishing in contact with an open flame and does not support combustion. US FDA regulations permit the use of ADC for the manufacturing of articles intended to come in contact with food. However, according to EU27, since 2005 ADC is prohibited for being used in sealing gaskets of metal lids intended for glass containers (e.g., baby food jars). The reason for this limitation is a possible formation of the semicarbazide, a compound shown to be weakly carcinogenic while detectable in food. Nevertheless, the industrial uses of it as a blowing agent for plastics remains permitted.

ADC is an exothermic blowing agent. The chemical reactions describing its decomposition are quite complex, and are also dependent on the polymer formulation. In addition to generating blowing gas during the decomposition, a certain part of the original material sublimes (as cyanuric acid, cyamelide, and urea), and the rest forms a solid residue (containing cyanuric acid, urazole, and hydrazodicarbonamide/semicarbazide). All decomposition products are colorless. ADCcan be activated by kickers (see Section 13.6.1).

The decomposition reaction of ADC is affected by the presence of moisture, which usually leads to an increased yield of blowing gas.

13.5 Effect of Blowing Agent Level

CBAs are usually added to plastic-containing compounds and matrixes at 0.5–5.0% ("parts per hundred parts of resin" (phr) units are also used to express the amount of blowing agents).

Figure 13.2 shows the effect of ADC concentration on foam density. As can be seen in this example, increasing the blowing agent level beyond 0.6 phr has little effect on reducing the density of foamed rigid PVC. It is interesting to note that achieving a minimum density of the foamed plastisol requires increasing the ADC level up to 5–6 phr, which is about a 10-fold increase compared to the foamed rigid compounds.

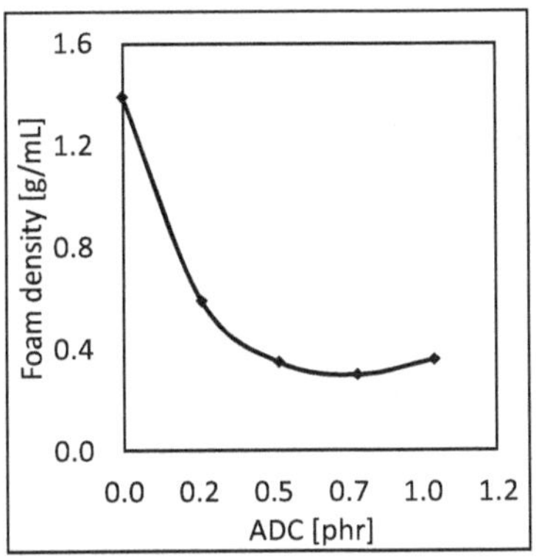

Figure 13.2
Foam density of rigid PVC prepared with ADC

13.6 Use of Kickers

13.6.1 Activation of ADC

Although the decomposition temperature of technically pure ADC is comparatively high (205–215 °C), it can be made suitable for the foaming of PVC plastisols and other polymers at lower processing temperatures. With the use of auxiliary additives, the decomposition temperature can be reduced to as low as 150 °C. Such additives are known as "kickers" or "activators".

Metal-containing compounds, where the metal is zinc, lead, or potassium – but preferably zinc and zinc/potassium combinations – are effective kickers for CBAs. Other types of chemicals that have been shown to function as activators include polyols, urea, amines, certain organic acids etc. Lowering the decomposition temperature may extend the time needed to complete the decomposition of the same amount of the blowing agent. Activator levels vary from 1% of the blowing agent level, to levels equal to that of the blowing agent, depending on the desired effect. It is always advisable to pre-blend the blowing agent and the activator before adding them to the polymer system to ensure the maximum effect.

Reduction of the decomposition temperature by zinc-containing compounds is explained by the formation of the thermally unstable zinc salts of azodicarboxylic acid intermediate. Activators for ADC include zinc oxide, zinc stearate (Figure 13.3), potassium stearate, other salts of zinc, salts of lead, and urea, among others. Depending on the amount of kicker in mixtures with ADC, its gas yield can be reduced to 150–220 mL/g. Activated ADC is used in PVC, PE, PP, EVA, PS, ABS, EPDM, NBR, SBR, and NBR/PVC.

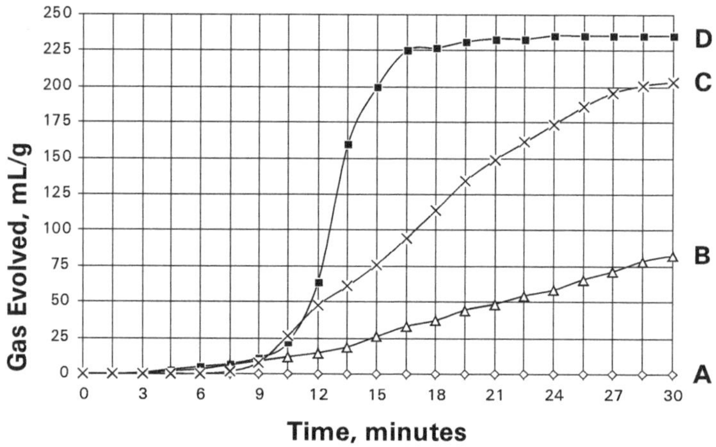

Figure 13.3 Effect of activators on decomposition of ADC-based CBA at 150 °C: (A) ADC; (B) 9:1 ADC/zinc stearate; (C) 32:1 ADC/zinc oxide; (D) 3:2 ADC/zinc oxide. Image reproduced courtesy of Galata Chemicals, LLC

13.6.2 Effect of CBA Particle Size and Kicker Concentration

The decomposition characteristics of ADC are also affected by the particle size of ADC and the kicker concentration. Specifically, smaller ADC particles have a greater specific surface area, increasing the rate at which it reacts with the kicker.

> The smaller the CBA particle size, the more pronounced the "kicking effect" of the activator.

CBAs with a smaller particle size provide a faster rate of decomposition and gas evolution.

In polymer systems where the process involves curing or fusion, the rate of gas evolution from the blowing agent must be balanced with the viscosity of the polymer system. If gas is generated when the polymer viscosity is too low, much of the gas escapes, resulting in little to no density reduction and a larger and more irregular cell structure. If gas is generated when the polymer viscosity is too high, less expansion may take place, and cracking occurs in extreme cases. Although blowing agent particle size and activator type and level may be varied to achieve this balance, it is sometimes necessary to modify the curing or fusion behavior of the polymer system as well.

The optimum specific surface for the foaming of PVC plastisols is >2 m^2/mL. For the manufacture of crosslinked polyolefin foams, coarser grades are preferred, and the ADC grades of choice for that have a specific surface of <1 m^2/mL. The decomposition behavior of three ADC grades with different particle size and amount of kicker are shown in Figure 13.4 and Figure 13.5, respectively.

The higher the kicker concentration, the more pronounced the "kicking effect".

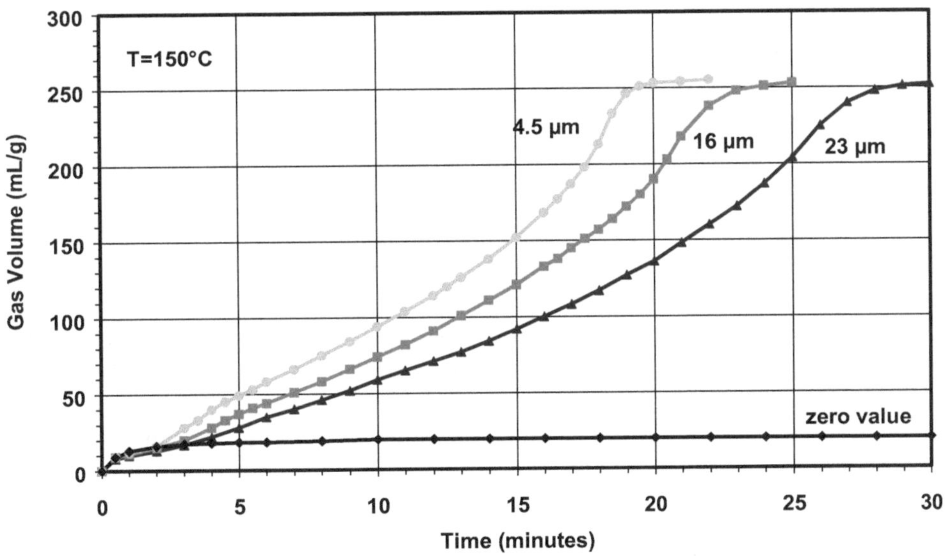

Figure 13.4 Effect of ADC particle size on isothermal (150 °C) decomposition of ADC; all formulations contained 1 g ADC and 1 g K/Zn kicker, dispersed in 15 mL DOP

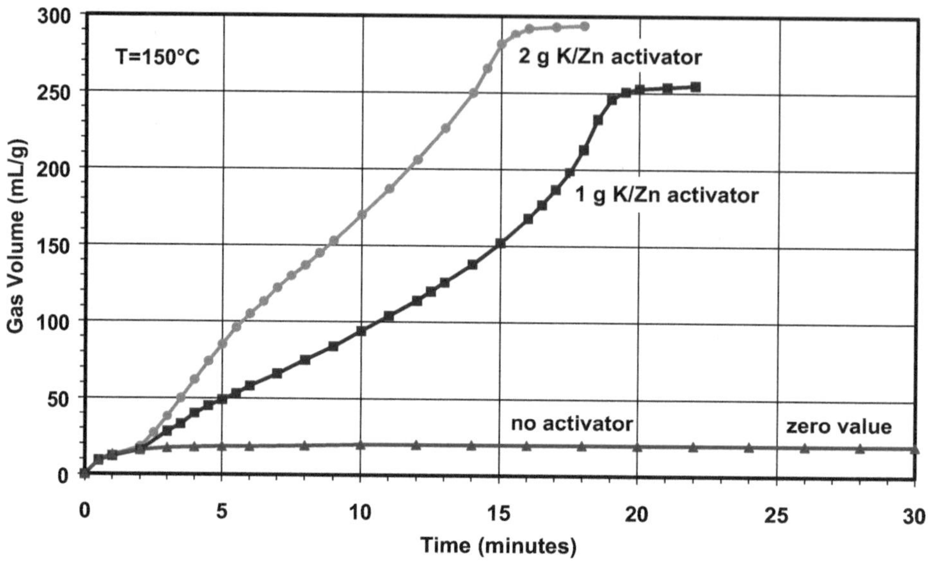

Figure 13.5 Effect of kicker amount on isothermal (150 °C) decomposition of ADC; all formulations contained 1 g ADC and 1 g K/Zn kicker, dispersed in 15 mL DOP

13.6.3 Inhibition of ADC Decomposition

As well as kickers, which accelerate the thermal decomposition of ADC, certain inhibitors can slow down the decomposition rate of kicker-containing ADCs. This effect is used in "chemical embossing" of flooring materials, wall coverings, and foamed PVC wallpaper. Effective inhibitors include di- and tricarboxylic acids and their anhydrides, as well as certain benzotriazole derivatives (see Chapter 3).

13.6.4 Activation of Other Organic CBAs

Sulfonylhydrazides (both TSH and OBSH) can be activated with either oxidizing agents, or solvents containing hydroxy groups, such as aliphatic alcohols or water. US FDA regulations permit the use of OBSH for the manufacture of articles intended to come in contact with food. OBSH can be activated with urea, alcohol amines, organic peroxides, and zinc- and lead-containing salts.

Effective kickers for **TSSC** include glycols, stearic acid, urea, zinc compounds, dibasic lead phthalate, and dibasic lead phosphite. TSSC may be combined with OBSH or ADC to achieve optimal foaming effects.

DNPT is flammable, and should be kept in refrigerated storage. For safe handling, DNPT is marketed exclusively in blends with a maximum active substance concentration of 80 wt%. The most common kickers for it are urea, benzoic acid, salicylic acid, phthalic anhydride, and glycols. For toxicological reasons and potential for it to impart discoloration on the end products, DNPT commercial use has been decreasing rapidly.

13.7 Sodium Bicarbonate

Inorganic blowing agents, containing bicarbonates and carbonates, are some of the oldest additives of this group, and are readily available commercially. However, high-quality foams are difficult to produce with these CBAs because they result in slow, inconsistent, and unsteady gas generation during the process. When carbonates are used as blowing agents, the foam obtained can be coarse and have an inconsistent cell structure.

Mechanisms of thermal decomposition of sodium bicarbonate (SBC) as well as its blend with citric acid (commonly used as a co-reagent for SBC) are shown in Figure 13.6 and Figure 13.7, respectively. These thermal decomposition reactions are endothermal and irreversible.

$$2\ NaHCO_3 \rightleftharpoons Na_2CO_3 + CO_2 + H_2O$$

Figure 13.6 Thermal Decomposition of SBC

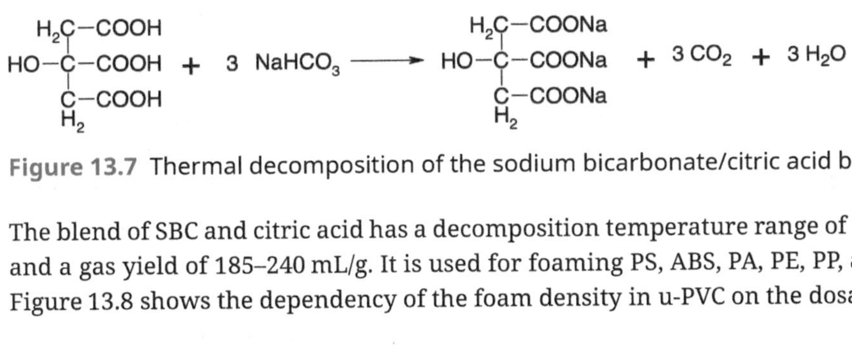

Figure 13.7 Thermal decomposition of the sodium bicarbonate/citric acid blend

The blend of SBC and citric acid has a decomposition temperature range of 150–230 °C and a gas yield of 185–240 mL/g. It is used for foaming PS, ABS, PA, PE, PP, and u-PVC. Figure 13.8 shows the dependency of the foam density in u-PVC on the dosage of SBC.

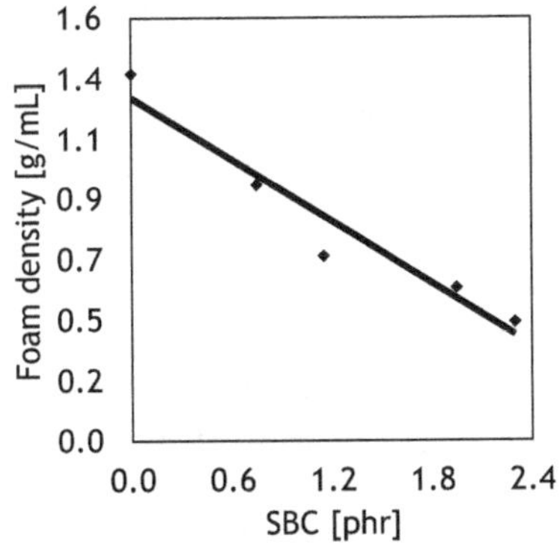

Figure 13.8
Dependence of PVC foam density on dosage of SBC

13.8 Combinations of Blowing Agents

Due to its versatility, many blends of ADC-based blowing agents are commercially viable.

For the combination of ADC with a blowing agent of another chemical group, the sulfonylhydrazides are usually favored, especially in combination with a kicker.

ADC/TSH and ADC/DNPT are combinations typically used in elastomers. Besides lowering the decomposition temperature, this mixture results in foamed materials with improved cell structures.

Combining SBC with ADC results in foams with distinctly reduced cell size compared with the use SBC as the sole blowing agent. This effect is explained by the slower diffusion of nitrogen generated from ADC in the ADC/SBC blend compared to only carbon dioxide generated from SBC when used alone to foam plastics. It has been shown that the lowest foam density is achieved in PVC with SBC added at 2.4 phr and ADC added at 0.2–0.4 phr. The synergism between ADC and SBC in the extrusion of rigid PVC foam is largely attributable to the complementary properties of the gaseous decomposition components, such as nitrogen and carbon dioxide.

Combinations of TSSC with ADC and OBSH are also known to be used for foaming plastics.

13.9 Foaming Polymers

For manufacturing foamed profiles, sheets, pipes, and cable jacketing, extrusion processes are used. The CBAs in a powder form are added as a component of the dry blend. Another option is to dose the blowing agents in powder concentrates and/or polymer masterbatches in a granular form directly into the extruder; the same applies to injection molding processes.

For the foam extrusion of thermoplastic materials other than unplasticized PVC, foaming typically starts with the granulated thermoplastic materials. The blowing agents may be added in various forms, including surface-oiled, dust-free powders, liquid pastes, high concentration masterbatches, etc.

Uniform mixing and dispersion of the blowing agent within the polymer are critical for obtaining a good-quality foam. It also helps to minimize the process variability. Pre-blending of blowing agents with the resin also enables ensuring that the best results are achieved.

Blowing agent selection requires matching the gas release temperature with the processing temperature used for the polymer. Chemical compatibility between the blowing gas components and the polymer composition should also be considered.

For example, combining ADC (which generates alkaline gases, such as ammonia) with hydrolyzable polymers, such as PC or thermoplastic polyesters (PET and/or PBT), should never be conducted; severe chemical degradation of the polymer and a significant loss in mechanical properties of the finished material would result.

Processes have been developed for the manufacture of many foamed plastics. The foamed plastics are produced from the standard polymers, such as PVC, PE, PP, and PET. For the production of integral skin foams, the applicable polymer resins and resin blends are: ABS, PBT, PC, (PC/ABS), (PBT/PC), (PPE/PS), PA, and TPU. Solid rubbers used to produce foams are EPDM, NBR, (NBR/PVC), and SBR.

13.10 Applications of Foamed Polymers

Of foamed materials produced globally, the largest volumes are based on PVC. Typical examples of articles made from foamed p-PVC are flooring and wall coverings, embossed tapestry, bathmats and gymnastics mats, foam coatings for handbags, upholstery and clothing, foamed gaskets for the construction industry, and many others.

The calendering process is widely used for the manufacture of p-PVC sheets. Foamed polymeric materials are also manufactured by extrusion. u-PVC foams have gained significant industrial importance, with foamed extruded profiles being used in the construction industry, windowsills, walls and wall linings. u-PVC pipes with foam cores are used for ambient-pressure applications, such as wastewater drainage. Foamed u-PVC sheets are also used in temporary housing, signage, displays, and advertising in the construction industry, among others. Large volumes of foamed PVC-coated fabrics have been produced using ADC and OBSH.

Since heat stabilizers not only protect PVC against thermal decomposition but also activate the decomposition of the blowing agents due to the presence of zinc-containing components, proper selection of the stabilizers is of great importance. Zinc-containing stabilizers yield good foams within the 166–204 °C temperature range. Calcium–zinc-containing stabilizers are generally considered to be milder activators compared to barium–cadmium–zinc stabilizer systems or zinc carboxylates alone. Additionally, if ADC is used as a CBA in PVC in combination with a calcium–zinc stabilizer, the stabilizer should not contain β-diketones, calcium acetylacetonate, or zinc acetylacetonate. The reason for this is that ammonia formed from decomposing ADC would react with these stabilizer components, forming yellow discoloration as soon as heat is applied to the formulation during its processing.

Inorganic pigments, such as titanium dioxide, have no measurable effect on the decomposition of ADC and/or OBSH. Organic pigments containing zinc or lead would also likely activate the CBA. Fillers, if used at high levels, can also affect foam quality and density.

Foamed PP films and sheets produced by extrusion are used for manufacturing trays for packaging fruits, vegetables and meat, as well as packaging materials for eggs, among other applications.

Foamed polyolefins are excellent insulation materials for all types of cable applications produced from HDPE, PP, and LDPE plastics, using TSSC and/or ADC as blowing agents.

Foamed cross-linked LDPE or NBR/PVC blends are traditionally used in the construction industry. Extruded and cross-linked polyolefin foams have become popular for manufacturing trunk and door linings, in sports and leisure equipment for gym mats, crash helmets, knee and elbow pads, and sport shoe insoles.

13.11 Technological Trends

It is likely that the number of commonly used blowing agents will continue to decrease. Due to its versatility, performance, and cost-in-use, ADC will likely remain to be the most commonly used CBA, especially for PVC.

Among the standard plastics for food packaging applications, it is likely that the use of PP foams will continue to grow. It is also expected that a larger share of recycled thermoplastics will be used more in manufacturing foamed materials.

13.12 Questions

1. What are the two main types of blowing agents?
2. What is the definition of a physical blowing agent? Give examples.
3. What is the definition of a chemical blowing agent? Give examples.
4. What are the functions of blowing agents in plastics?
5. What are the three modes of liberating blowing gas?
6. What are the general requirements for chemical blowing agents (CBAs) used in plastics?
7. What are the units for measuring gas yield?
8. What is the highest production volume CBA available commercially?
9. What is the role of activators (kickers) for CBAs?
10. What is the effect of particle size on the efficiency of CBAs?
11. What is the typical use level range (wt%) of CBAs in plastics?

12. What are the applications of CBAs in u-PVC?
13. What are the applications of CBAs in p-PVC?
14. What are the applications of CBAs in polyolefins?
15. What are the applications of CBAs in polyurethanes?

14 Bio-attributed Additives

Content of this chapter was adapted from:

- P. Frenkel, J. Leadbitter, and D. DeCaria, "Decarbonization of PVC" (Chapter 22), in: *PVC Handbook* (2nd edition), ed. R. P. Krock, M. Schiller, P. Frenkel, J. W. Summers, and C. A. Daniels, Hanser, 2025, pp. 1049–1074.

14.1 Terms and Definitions

Bio-based content	The proportion of carbon in a product derived from biological or renewable sources, typically expressed as a percentage of total carbon content.
Bio-attributed carbon	Carbon content in a product that is credited to biomass through a mass balance approach.
Renewable raw materials	Natural resources that can be replenished within a short time and are used as raw materials in production processes.

14.2 Transition to Bio-sourced Feedstocks in the Value Chain

There is now widespread consensus that the human contribution to climate change is no longer an issue up for debate. Accordingly, one of the main priorities in the sustainable development of chemicals and materials is the transition from fossil-based feedstocks towards bio-sourced or renewable feedstocks – and indeed, this is one of the 12 principles of "green chemistry".[1] Attaining decarbonization with renewable or bio-sourced feedstocks for manufacturing plastics and their additives is reviewed below.

Raw materials are considered to be bio-renewable if the carbon acquisition (as a result of CO_2 sequestration via photosynthesis that occurs within those raw materials) takes place within a 10-year period and positively impacts (reduces) the global warming potential (GWP) (Table 14.1).

Table 14.1 Comparison of Bio-based (Renewable) and Petrochemical-based (Fossil) Carbon

Parameter	Type of carbon source	
	Bio-based carbon	Fossil carbon
Period of carbon acquisition for manufacturing raw materials	1–10 years	>100 (106) years
Typical sources	Agricultural crops, algae, short-rotation wood/trees	Crude oil, coal, natural gas

Environmental impacts are assessed with the use of standardized life-cycle assessment (LCA) methodology (ISO 14040), which is conducted based on establishing detailed and comprehensive material and energy balances for each studied product/process/manufacturing site.

Over the years, a check-list and criteria have been developed for successful market introduction of bio-sourced chemicals, polymers, and polymeric materials. Compared to petroleum-based incumbent materials, commercially viable bio-sourced products should meet or exceed the following expectations:

- Competitive price
- Superior or comparable performance
- Near drop-in replacement in terms of processability
  - Same or comparable processing equipment and its configuration
  - No (or only minor) product reformulation
- Use of renewable feedstocks (by design)
  - Faster replenishment of renewable feedstocks
  - Better-controlled raw material costs of the renewable feedstocks
- Positive environmental impacts (according to LCA)
  - Lower process energy consumption
  - Reduced carbon footprint according to LCA (expressed in kg CO_2 equivalent per kg material).

[1] P. T. Anastas and J. C. Warner, *Green Chemistry: Theory and Practice*, Oxford University Press, 1998.

Several countries (the USA, Germany, and Belgium) have launched product labelling programs to identify commercial bio-based products manufactured from renewable feedstocks and incentivize manufacturers to certify their products and intermediates for bio-based product labels. Affixing those labels on consumer products was shown to be very effective to raise public awareness and promote the use of articles produced from raw materials derived from renewable feedstocks. The minimum bio-based content required to qualify for the USDA certification was set at 25% in accordance with ASTM D6866. The method uses ^{14}C decay to measure the amount of organic carbon in a sample that is derived from renewable sources as a percentage of the total organic carbon (which includes both renewable carbon and fossil carbon).

The bio-based content determined according to ASTM D6866 is not affected by the presence of other chemical elements and inorganic (mineral) carbon.

The bio-based content determined by ASTM D6866 depends on bio-based content of its components and the concentration of these components in the final formulation or an article. Bio-based raw materials can be derived from plant- or animal-sourced feedstocks.

Most typical manufacturing approaches for producing bio-sourced plastic materials using conventional petroleum-based polymer resins use either:

- Plastic composites composed of conventional polymers and biogenic (natural but not mineral) fillers
- Plastics compounds produced with conventional polymers and bio-based plasticizers (the plasticized compounds can be called *plasticizates*); this approach is primarily applicable to PVC.

Using these approaches, >25% bio-based content of plastic materials can be attained. Additionally, the plasticized PVC compounds may also contain biogenic fillers.

14.3 Bio-based PVC Additives

14.3.1 Bio-based Plasticizers for PVC Plasticizates

Information about commercial and nearly commercial bio-based plasticizers for PVC, as well as the plasticizates, is summarized in Chapter 5. Selected commercially available primary bio-based plasticizers, their applications, chemical functionality, and raw materials, and the renewable feedstocks used for manufacturing them, are shown in Table 14.2.

Table 14.2 Selected Primary Bio-based Plasticizers

Ester type	Ester plasticizers	Renewable raw materials	Feedstocks	Performance attributes	Applications
Di-esters	Azelates	Azelaic acid	Vegetable oils/fats; castor oil; sugars	Low-temperature flexibility; high efficiency	Automotive; wire & cable
	Sebacates	Sebacic acid			
	Succinates	Succinic acid			
	Furoates	2,5-Furan dicarboxylic acid			
Tetra-esters	Citrates	Citric acid	Sugars	Food contact approval	Packaging film, toys; medical devices; blood bags; textiles
	Valerates	Valeric acid			
Epoxidized mono-esters	Epoxidized ethylhexyl esters of fatty acids (EEFAs)	Fatty acids	Plant oils; fats; rosin	Low temp. flexibility; low viscosity; heat co-stabilizers	Coatings; moldings; castings
Epoxidized tri-esters	Epoxidized soybean oil (ESO or ESBO)	Fatty acids; glycerin	Vegetable oils	Suitable for food contact; heat co-stabilizers	Secondary plasticizers

As can be seen, a broad range of renewable feedstocks can be used for manufacturing primary plasticizers with high bio-based content for use in flexible PVC compounds and their applications. A typical bio-based plasticizer/PVC composition would be 30:70 by weight.

Among all of the additives, bio-based plasticizers are the main source of renewable carbon in flexible PVC compounds, if the resin type used is petroleum-based. Incorporation of 30 phr (and more) of primary plasticizers with high bio-based content in PVC compounds may enable achieving the >25% bio-based content required for qualifying the compounds for bio-based product labeling programs.

14.3.2 Biogenic Fillers for Plastic Composites

PVC is one of three major polymers used for manufacturing bio-based composites. The other two are PE and PP (see also Section 12.5.6). The preparation of plastic composites with biogenic fillers is very effective in reducing the plastics content and incorporating renewable carbon into typical rigid articles. Compared to conventional plastics, its composites may have a different feel, appearance, and aesthetics with somewhat changed mechanical properties.

Wood is used as a biogenic filler in manufacturing wood–plastic composite decking, window profiles, railing, fencing, furniture, and automotive interior parts. Other natural biogenic fillers containing organic carbon include starch, cellulose, chitin, chitosan, lignin, jute, rice straw, sisal, hemp, flax, kenaf, and bamboo. Wood flour and several other fibers act as reinforcing additives that also reduce the cost and weight of the plastic composites containing them. Typically, compositions of these composites consist of 40:60 wood flour/PVC combinations and 50:50 wood flour/PE or PP.

Using biogenic fillers and bio-based plasticizers allows achieving bio-based content of PVC compounds and materials to greatly exceed the minimum required in accordance with the USDA BioPreferred® voluntary labeling and other similar programs.

14.3.3 Bio-based Heat Stabilizers

Most mixed-metal carboxylate-containing heat stabilizers and all alkyltin reverse-ester heat stabilizers (Chapter 4) for PVC are derived from fatty acids, and therefore have high bio-based content (Table 14.3).

Table 14.3 Selected Bio-based Heat Stabilizer Components

Stabilizer type	Examples of bio-based stabilizer components	Calculated total carbon content (%)	Calculated bio-based content (%)	Typical loading in PVC compounds (%)	Calculated bio-based content contribution to PVC compounds (%)[a]
Mixed-metal carboxylates	Calcium stearate	71	100	1.5	2.9
	Calcium oleate	72	100	1.0	1.9
	Calcium laurate	66	100	1.0	1.7
	Barium oleate	62	100	1.0	1.6
	Barium stearate	61	100	1.0	1.6
	Zinc stearate	68	100	1.2	2.1
	Zinc oleate	69	100	1.0	1.7
	Zinc laurate	52	100	1.0	1.4
Alkyltin "reverse-esters"	Monomethyltin tris(2-mercaptoethyl oleate)	63	89	1.0	1.6
	Dimethyltin bis(2-mercaptoethyl oleate)	61	86	0.5	0.8
	Monobutyltin tris(2-mercaptoethyl oleate)	64	84	1.0	1.3

[a] For the calculations, the total carbon content of PVC was considered to be 38%.

As can be seen from Table 14.3, the contribution made by bio-based stabilizer components and fully formulated stabilizers – when added at typical, relatively low loadings – towards the total bio-based content of PVC compounds is relatively small, being about 1–3% and 3–6%, respectively.

14.3.4 Bio-based Lubricants

Table 14.4 Selected Bio-based Lubricants

Lubricant type	Examples of bio-based lubricants	Calculated total carbon content (%)	Calculated bio-based content (%)	Typical loading in PVC compounds (%)	Calculated bio-based content contribution in PVC compounds (%)[a]
Fatty acid	Stearic acid	76	100	0.1–0.5	0.2–1.0
	2-Hydroxystearic acid	72	100	0.1–0.5	0.4–0.9
Fatty alcohol	Stearyl alcohol	80	100	0.3–1.5	0.6–3.2
Fatty acid ester	Distearyl phthalate	78	82	0.2–1.2	0.3–2.0
	Stearyl stearate	81	100	0.2–1.2	0.4–2.6
	Sucrose octastearate	76	100	0.2–1.2	0.4–2.4
	Pentaerythritol monostearate	69	78	0.2–1.2	0.2–1.7
	Pentaerythritol distearate	74	88	0.2–1.2	0.3–2.1
	Pentaerythritol tetrastearate	77	94	0.2–1.2	0.4–2.3
	Glycerol monostearate	70	100	0.2–1.2	0.4–2.2
	Glycerol tristearate	77	100	0.2–1.2	0.4–2.4
	Glycerol monooleate	71	100	0.2–1.2	0.4–2.2
	Glycerol trioleate	77	100	0.2–1.2	0.4–2.4
	Hydrogenated castor oil	72	100	0.2–1.2	0.4–2.3
Fatty acid amide	*N,N'*-Ethylene bis(stearamide)	77	95	0.5–2.0	1.0–3.9

[a] For the calculations, the total carbon content of PVC was considered to be 38%.

Many lubricants used for processing plastics are bio-based. They include fatty alcohols, fatty acids, metal soaps, fatty acid amides, and esters of fatty alcohols and/or fatty acids. The bio-based portions of these lubricants are derived mostly from fatty acids sourced from fats and oils. Examples of bio-based lubricants used in processing PVC compounds are shown in Table 14.4.

As can be seen from Table 14.4, the contribution made by bio-based lubricant components and fully formulated lubricant packages – when added at typical, relatively low loadings – towards the total bio-based content of PVC compounds is relatively small, being about 0.2–3% and 0.2–4%, respectively.

Although the bio-based content of the described heat stabilizers and lubricants is 80–100%, and their contribution to the total bio-based content of PVC compounds is only 2–6%, the replacement of petroleum-based additives with bio-based additives at any PVC manufacturing site (as opposed to a specific PVC compound) may amount to a substantial annual reduction in carbon footprint expressed in kg CO_2 equivalents per kg material.

It should be noted that metal stearates and other salts of fatty acids are broadly used as lubricants for other plastics (Chapter 7). Most other additives, such as flame retardants, processing aids, impact modifiers, antimicrobials, colorants, and mineral fillers such as calcium carbonate, do not have high bio-based content, and therefore do not, in any meaningful way, contribute towards increasing the levels of renewable carbon in plastics, or reducing their carbon footprint.

14.4 Questions

1. What are the main sources of renewable carbon in flexible PVC compounds?
2. What are the main sources of renewable carbon in rigid PVC compounds?
3. What are the main chemical types of bio-based heat stabilizer components in PVC?
4. What are the main chemical types of bio-based lubricants used in PVC?

Index